Comparison Geometry asks: What can we say about a Riemannian manifold if we know a (lower or upper) bound for its curvature, and perhaps something about its topology? Powerful results that allow the exploration of this question were first obtained in the 1950s by Rauch, Alexandrov, Toponogov, and Bishop, with some ideas going back to Hopf, Morse, Schoenberg, Myers, and Synge in the 1930s.

In the last decade the field has witnessed many important advances: first in conjunction with Morse theory and convexity, then with critical point theory for distance functions, and most recently with the Gromov–Hausdorff topology on spaces of Riemannian manifolds, and the geometry of singular spaces. As a result, our understanding of relations between the geometry and topology of Riemannian manifolds has expanded, and no longer consists of small unrelated pieces of scholarship.

This volume, arising from a 1994 MSRI program, is an up-to-date panorama of Comparison Geometry, featuring surveys and new research. Surveys present classical and recent results, and often include complete proofs, in some cases involving a new and unified approach. The historical evolution of the subject is summarized in charts and tables of examples.

This volume will be a valuable source for researchers and graduate students in Riemannian Geometry.

Mathematical Sciences Research Institute
Publications

30

Comparison Geometry

Mathematical Sciences Research Institute
Publications

Volumes 1 through 27 are available from Springer-Verlag

Comparison Geometry

Edited by

Karsten Grove
University of Maryland

Peter Petersen
University of California, Los Angeles

Karsten Grove
Department of Mathematics
University of Maryland
College Park, MD 20742-0001
kng@math.umd.edu

Peter Petersen
Department of Mathematics
University of California, Los Angeles
Los Angeles, CA 90024-1555
petersen@math.ucla.edu

Mathematical Sciences Research
  Institute
1000 Centennial Drive
Berkeley, CA 94720

MSRI Book Series Editor
Silvio Levy

MSRI Editorial Committee
T. Y. Lam (chair)
Alexandre Chorin
Robert Osserman
Peter Sarnak

The Mathematical Sciences Research Institute wishes to acknowledge
support by the National Science Foundation.

CAMBRIDGE UNIVERSITY PRESS
Cambridge, New York, Melbourne, Madrid, Cape Town, Singapore, São Paulo, Delhi

Cambridge University Press
The Edinburgh Building, Cambridge CB2 8RU, UK

Published in the United States of America by Cambridge University Press, New York

www.cambridge.org
Information on this title: www.cambridge.org/9780521592222

First published 1997
This digitally printed version 2008

A catalogue record for this publication is available from the British Library

ISBN 978-0-521-59222-2 hardback
ISBN 978-0-521-08945-6 paperback

Comparison Geometry
MSRI Publications
Volume 30, 1997

Contents

Comparison Geometry
MSRI Publications
Volume 30, 1997

Preface

One of the main activities during the 1993–94 special year in differential geometry at MSRI was focused on the subject baptized "Comparison Geometry" during the planning phase of the workshops.

Although a name has been lacking for this beautiful and most geometric branch of riemannian geometry, its history can be traced back to the nineteenth century. It did not take root, however, until the 1930's, through the work of H. Hopf, Morse and Schoenberg, Myers, and Synge. The real breakthrough came in the 1950's with the pioneering work of Rauch and the foundational work of Alexandrov, Toponogov and Bishop. Since then, the simple idea of comparing the geometry of an arbitrary riemannian manifold with the geometries of constant curvature spaces has seen a tremendous evolution: first in conjunction with Morse theory and convexity, then with critical point theory for distance functions, and most recently with the Gromov–Hausdorff topology on spaces of riemannian manifolds, and the geometry of singular spaces. As a result, our understanding of relations between the geometry and topology of riemannian manifolds has gained tremendous breadth and consists no longer of just a short string of pearls.

At the outset it is worth mentioning that the flavor and character of problems and techniques related to upper rather than lower curvature bounds to a large extent are remarkably different. This volume is an up-to-date reflection of the above mentioned development regarding spaces with lower, or two-sided, curvature bounds. The subject of manifolds with negative or nonpositive curvature, with its ramifications to dynamics and number theory, is not represented here.

The content of the volume reflects some of the most exciting activities on Comparison Geometry during the year, and especially of the workshop devoted to the subject. As a consequence, the book features survey articles (by Abresch and Meyer, Anderson, Colding, Greene, Otsu, Petersen, and Zhu) and research articles (by Perelman and Petrunin). Each of the survey articles stems from recent interesting developments concerning either classical or more recent important problems. Complete proofs are often provided, and in one case (Petersen) a new unified strategy is presented and new proofs are offered.

We hope that this volume will be a valuable source for those who wish to learn about this beautiful subject and contribute to it.

Karsten Grove
Peter Petersen

Comparison Geometry
MSRI Publications
Volume **30**, 1997

Injectivity Radius Estimates and Sphere Theorems

UWE ABRESCH AND WOLFGANG T. MEYER

ABSTRACT. We survey results about the injectivity radius and sphere theorems, from the early versions of the topological sphere theorem to the authors' most recent pinching below-$\frac{1}{4}$ theorems, explaining at each stage the new ideas involved.

Introduction

Injectivity radius estimates and sphere theorems have always been a central theme in global differential geometry. Many tools and concepts that are now fundamental for comparison geometry have been developed in this context.

This survey of results of this type reaches from the early versions of the topological sphere theorem to the most recent pinching below-$\frac{1}{4}$ theorems. Our main concern is to explain the new ideas that enter at each stage. We do not cover the differentiable sphere theorem and sphere theorems based on Ricci curvature.

In Sections 1–3 we give an account of the entire development from the first sphere theorem of H. E. Rauch to M. Berger's rigidity theorem and his pinching below-$\frac{1}{4}$ theorem. Many of the main results depend on subtle injectivity radius estimates for compact, simply connected manifolds.

In Section 4 we present our recent injectivity radius estimate for odd-dimensional manifolds M^n with a pinching constant below $\frac{1}{4}$ that is independent of n [Abresch and Meyer 1994]. With this estimate the restriction to even-dimensional manifolds can be removed from the hypotheses of Berger's pinching below-$\frac{1}{4}$ theorem.

Additional work is required in order to get a sphere theorem for odd-dimensional manifolds M^n with a pinching constant $< \frac{1}{4}$ independent of n. This result and the basic steps involved in its proof are presented in Sections 5–7; details can

1991 *Mathematics Subject Classification.* Primary 53C20; Secondary 53C21, 53C23.

Key words and phrases. Curvature and Topology, Pinching, Sphere Theorems.

be found in [Abresch and Meyer a]. The essential step in the geometric part of the argument is to establish Berger's horseshoe conjecture for simply connected manifolds. For this purpose we need the new Jacobi field estimates in Section 6.

1. On the Topological Sphere Theorem

The topological sphere theorem was one of the first results in Riemannian geometry where the topological type of a manifold M^n is determined by inequalities for its sectional curvature K_M and some mild global assumptions. Building on earlier work of Rauch and Berger, the final version of this theorem was obtained by W. Klingenberg [1961]:

THEOREM 1.1 (TOPOLOGICAL SPHERE THEOREM). *Let M^n be a complete, simply connected Riemannian manifold with strictly $\frac{1}{4}$-pinched sectional curvature. Then M^n is homeomorphic to the sphere $\mathbb{S}^n$.*

Note that a positively curved manifold M^n is said to be *strictly δ-pinched* if and only if $\inf K_M > \delta \sup K_M$.

The first version of the theorem, for manifolds that are approximately 0.74-pinched, had been obtained by Rauch [1951]. Several years passed until Berger and Klingenberg managed to improve the result. The proofs of all these theorems are based on direct methods in comparison geometry. We shall describe the basic ideas later in this section.

First we discuss the hypotheses of the topological sphere theorem and explain in what sense the theorem is optimal.

REMARKS 1.2. (i) Strict $\frac{1}{4}$-pinching implies in particular that $\inf K_M \geq \lambda > 0$, and by Myers' theorem [1935; 1941] M^n is a compact manifold with diameter $\leq \pi/\sqrt{\lambda}$. For this reason the theorem can equivalently be stated for compact rather than complete manifolds.

(ii) By Synge's lemma [1936] any compact, oriented, even-dimensional manifold with $K_M > 0$ is simply connected. Thus, in the even-dimensional case it is enough to assume that the manifold M^n is orientable rather than simply connected.

(iii) The hypothesis $\min K_M > \frac{1}{4} \max K_M$ is optimal provided that the dimension of M^n is even and ≥ 4. In fact, the sectional curvature of the Fubini–Study metric on the complex and quaternionic projective spaces $\mathbb{CP}^m$ and $\mathbb{HP}^m$, and on the Cayley plane $\mathbb{CaP}^2$, is weakly quarter-pinched. The dimensions of these spaces are $2m$, $4m$, and 16, and except for $\mathbb{CP}^1$ and $\mathbb{HP}^1$ these spaces are not homotopically equivalent to spheres.

(iv) Nevertheless, it is possible to write more adapted curvature inequalities and relax the hypotheses of the topological sphere theorem accordingly. Sphere theorems based on pointwise pinching conditions were established in the mid seventies [Im Hof and Ruh 1975]. The most advanced result in this direction was obtained by M. Micallef and J. D. Moore in 1988, and will be discussed at the end of this section.

For two-, three-, and four-dimensional manifolds stronger results are known. Any compact, simply connected surface is diffeomorphic to $\mathbb{S}^2$, and by the Gauss–Bonnet theorem any compact, orientable surface with strictly positive curvature is also diffeomorphic to $\mathbb{S}^2$.

Furthermore, any compact, simply connected three-manifold M^3 with strictly positive Ricci curvature is diffeomorphic to the standard three-sphere. This assertion follows from a more general result about the Ricci flow:

THEOREM 1.3 [Hamilton 1982]. *Let (M^3, g) be a compact, connected, three-dimensional Riemannian manifold with Ricci curvature* ric > 0 *everywhere. Then g can be deformed in the class of metrics with* ric > 0 *into a metric with constant sectional curvature K_M.*

Finally, in dimension 4 one can determine all homeomorphism types under some weaker pinching condition by combining Bochner techniques with M. Freedman's classification of simply connected, topological four-manifolds [1982]:

THEOREM 1.4 [Seaman 1989]. *Let M^4 be a compact, connected, oriented Riemannian four-manifold without boundary. Suppose that the sectional curvature K_M of M^4 satisfies $0.188 \approx (1 + 3\sqrt{1 + 2^{5/4} \cdot 5^{-1/2}})^{-1} \leq K_M \leq 1$. Then M^4 is homeomorphic to $\mathbb{S}^4$ or $\mathbb{CP}^2$.*

Beginning with D. Gromoll's thesis [1966], various attempts have been made to prove that M^n is diffeomorphic to the sphere $\mathbb{S}^n$ with its standard differentiable structure. The optimal pinching constant for a *differentiable sphere theorem* is not known. Except for low-dimensional special cases, the best constant obtained so far is approximately 0.68 [Grove et al. 1974a; 1974b; Im Hof and Ruh 1975]. However, not a single exotic sphere is known to carry a metric with $K_M > 0$. An exotic Milnor seven-sphere that comes with a metric of nonnegative sectional curvature has been described by Gromoll and W. T. Meyer [1974].

The most recent results in this direction are due to M. Weiss [1993]. Using sophisticated topological arguments, he has shown that any exotic sphere Σ^n that bounds a compact, smooth, parallelizable $4m$-manifold with $m \geq 2$ does not admit a strictly quarter-pinched Riemannian metric, provided that Σ^n represents an element of even order in the group Γ_n of differentiable structures on $\mathbb{S}^n$.

For the sequel it will be convenient to introduce *generalized trigonometric functions*, which interpolate analytically between the usual trigonometric and hyperbolic functions. The generalized sine $\operatorname{sn}_\lambda$ is defined as the solution of $y'' + \lambda y = 0$ with initial data $y(0) = 0$ and $y'(0) = 1$: explicitly,

$$\operatorname{sn}_\lambda(\varrho) := \begin{cases} \lambda^{-1/2}\sin(\sqrt{\lambda}\,\varrho) & \text{if } \lambda > 0, \\ \varrho & \text{if } \lambda = 0, \\ |\lambda|^{-1/2}\sinh(\sqrt{|\lambda|}\,\varrho) & \text{if } \lambda < 0. \end{cases}$$

The generalized cosine and cotangent are given by $\mathrm{cn}_\lambda := \mathrm{sn}'_\lambda$ and $\mathrm{ct}_\lambda := \mathrm{cn}_\lambda/\mathrm{sn}_\lambda$.

ON RAUCH'S PROOF [1951] OF THE SPHERE THEOREM. The argument dispenses with the two-dimensional case by referring to the Gauss–Bonnet theorem as explained above. His basic idea for proving the theorem in dimensions $n \geq 3$ was to recover the structure of a *twisted sphere* on the manifold M^n under consideration. For this purpose he studies the exponential maps $\exp_p$ and $\exp_q$ at two points $p, q \in M^n$. This approach requires a pinching constant δ equal to the positive root of $\sin(\pi\sqrt{\delta}) = \frac{1}{2}\sqrt{\delta}$, or approximately 0.74. (Note: We always report approximate bounds by rounding to the safe side, so they sometimes differ from occurrences in the literature that are rounded to the nearest value either way.)

Without loss of generality Rauch scales the Riemannian metric on M^n to ensure that $\delta < K_M < 1$. He picks the point $p \in M^n$ and some unit tangent vector $v_0 \in T_pM$ arbitrarily and sets $q := \exp_p \pi v_0$. In order to control the geometry of the exponential maps $\exp_p$ and $\exp_q$, he then develops the by now well-known *Rauch comparison theorems* for Jacobi fields. These estimates imply in particular that

(i) the conjugate radius $\mathrm{conj}\,M^n$ is $> \pi$;
(ii) the image of the sphere $S(0,\pi) \subset T_pM$ under $\exp_p$ has diameter $< \pi\,\mathrm{sn}_\delta(\pi)$;
(iii) for any $\hat{\varrho} < \frac{1}{2}\pi$ the ball $B(0,\hat{\varrho}) \subset \tilde{B}_q$, where $\tilde{B}_q$ is the ball $B(0,\pi) \subset T_qM$ equipped with the metric $\exp_q^* g$, is strictly convex. Its boundary has strictly positive second fundamental form.

Specializing to $\delta \approx 0.74$, Rauch concludes that the diameter of $\exp_p\big(S(0,\pi)\big)$ is bounded by $\frac{1}{2}\pi - 2\varrho$ for $\varrho > 0$ sufficiently small. Lifting the restriction $\exp_p|_{S(0,\pi)}$ under $\exp_q$, he thus obtains an immersion ϕ of $S(0,\pi) \subset T_pM$ into the closed ball of radius $\frac{1}{2}\pi - 2\varrho$ centered at the origin in T_qM, mapping πv_0 to the origin. For this construction it is crucial that the sphere $S(0,\pi) \subset T_pM$ be simply connected, that is, that the dimension of M^n be ≥ 3. (Rauch describes the lifting under the local diffeomorphism $\exp_q$ in a more classical terminology, widely used in complex analysis when dealing with monodromy. He speaks of a "c-process" based on a "purse-string construction".)

The next step is to find for each $w \in S(0,\pi) \subset T_pM$ the smallest number $t_w \in (0,1)$ such that the lift of the geodesic $t \mapsto \exp_p((1-t)w)$ under $\exp_q$ starting at $\phi(w)$ leaves the ball $B(0, \frac{1}{2}\pi - \varrho) \subset T_qM$. It is not a priori clear that such a number t_w exists unless $w = \pi v_0$. On the other hand, if t_w exists, it follows from (iii) that the lifted geodesic must intersect the sphere $S(0, \frac{1}{2}\pi - \varrho) \subset T_qM$ transversally in $t = t_w$. This observation is the basis for an elaborate continuity argument that shows that t_w exists for all $w \in S(0,\pi) \subset T_pM$. Moreover, this continuity argument provides a homeomorphism $\theta : S(0, \frac{1}{2}\pi - \varrho) \subset T_qM \to \partial\Omega$, where $\partial\Omega$ denotes the boundary of the star-shaped set

$$\Omega := \{\tau w \mid 0 \leq \tau < 1 - t_w,\ w \in S(0,\pi) \subset T_pM\}.$$

When gluing $\bar{\Omega}$ and $\bar{B}(0, \frac{1}{2}\pi - \varrho) \subset T_q M$ by means of θ, one obtains a twisted sphere Σ^n. The exponential maps $\exp_p$ and $\exp_q$ fit together. They induce a local homeomorphism $\Sigma^n \to M^n$. Since M^n is simply connected this local homeomorphism must actually be a global homeomorphism. $\qquad\square$

The starting point for improving Rauch's sphere theorem was the following injectivity radius estimate:

THEOREM 1.5 [Klingenberg 1959]. *Let M^n be an even-dimensional, compact, simply connected Riemannian manifold with strictly positive sectional curvature K_M. Then the injectivity radius $\operatorname{inj} M^n$ is controlled in terms of the conjugate radius $\operatorname{conj} M^n$:*

$$\operatorname{inj} M^n = \operatorname{conj} M^n \geq \pi / \sqrt{\max K_M}\,.$$

We shall discuss this result together with Theorem 1.6 below. For the moment our issue is to explain how such an estimate can be employed for the proof of the sphere theorem. Actually, as a first application of Theorem 1.5, Klingenberg obtained a sphere theorem for even-dimensional manifolds requiring only a pinching constant $\delta \approx 0.55$, the positive solution of $\sin(\pi\sqrt{\delta}) = \sqrt{\delta}$ [Klingenberg 1959, Theorem 2].

The basic advantage provided by the injectivity radius estimate is the fact that the immersions $\exp_p : B(0, \pi) \to B(p, \pi)$ and $\exp_q : B(0, \pi) \to B(q, \pi)$ are recognized as diffeomorphisms onto their images in M^n. Since the points p and $q \in M^n$ are chosen in the same way as in Rauch's original approach, a slightly modified continuity argument implies that the cut locus C_q of q lies in $B(p, \pi)$ and vice versa. Thus the open balls $B(p, \pi)$ and $B(q, \pi)$ cover the manifold. It follows that M^n is the union of two closed cells with a common boundary homeomorphic to $\mathbb{S}^{n-1}$. In other words, M^n itself is recognized as a twisted sphere. This construction avoids several lifting arguments from Rauch's proof, and thus it eliminates some constraints on the pinching constant.

Theorem 1.5 was also the starting point of Berger's work on the topological sphere theorem. Combining Klingenberg's injectivity radius estimate with Toponogov's triangle comparison theorem, which had just appeared in the literature, Berger [1960b, Théorème 1] established Theorem 1.1 for even-dimensional manifolds with the optimal pinching constant. Subsequently, he published an independent proof [Berger 1962b, Theorem 3] of the triangle comparison theorem, based on an extension of Rauch's comparison theorems for Jacobi fields rather than on Alexandrov's ideas for surfaces.

ON BERGER'S PROOF OF THE SPHERE THEOREM. Here the starting point is the observation that the choice of the points $p, q \in M^n$ in the preceding work of Rauch and Klingenberg was not optimal. Berger suggested picking p and q in such a way that the distance $d(p, q)$ in M^n is maximal. The key property of such a pair of antipodal points is the fact that for any unit tangent vector $v \in T_q M$

there exists a minimizing geodesic $c : [0, 1] \to M^n$ from q to p making an acute angle with v, that is, such that $g(c'(0), v) \geq 0$. Assuming that $\frac{1}{4} < K_M \leq 1$, he can thus apply Toponogov's triangle comparison theorem in order to conclude that the metric balls $B(p, \pi)$ and $B(q, \pi)$ cover M^n.

Then he can proceed with some arguments from Klingenberg's proof: Theorem 1.5 reveals that $B(p, \pi)$ and $B(q, \pi)$ are diffeomorphic to the balls $B(0, \pi)$ in the tangent spaces $T_p M$ and $T_q M$, respectively, and again M^n can be recognized as a twisted sphere.   □

Berger [1960b, Théorème 2] also succeeded in analyzing the additional phenomena that occur for simply connected, weakly quarter-pinched manifolds. We shall review this result in Theorem 2.1.

With Berger's proof, the only missing ingredient for the final version of the topological sphere theorem as stated in 1.1 was a suitable injectivity radius estimate in the odd-dimensional case. Such an estimate was established shortly afterwards:

THEOREM 1.6 [Klingenberg 1961]. *Let M^n be a compact, simply connected Riemannian manifold with strictly $\frac{1}{4}$-pinched sectional curvature. Then the injectivity radius* inj M^n *and the conjugate radius* conj M^n *coincide:*

$$\operatorname{inj} M^n = \operatorname{conj} M^n \geq \pi / \sqrt{\max K_M}\,.$$

REMARKS 1.7. (i) In Theorem 1.5 it is crucial to assume that the manifold has nonnegative sectional curvature. Otherwise, there is not even a uniform lower bound on the injectivity radius for simply connected surfaces. In fact, it is easy to construct surfaces of revolution with $-1 \leq K_M \leq 1$ and arbitrarily small injectivity radius. The diameter of these surfaces, which look like hourglasses, increases without bound as the injectivity radius approaches zero.

(ii) The most significant difference between Theorems 1.5 and 1.6 is the pinching condition that appears in the hypothesis of the latter. Such a condition is necessary to get a result for odd-dimensional manifolds at all. The optimal value for the pinching constant in Theorem 1.6 is not known. Berger has shown [1962a] that a constant $< \frac{1}{9}$ is not sufficient in order to obtain even the slightly weaker inequality inj $M^n \geq \pi / \sqrt{\max K_M}$. For this purpose he considers a family of Riemannian metrics g_ε on the odd-dimensional spheres $\mathbb{S}^{2n+1} \subset \mathbb{C}^{n+1}$. These metrics are defined by shrinking the standard metric in the direction of the Hopf circles $\{e^{it} p \mid t \in \mathbb{R}/2\pi\mathbb{Z}\} \subset \mathbb{S}^{2n+1}$ in such a way that their lengths with respect to g_ε become $2\pi\varepsilon$. The range of the sectional curvature of g_ε is the interval $[\varepsilon^2, 4 - 3\varepsilon^2]$, provided of course that $0 < \varepsilon \leq 1$. Clearly, $\pi\varepsilon < \pi/\sqrt{4 - 3\varepsilon^2}$ for $\varepsilon^2 < \frac{1}{3}$. This means that for any $\delta \in (0, \frac{1}{9})$ there exists a Berger metric g_ε whose sectional curvature is δ-pinched and whose injectivity radius is strictly less than $\pi/\sqrt{\max K_M}$. Unless $\varepsilon = 1$ there does not exist any pair consisting of a (horizontal) geodesic $\gamma : \mathbb{R} \to (\mathbb{S}^{2n+1}, g_\varepsilon)$ and a parallel unit normal field v along γ such that the sectional curvature on each plane span$\{\gamma'(s), v(s)\}$ equals

$4 - 3\varepsilon^2$. Hence for any $\varepsilon \in (0,1)$ the conjugate radius of g_ε is strictly greater than $\pi/\sqrt{\max K_M}$. A direct computation reveals that the conjugate radius of g_ε arises as the first positive zero of the map $s \mapsto \varepsilon^2 + (1-\varepsilon^2)s \cot s$, and thus the injectivity radius of g_ε ceases to be equal to its conjugate radius as ε becomes less than 0.589. This means that the optimal value for the pinching constant in Theorem 1.6 is necessarily at least $0.117 > \frac{1}{9}$.

(iii) Even worse, in dimension 7, the Aloff–Wallach examples [1975] described in the appendix contain a sequence of simply connected, homogeneous Einstein spaces whose pinching constants converge to $\frac{1}{37}$ and whose injectivity radii converge to zero. In other words, if $\delta < \frac{1}{37}$ there does not exist any a priori lower bound for the injectivity radius of a seven-dimensional, simply connected, δ-pinched Riemannian manifold.

The proofs of Theorems 1.5 and 1.6, and of essentially any other injectivity radius estimate, begin with the observation that the injectivity radius of a compact Riemannian manifold can be related to its conjugate radius and to the length of a shortest geodesic loop. We give a refined version of [Klingenberg 1959, Lemma 4]:

LEMMA 1.8 [Cheeger and Ebin 1975, Lemma 5.6]. *Let M^n be a complete Riemannian manifold, and let $p \in M^n$. Let $\ell_M(p)$ denote the minimal length of a nontrivial geodesic loop $c_0 : [0,1] \to M^n$ starting and ending at p. Then the injectivity radius of M^n at p is* $\mathrm{inj}_M(p) = \min\{\mathrm{conj}\, M^n, \frac{1}{2}\ell_M(p)\}$.

By definition, $\mathrm{inj}\, M^n = \inf_p \mathrm{inj}_M(p)$. For compact manifolds M^n the infimum is always achieved at some point $p_0 \in M^n$. Furthermore, if $\ell_M(p_0) < 2\,\mathrm{conj}\, M^n$, it is easy to see that the geodesic loop $c_0 : [0,1] \to M^n$ of length $2\,\mathrm{inj}\, M^n$ with $c_0(0) = c_0(1) = p_0$ is actually a closed geodesic $c_0 : \mathbb{R}/\mathbb{Z} \to M^n$. This means that for compact Riemannian manifolds one has $\mathrm{inj}\, M^n = \min\{\mathrm{conj}\, M^n, \frac{1}{2}\ell(M^n)\}$, where $\ell(M^n) = \inf_p \ell_M(p)$ is the minimal length of a nontrivial closed geodesic $c_0 : \mathbb{R}/\mathbb{Z} \to M^n$.

ON THE PROOF OF THEOREM 1.5. It is a standard fact that a geodesic is not minimizing beyond the first conjugate point. Hence $\mathrm{inj}\, M^n \le \mathrm{conj}\, M^n$, and it is possible to proceed indirectly.

Assuming that $\mathrm{inj}\, M^n < \mathrm{conj}\, M^n$, Lemma 1.8 asserts that there is a closed geodesic $c_0 : \mathbb{R}/\mathbb{Z} \to M^n$ of length $L(c_0) = 2\,\mathrm{inj}\, M^n$. As in the proof of Synge's lemma [1936], the canonical form theorem for the orthogonal group $SO(n-1)$ leads to a closed, parallel, unit normal field v_0 along c_0. Since $K_M > 0$, the second variation formula reveals that the nearby curves $c_t : \mathbb{R}/\mathbb{Z} \to M^n$, $s \mapsto \exp_{c_0(s)} t v_0(s)$, are strictly shorter than c_0, provided that t is nonzero and sufficiently small.

At this point Klingenberg observes that the image of such a curve c_t is contained in the closed ball with radius $\frac{1}{2}L(c_t) < \mathrm{inj}\, M^n$ centered at $c_t(0)$, and thus he can lift c_t under $\exp_{c_t(0)}$ to a map $\tilde{c}_t : \mathbb{R}/\mathbb{Z} \to T_{c_t(0)}M$ such that $\tilde{c}_t(0) = 0$.

Notice that the argument that guarantees the existence of $\tilde{c}_t$ is very different from the arguments that justify taking local lifts under $\exp_p$ up to the conjugate radius, encountered in Rauch's proof of the sphere theorem.

By assumption, $\operatorname{inj} M^n < \operatorname{conj} M^n$, and thus the curves $\tilde{c}_t : \mathbb{R}/\mathbb{Z} \to T_{c_t(0)}M$ define an equicontinuous map $\tilde{c} : \mathbb{R}/\mathbb{Z} \times (0, \varepsilon) \to TM$. Extending this map continuously to $\mathbb{R}/\mathbb{Z} \times [0, \varepsilon)$, one obtains a lift $\tilde{c}_0 : \mathbb{R}/\mathbb{Z} \to T_{c_0(0)}M$ of c_0 under $\exp_{c_0(0)}$, satisfying $\tilde{c}_0(0) = 0$. On the other hand the lift of the geodesic c_0 must be the line $s \mapsto s c'(0)$, contradicting the periodicity of $\tilde{c}_0$. $\square$

The proof of Theorem 1.6 is much more subtle. It will be explained in Section 4 together with our recent extension of this injectivity radius estimate (Theorem 4.1).

Further progress in understanding the topological sphere theorem was made by K. Grove and K. Shiohama in the late seventies. Observing that the twisted sphere construction in the proof of Theorem 1.1 resembles the proof of Reeb's theorem in Morse theory, Grove and Shiohama investigated under what conditions the function $f_{pq} : x \mapsto \operatorname{dist}(x, p) - \operatorname{dist}(x, q)$ has only two critical points, an absolute minimum at p and an absolute maximum at q. Persuing this idea, they proved the following result:

THEOREM 1.9 (DIAMETER SPHERE THEOREM [Grove and Shiohama 1977]). *Let M^n be a connected, complete Riemannian manifold with sectional curvature $K_M \geq \lambda > 0$ and diameter $\operatorname{diam} M^n > \pi/(2\sqrt{\lambda})$. Then M^n is homeomorphic to the sphere $\mathbb{S}^n$.*

REMARK 1.10. The Fubini–Study metrics on the projective spaces $\mathbb{CP}^m$, $\mathbb{HP}^m$, and $\mathrm{Ca}\mathbb{P}^2$ have diameter $\pi/(2\sqrt{\min K_M})$ if $m \geq 2$. Thus in Theorem 1.9 the hypothesis on the diameter is optimal if the dimension of the manifold is even and ≥ 4.

Notice that the distance functions to fixed points p and q are only Lipschitz functions. They are not differentiable at the cut loci C_p and C_q. In fact, the proof of Theorem 1.9 has led to a fruitful definition of what a critical point of a distance function should be. Eventually, an elaborate *critical point theory for distance functions* was developed and successfully applied to many other problems [Grove 1993]. For instance, the proof of M. Gromov's Betti numbers theorem uses critical point theory for distance functions in a substantial way [Abresch 1985; 1987; Gromov 1981b].

With this particular critical point theory it is possible to prove the diameter sphere theorem in a conceptually straightforward way by applying Toponogov's triangle comparison theorem twice [Grove 1987; 1993; Karcher 1989; Meyer 1989]. Moreover, Theorem 1.9 ties in nicely with the injectivity radius estimate from Theorem 1.6. Thinking of the injectivity radius as a lower bound for the diameter, one obtains an alternate, and structurally more appealing, proof of the topological sphere theorem.

The pointwise pinching problem requires completely different techniques. Results of this type are typically not based on direct comparison methods but on partial differential equations on the manifold M^n [Ruh 1982] or on minimal surfaces in M^n [Micallef and Moore 1988]. Here we shall focus on this theorem:

THEOREM 1.11 [Micallef and Moore 1988]. *Let M^n be a compact, simply connected Riemannian manifold of dimension $n \geq 4$. Suppose that M^n has positive curvature on totally isotropic two-planes. Then M^n is homeomorphic to $\mathbb{S}^n$.*

REMARKS 1.12. (i) The notion of positive curvature on totally isotropic two-planes arises naturally when one looks for a condition determining the sign of the difference between the Hodge Laplacian and the rough Laplacian on two-forms.

(ii) The projective spaces $\mathbb{CP}^m$, $\mathbb{HP}^m$, and $\mathrm{Ca}\mathbb{P}^2$ have nonnegative curvature on totally isotropic two-planes.

(iii) A direct computation shows that pointwise strict $\frac{1}{4}$-pinching implies positive curvature on totally isotropic two-planes. This means that the topological sphere theorem holds in dimensions $n \geq 4$ already for compact, simply connected, *pointwise* strictly $\frac{1}{4}$-pinched manifolds.

ON THE PROOF OF THEOREM 1.11. Note that any conformal harmonic map $f : \mathbb{S}^2 \to M^n$ is a common critical point for the Dirichlet functional D and the area functional and, moreover, area$(f) = D(f)$. In general, area $\leq D$, so the Dirichlet functional is an upper barrier for the area functional at the surface represented by f.

The key step for the proof of the theorem is to show that any nonconstant, branched, minimal two-sphere $f : \mathbb{S}^2 \to M^n$ in a Riemannian manifold of dimension ≥ 4 with positive curvature on totally isotropic two-planes has index $\mathrm{ind}_D(f) \geq \frac{1}{2}(n - 3)$. On the other hand, Micallef and Moore prove that any compact Riemannian manifold M^n with $\pi_k(M^n) \neq 0$ for some $k \geq 2$ contains a nonconstant harmonic two-sphere $f : \mathbb{S}^2 \to M^n$ with index $\mathrm{ind}_D(f) \leq k - 2$. Combining these two facts, it follows that $\pi_1(M^n) = \cdots = \pi_{\lfloor n/2 \rfloor}(M^n) = 0$. Hence the Hurewicz isomorphism theorem implies that $H_1(M^n; \mathbb{Z}) = \cdots = H_{\lfloor n/2 \rfloor}(M^n; \mathbb{Z}) = 0$, and by the Poincaré duality theorem M^n must be a homology sphere. Thus the result follows using S. Smale's solution of the generalized Poincaré conjecture in dimensions $n \geq 5$ [Milnor 1965, p. 109; Smale 1961] and Freedman's classification of compact, simply connected four-manifolds [1982].

The existence of nonconstant harmonic two-spheres $f : \mathbb{S}^2 \to M^n$ with index $\mathrm{ind}_D(f) \leq k - 2$ is established by means of the standard saddle point arguments from Morse theory. Micallef and Moore work with a perturbed version of the α-energy introduced by S. Sacks and K. Uhlenbeck, in order to have a nondegenerate Morse functional that satisfies Condition C of Palais and Smale, and study the limit as $\alpha \to 1$.

In order to obtain the lower bound for the index of such a nonconstant, conformal, branched, minimal two-sphere $f : \mathbb{S}^2 \to M^n$, Micallef and Moore express the Hessian of the Dirichlet functional D in terms of the squared norm of the $\bar{\partial}$-operator of the complexified bundle $f^*TM \otimes_{\mathbb{R}} \mathbb{C}$, rather than in terms of the squared norm of the full covariant derivative ∇. With this modification the zero-order term in the Hessian becomes an expression in the curvatures of totally isotropic two-planes containing ∂f. In particular, the claimed estimate for the index $\mathrm{ind}_D(f)$ follows upon constructing sufficiently many isotropic, holomorphic sections in $f^*TM \otimes_{\mathbb{R}} \mathbb{C}$ whose exterior product with ∂f is nontrivial. The appropriate tool for this purpose is Grothendieck's theorem on the decomposition of holomorphic vector bundles over $\mathbb{CP}^1$. Combining this theorem with the fact that the first Chern class of $f^*TM \otimes_{\mathbb{R}} \mathbb{C}$ vanishes, the authors construct a complex linear space of dimension $\geq \frac{1}{2}(n-1)$ of isotropic holomorphic sections. $\qquad\square$

2. Berger's Rigidity Theorem and Related Results

In this section the principal goal is to present the extensions of the topological sphere theorem and the diameter sphere theorem that hold when all strict inequalities in the hypotheses of these theorems are replaced by their weak counterparts. In particular, we shall see that the projective spaces mentioned in Remarks 1.2(iii) and 1.10 are the only other possibilities for the topological type of M^n. It is not known whether the sphere theorem by Micallef and Moore can be extended correspondingly or not.

When working on the topological sphere theorem, Berger actually studied the limiting case of simply connected, weakly quarter-pinched, even-dimensional manifolds, too:

THEOREM 2.1 (BERGER'S RIGIDITY THEOREM [Berger 1960b, Théorème 2]). *Let M^n be an even-dimensional, complete, simply connected Riemannian manifold with $\frac{1}{4} \leq K_M \leq 1$. Then either*

(i) *M^n is homeomorphic to the sphere $\mathbb{S}^n$, or*
(ii) *M^n is isometric to one of the other rank-one symmetric spaces, namely $\mathbb{CP}^{n/2}$, $\mathbb{HP}^{n/4}$, or $\mathrm{Ca}\mathbb{P}^2$.*

The first step in Berger's proof of the rigidity theorem was in a sense a predecessor of the diameter sphere theorem. Theorem 1.5 asserts that the manifold has injectivity radius $\mathrm{inj}\, M^n \geq \pi$. Refining the comparison argument from the proof of the topological sphere theorem, Berger proves that a Riemannian manifold M^n with $\frac{1}{4} \leq K_M \leq 1$, $\mathrm{inj}\, M^n \geq \pi$, and $\mathrm{diam}\, M^n > \pi$ is homeomorphic to $\mathbb{S}^n$. There remains the case where $\frac{1}{4} \leq K_M \leq 1$ and $\mathrm{diam}\, M^n = \mathrm{inj}\, M^n = \pi$; here he shows that any point $p \in M^n$ lies on a closed geodesic $c : \mathbb{R}/\mathbb{Z} \to M^n$ of length $L(c) = 2\pi$. With this additional information M^n can be recognized as a rank-one symmetric space by means of an argument that Berger had used shortly before in the proof of a weaker rigidity result [Berger 1960a, Théorème 1].

The assertion that a compact, simply connected Riemannian manifold with $\frac{1}{4} \leq K_M \leq 1$ and diam $M^n = \pi$ is isometric to a symmetric space is known as Berger's minimal diameter theorem [Cheeger and Ebin 1975, Theorem 6.6(2)]. J. Cheeger and D. Ebin also give a geometrically more direct proof for this result. Their idea is to study the metric properties of the geodesic reflections $\phi_p : B(p, \pi) \to B(p, \pi)$ and prove directly that these maps are isometries that can be extended continuously to M^n. Thus M^n is recognized as a symmetric space. It is an elementary fact that any symmetric space with $K_M > 0$ has rank one.

REMARK 2.2. Berger's rigidity theorem has been extended to cover odd-dimensional manifolds as well, asserting that any complete, simply connected, odd-dimensional manifold with weakly quarter-pinched sectional curvature is homeomorphic to the sphere. It should be clear from our sketch of Berger's original proof that such an extension follows immediately once the injectivity radius estimate from Theorem 1.6 has been generalized to the class of simply connected, weakly quarter-pinched manifolds. Such a generalization appeared shortly afterwards in the work of Klingenberg [1962]. However, the argument is technically extremely·subtle, and complete proofs were only given much later in two independent papers by Cheeger and Gromoll [1980] and by Klingenberg and T. Sakai [1980].

The diameter sphere theorem due to Grove and Shiohama can be generalized as follows:

THEOREM 2.3 (DIAMETER RIGIDITY THEOREM [Gromoll and Grove 1987]). *Let M^n be a connected, complete Riemannian manifold with sectional curvature $K_M \geq \lambda > 0$ and diameter* diam $M^n \geq \pi/(2\sqrt{\lambda})$. *Then*

(i) *M^n is homeomorphic to the sphere $\mathbb{S}^n$, or*
(ii) *the universal covering $\tilde{M}^n$ of M^n is isometric to $\mathbb{S}^n$, $\mathbb{CP}^{n/2}$, or $\mathbb{HP}^{n/4}$, or*
(iii) *the integral cohomology ring of $\tilde{M}^n$ is isomorphic to that of $\mathrm{Ca}\mathbb{P}^2$.*

We are discussing this theorem mainly because its proof has required an entirely new approach for recognizing rank-one symmetric spaces. The details are technically quite subtle, but the basic ideas are geometrically nice and simple.

Beforehand we mention that there are only a few possibilities for the covering maps $\tilde{M}^n \to M^n$ in assertions (ii) and (iii), since by Synge's lemma any orientable even-dimensional manifold M^n with $K_M > 0$ is simply connected. The only nontrivial quotients that can arise are the real projective spaces $\mathbb{RP}^n$, the space forms $\mathbb{S}^{2m+1}/\Gamma$ where the action of Γ preserves some proper orthogonal decomposition of $\mathbb{R}^{2m+2}$, and the spaces $\mathbb{CP}^{2m+1}/\mathbb{Z}_2$ where the $\mathbb{Z}_2$-action is given by the antipodal maps in the fibers of some standard projection $\mathbb{CP}^{2m+1} \to \mathbb{HP}^m$.

Because of the structure of their cohomology rings, the spaces $\mathbb{CP}^{2m}$, $\mathbb{HP}^{2m}$, and $\mathrm{Ca}\mathbb{P}^2$ do not admit any orientation-reversing homeomorphisms. With a little more effort one can also show that none of the spaces $\mathbb{HP}^{2m+1}$, where $m \geq 1$, admits a smooth, orientation-reversing, fixed-point-free $\mathbb{Z}_2$-action. For this purpose one verifies that the projection $\pi : \mathbb{HP}^{2m+1} \to \mathbb{HP}^{2m+1}/\mathbb{Z}_2$ onto the hypothetical quotient space induces the zero-map on the fourth integral cohomology groups. Since $T\,\mathbb{HP}^{2m+1} = \pi^*T(\mathbb{HP}^{2m+1}/\mathbb{Z}_2)$, it follows that the first Pontrjagin class $p_1(\mathbb{HP}^{2m+1})$ should vanish, contradicting the fact that $p_1(\mathbb{HP}^k) = 2(k-1)\xi$ where ξ is a generator of $H^4(\mathbb{HP}^k) \cong \mathbb{Z}$ [Greub et al. 1973, Chapter IX, Problem 31(ii)]. In the quaternionic case, the existence of quotients can alternatively be ruled out by observing that the full group of isometries $\operatorname{Isom}\mathbb{HP}^k$ is connected for $k \geq 2$ [Wolf 1977, p. 381].

ON THE PROOF OF THE DIAMETER RIGIDITY THEOREM. The starting point is to observe that the cut locus of a subspace $\mathbb{CP}^k \subset \mathbb{CP}^m$ is the dual subspace $\mathbb{CP}^{m-k-1} \subset \mathbb{CP}^m$. Moreover, the pairs consisting of such a $\mathbb{CP}^k \subset \mathbb{CP}^m$ and its cut locus can be characterized geometrically as pairs of dual convex sets $A, A' \subset \mathbb{CP}^n$. This means that $\operatorname{dist}(p, p') = \operatorname{diam}\mathbb{CP}^n$ for any pair of points $(p, p') \in A \times A'$.

If $k = 0$, that is, if A' consists of a single point, it is possible to recover the total space $\mathbb{CP}^m$ as the Thom space of the normal bundle of $A \subset \mathbb{CP}^m$. In particular, the total space of the corresponding unit sphere bundle is an $\mathbb{S}^{2m-1}$ that is foliated by equidistant circles. Similar structures can be found on the quaternionic projective spaces $\mathbb{HP}^m$ and on the Cayley plane $\mathbb{CaP}^2$. The only differences are that the total spaces of the unit normal bundles of A are spheres $\mathbb{S}^{4m-1}$ foliated by three-spheres and an $\mathbb{S}^{15}$ foliated by seven-spheres, respectively.

In fact, when we normalize λ to 1, the diameter sphere theorem reduces the proof of Theorem 2.3 to the study of compact Riemannian manifolds with $K_M \geq 1$ and $\operatorname{diam} M^n = \frac{1}{2}\pi$. The first step in investigating this setup is to analyze the structure of dual convex sets $A, A' \subset M^n$ by means of critical point theory, establishing more and more of the properties described above. In particular, A and A' are totally geodesic submanifolds, and for any point $p' \in A'$ the exponential map defines a Riemannian submersion π_A from the unit normal sphere in p' to A. It also follows that A is a deformation retract of $M^n \smallsetminus A'$ and vice versa. By the latter assertion, at least one of the sets A or A' is not contractible unless M^n is homeomorphic to a sphere.

On the other hand, if A' were contractible, it would consist of a single point $p' \in M^n$. In this case the manifold M^n is the mapping cone of π_A, and the fibers of this submersion are homotopy spheres of dimensions 1, 3, or 7. Now the key point is to resort to the results about low-dimensional metric foliations of Euclidean spheres in [Gromoll and Grove 1988] to conclude that π_A is isometric to a standard Hopf fibration, except possibly when $n - 1 = 15$ and $\dim A = 8$. Clearly, the isometry between π_A and such a standard Hopf fibration induces a continuous map between the corresponding mapping cones, that is, it gives rise to a map from M^n to $\mathbb{CP}^{n/2}$ or $\mathbb{HP}^{n/4}$.

In order to prove the result for simply connected manifolds M^n, it remains to show that the latter map remains an isometry and, moreover, that there exists a pair of dual convex sets $A, A' \subset M^n$ such that A' is contractible. Both steps are accomplished by an argument that uses recursively the concepts presented so far. Finally, the case where M^n is not simply connected is reduced to the preceding one by means of covering theory.   □

The problems in recovering the Cayley plane in Theorem 2.3 up to isometry are due to some shortcomings in understanding metrical foliations of Euclidean spheres. Recently, F. Wilhelm [1995] has treated the case of the Cayley plane under more restrictive geometric conditions that yield better information about the structure of the family of dual convex sets in M^{16}:

THEOREM 2.4 (RADIUS RIGIDITY THEOREM). *Let M^n be a connected, complete Riemannian manifold with sectional curvature $K_M \geq \lambda > 0$ and radius* $\operatorname{rad} M^n \geq \pi/(2\sqrt{\lambda})$. *Then*

(i) *M^n is homeomorphic to the sphere $\mathbb{S}^n$, or*
(ii) *the universal covering $\tilde{M}^n$ of M^n is isometric to $\mathbb{S}^n$, $\mathbb{CP}^{n/2}$, $\mathbb{HP}^{n/4}$, or $\mathbb{CaP}^2$.*

Recall that the radius $\operatorname{rad} M^n$ of a compact, connected Riemannian manifold is defined as the infimum of the function $p \mapsto \operatorname{rad}_M(p) := \max_{q \in M^n} \operatorname{dist}(p, q)$. Clearly, $\operatorname{inj} M^n < \operatorname{rad} M^n \leq \operatorname{diam} M^n$. If M^n is a compact, simply connected, rank-one symmetric space, all three quantities coincide.

3. On Berger's Pinching Below-$\frac{1}{4}$ Theorem

Since the early sixties it had been a challenging problem to find out whether there is a stability result extending Berger's rigidity theorem. An affirmative answer was only found in 1983.

THEOREM 3.1 [Berger 1983]. *For any even number n there exists a constant $\delta_n < \frac{1}{4}$ such that any n-dimensional, complete, simply connected Riemannian manifold M^n with $\delta_n \leq K_M \leq 1$ is either*

(i) *homeomorphic to the sphere $\mathbb{S}^n$, or*
(ii) *diffeomorphic to $\mathbb{CP}^{n/2}$, $\mathbb{HP}^{n/4}$, or $\mathbb{CaP}^2$.*

Up to now there has been no analogous theorem where the upper curvature bound is replaced by a corresponding lower bound for the diameter. Results in this direction by O. Durumeric [1987] involve some additional hypotheses.

In contrast to the pinching theorems discussed so far, the proof of Theorem 3.1 does not provide an explicit constant δ_n, because it relies on the *precompactness* of certain spaces of isometry classes of Riemannian manifolds.

Our plan is to summarize the precompactness result, explaining in particular the injectivity radius estimates that are required in this context. We conclude this section by describing how Theorem 3.1 can be deduced from a rigidity theorem that extends Theorem 2.1.

In order to discuss precompactness, one needs a topology on the class of connected Riemannian manifolds. In this context there are actually two natural topologies, the Hausdorff topology and the Lipschitz topology. Both are defined in terms of appropriate distance functions on the class of inner metric spaces. Recall that a connected Riemannian manifold (M^n, g) can be considered as an inner metric space (M, d), where d denotes the Riemannian distance function corresponding to the metric g.

The Hausdorff distance between two inner metric spaces (M_μ, d_μ) is defined as an infimum over all isometric embeddings $\iota_\mu : (M_\mu, d_\mu) \to (X, d)$ into some bigger metric space:

$$\text{dist}_H\big((M_1, d_1), (M_2, d_2)\big) := \inf_{\iota_\mu : M_\mu \to X} d_H\big(\iota_1(M_1), \iota_2(M_2)\big),$$

where d_H denotes the Hausdorff distance of the closed subsets $\iota_1(M_1)$ and $\iota_2(M_2)$ within the metric space (X, d):

$$d_H\big(\iota_1(M_1), \iota_2(M_2)\big) = \inf\big\{\varepsilon > 0 \mid \iota_1(M_1) \subset U_\varepsilon(\iota_2(M_2)), \ \iota_2(M_2) \subset U_\varepsilon(\iota_1(M_1))\big\}.$$

The Lipschitz distance of (M_1, d_1) and (M_2, d_2), on the other hand, is defined as an infimum over the class of all bijective maps $f : M_1 \to M_2$:

$$\text{dist}_L\big((M_1, d_1), (M_2, d_2)\big) := \inf_{f : M_1 \to M_2} \log_+(\text{dil} f) + \log_+(\text{dil} f^{-1}),$$

where $\log_+(x) := \sup\{0, \log(x)\}$ and

$$\text{dil}(f) := \sup_{p \neq q} \frac{d_2(f(p), f(q))}{d_1(p, q)}.$$

Note that $\text{dist}_L((M_1, d_1), (M_2, d_2)) = +\infty$ unless M_1 and M_2 are homeomorphic.

In the presence of a uniform bound for the diameter it is not hard to show that any two inner metric spaces that are Lipschitz close have small Hausdorff distance, too. The converse is not true in such generality, as the example consisting of a finite graph $X \subset \mathbb{R}^3$ and its distance tubes $U_\varepsilon(X)$ shows. The graph X and its tubes $U_\varepsilon(X)$ are not homeomorphic, so $\text{dist}_L(X, U_\varepsilon(X)) = +\infty$ despite the fact that $\text{dist}_H(X, U_\varepsilon(X)) \to 0$ as $\varepsilon \to 0$.

However, Gromov [1981a] has obtained the following result for the space $\mathfrak{M}_{\lambda \ . \ v}^{\Lambda \ D \ .}(n)$ consisting of all isometry classes of compact, n-dimensional Riemannian manifolds (M^n, g) with $\lambda \leq K_M \leq \Lambda$, $\text{diam} M^n \leq D$, and $0 < v \leq \text{vol} M^n$ (compare also [Peters 1987]):

THEOREM 3.2 (GROMOV'S COMPACTNESS THEOREM). *Let $n \in \mathbb{N}$, $\lambda \leq \Lambda$, and $v, D > 0$. Then, on $\mathfrak{M} := \mathfrak{M}_{\lambda \ . \ v}^{\Lambda \ D \ .}(n)$, the Hausdorff and the Lipschitz topologies coincide. Furthermore, $\mathfrak{M}$ is relatively compact in the space of isometry classes of $C^{1,1}$-manifolds with C^0-metrics g.*

REMARK 3.3. An upper bound for the number of diffeomorphism types of the Riemannian manifolds corresponding to the points in $\mathfrak{M}$ had been obtained earlier by Cheeger. This result is known as *Cheeger's finiteness theorem*.

It is possible to recover Cheeger's finiteness theorem, which is a refinement of a finiteness theorem for homotopy types due to A. Weinstein [1967], from Gromov's compactness theorem, except for the precise value of the upper bound, of course. In fact, by Shikata's work [1966] on the differentiable sphere theorem, there exists for any $n \in \mathbb{N}$ a constant $\varepsilon_n > 0$ such that any two n-dimensional Riemannian manifolds (M_μ^n, g_μ) with $\operatorname{dist}_L\big((M_1^n, d_1), (M_2^n, d_2)\big) < \varepsilon_n$ are diffeomorphic. By Gromov's compactness theorem it is possible to cover the space $\mathfrak{M}$ with finitely many dist_L-balls of radius ε_n. Thus there are only finitely many diffeomorphism types among the compact, n-dimensional Riemannian manifolds with $\lambda \le K_M \le \Lambda$, $\operatorname{diam} M^n \le D$, and $0 < v \le \operatorname{vol} M^n$.

REMARKS 3.4. (i) The lens spaces $\mathbb{S}^{2m-1}/\mathbb{Z}_k$ show that the lower volume bound is crucial for Cheeger's finiteness theorem and hence also for Gromov's compactness theorem. Moreover, in dimension 7 the Aloff–Wallach examples [1975] discussed in the appendix provide a family of counterexamples that are not just coverings of each other.

(ii) It is even easier to see that the upper diameter bound is necessary. One simply considers ladders, that is, connected sums of an unbounded number of copies of the same topologically nontrivial manifold.

(iii) Similarly, the lower bound for the sectional curvature turns out to be essential, whereas the upper bound for K_M was discarded in a later finiteness theorem by Grove, P. Petersen, and J.-Y. Wu [Grove et al. 1990].

Cheeger's finiteness theorem can indeed be viewed as an immediate predecessor of Gromov's compactness theorem. The proofs of both rely on a particular injectivity radius estimate, which, in contrast to Theorems 1.5 and 1.6, must not impose any restriction on the sign of the sectional curvature K_M. As explained in Remark 1.7, such an injectivity radius estimate requires some further geometrical hypotheses in addition to the bounds for K_M. The first result in this direction was Cheeger's propeller lemma [Cheeger and Ebin 1975, Theorem 5.8]; we give a version with improved numerical constants, due to E. Heintze and H. Karcher [1978, Corollary 2.3.2]:

PROPOSITION 3.5 (PROPELLER LEMMA). *Let M^n be a complete Riemannian manifold with $\lambda \le K_M$, $\operatorname{diam} M^n \le D$, and $\operatorname{vol} M^n \ge v > 0$. Then the injectivity radius of M^n is bounded from below by*

$$\operatorname{inj} M^n \ge \inf\left\{ \operatorname{conj} M^n, \ \frac{\pi v}{\operatorname{vol} \mathbb{S}^n} \operatorname{sn}_\lambda\big(\min\{D, \pi/(2\sqrt{\lambda})\}\big)^{-(n-1)} \right\},$$

where $\operatorname{vol} \mathbb{S}^n$ is the volume of the unit sphere in $(n+1)$-dimensional Euclidean space.

PROOF OF PROPOSITION 3.5. The idea is to use Lemma 1.8 in order to reduce the assertion to a simple volume computation. One considers distance tubes $U_r(\gamma)$ around a closed geodesic γ of length $\ell(M^n) = \inf_p \ell_M(p)$. Their volumes can be estimated as follows:

$$\operatorname{vol} U_r(\gamma) \le \ell(M^n) \operatorname{vol}(\mathbb{S}^{n-2}) \int_0^r \operatorname{sn}_\lambda(\varrho)^{n-2} \operatorname{cn}_\lambda(\varrho) \, d\varrho$$

$$= \ell(M^n) \operatorname{vol}(\mathbb{S}^{n-2}) \frac{1}{n-1} \operatorname{sn}_\lambda(r)^{n-1} = \ell(M^n) \frac{1}{2\pi} \operatorname{vol}(\mathbb{S}^n) \operatorname{sn}_\lambda(r)^{n-1} .$$

Finally, one observes that the closed tube $\bar{U}_r(\gamma)$ covers M^n for $r \ge \operatorname{diam} M^n$ or for $r \ge \pi/(2\sqrt{\lambda})$ if $\lambda > 0$. $\qquad\square$

Roughly speaking, the propeller lemma asserts that, in the presence of a lower sectional curvature bound and an upper diameter bound, giving a lower volume bound is equivalent to giving a lower injectivity radius bound. We have followed the approach of Heintze and Karcher, since it invokes the lower volume bound in a more intuitive way than Cheeger's original approach, which gave rise to the name of the result. As a word of caution, the upper bound for $\operatorname{vol} U_r(\gamma)$ used in the proof does not remain valid when the lower bound for the sectional curvature K_M is replaced by the corresponding lower bound for the Ricci curvature [Anderson 1990].

ON THE PROOF OF CHEEGER'S FINITENESS THEOREM. The basic idea is to cover a Riemannian manifold M^n with balls $B_i = B(p_i, 2\varrho)$ such that the concentric balls of radius ϱ are disjoint, and to consider the nerve complex corresponding to this covering. Since the conjugate radius of $M^n \in \mathfrak{M}$ is bounded below by $\pi/\sqrt{\Lambda}$ if $\Lambda > 0$ and is $+\infty$ otherwise, Proposition 3.5 provides a uniform lower bound for the injectivity radius on the whole class of Riemannian manifolds.

The idea is to work with some radius ϱ that is a small fraction of the preceding injectivity radius bound. Thus the balls B_i are actually topological balls, and the edges of the nerve complex correspond to minimizing geodesics $\gamma_{ij} : [0, 1] \to M^n$ from p_i to p_j. Picking an orthogonal frame $(e_i^\nu)_{\nu=1}^n$ at the center p_i of each ball in our covering, the edges of the nerve complex can be labeled by the length of γ_{ij} and by the orthogonal transformation that maps the frame $(e_i^\nu)_{\nu=1}^n$ to the frame $(e_j^\nu)_{\nu=1}^n$, when the tangent spaces $T_{p_i}M$ and $T_{p_j}M$ are identified by means of parallel transport along γ_{ij}.

The combinatorial properties of the nerve complex can be controlled using the relative volume comparison theorem. More sophisticated arguments from comparison geometry show in addition that two Riemannian manifolds in $\mathfrak{M}$ that admit combinatorially equivalent nerve complexes are diffeomorphic if the labelings of these complexes are sufficiently close. $\qquad\square$

In some sense, Cheeger's finiteness theorem may be regarded as a vast generalization of the topological sphere theorem. For instance, the counterpart of the nerve complex appearing in the proof of the topological sphere theorem is a

complex consisting of two vertices joined by one edge. The vertices correspond
to the two balls in the description of a twisted sphere.

REMARK 3.6. S. Peters [1987] has given a proof of Gromov's compactness
theorem following the approach to Cheeger's finiteness theorem described above.
He used harmonic coordinates to construct the diffeomorphism between two
manifolds in $\mathfrak{M}$ whose labeled nerve complexes are sufficiently close. In this way
he actually proved that the spaces $\mathfrak{M}$ are relatively compact in the $C^{2,\alpha}$-topology.
As observed by I. Nikolaev [1983], the limiting objects are $C^{3,\alpha}$-manifolds with
Riemannian metrics of class $C^{1,\alpha}$. They have curvature bounds λ and Λ in
distance comparison sense.

Before we finish this section with an outline of the proof of Theorem 3.1, we
present another injectivity radius estimate, which extends the basic idea behind
Proposition 3.5. In fact, even for complete, noncompact manifolds, it is possible
to relate a lower injectivity radius bound to some lower volume bound, provided
one "localizes" the relevant geometric quantities appropriately:

THEOREM 3.7 [Cheeger et al. 1982, Theorem 4.7]. *Consider two points p_0 and p
in a connected, complete Riemannian manifold M^n with $\lambda \leq K_M \leq \Lambda$. Furthermore, let $r_0, r > 0$. Suppose that $r < \pi/(4\sqrt{\Lambda})$ if $\Lambda > 0$. Then the injectivity
radius at the point p can be bounded from below as follows*:

$$\operatorname{inj}_M(p) \geq r \, \frac{\operatorname{vol} B(p,r)}{\operatorname{vol} B(p,r) + V_\lambda^n(2r)}$$

$$\geq r \, \frac{V_\lambda^n(r) \operatorname{vol} B(p_0, r_0)}{V_\lambda^n(r) \operatorname{vol} B(p_0, r_0) + V_\lambda^n(2r) V_\lambda^n(\hat{r})}, \qquad (3.1)$$

*where $\hat{r} := \max\{r, r_0 + \operatorname{dist}(p_0, p)\}$, and where $V_\lambda^n(\varrho)$ denotes the volume of a
ball of radius ϱ in the n-dimensional model space M_λ^n with constant sectional
curvature λ.*

REMARK 3.8. In [Cheeger et al. 1982] one can find even more refined versions
of the preceding theorem. Here, however, we prefer to point out one important
special case: if we choose the parameter r_0 as the injectivity radius at the point
$p_0 \in M^n$, inequality (3.1) turns into the following *relative injectivity radius
estimate*:

$$\operatorname{inj}_M(p) \geq \sup_{0 < r < \pi/(4\sqrt{\Lambda})} r \, \frac{V_\lambda^n(r) V_\lambda^n(r_0)}{V_\lambda^n(r) V_\lambda^n(r_0) + V_\lambda^n(2r) V_\lambda^n(\hat{r})}.$$

In other words, $\operatorname{inj}_M(p) \geq \varphi_{n,\lambda,\Lambda}\big(\operatorname{inj}_M(p_0), \operatorname{dist}(p_0, p)\big)$, where $\varphi_{n,\lambda,\Lambda}$ is a universal, strictly positive function that depends only on the dimension n and on
the curvature bounds λ and Λ. This is the way the result is stated in [Gromov
1981a, Proposition 8.22]. Moreover, this is the version typically used when studying degenerate limits where the dimension of the Riemannian manifolds drops.

Extending Theorem 3.2, a whole theory of *collapsing Riemannian manifolds* has been developed.

ON THE PROOF OF THEOREM 3.7. As far as the proof is concerned, the first inequality in (3.1) is the central assertion of the theorem, whereas the second comes almost free, as a direct consequence of the relative volume comparison theorem. Yet it is the second inequality that is crucial for controlling the injectivity radius $\mathrm{inj}_M(p)$ at points $p \in M^n$ far away from the base point p_0.

The link between $\mathrm{inj}_M(p)$ and $\mathrm{vol}\, B(p, r)$ described in the first inequality in (3.1) is a purely local result. By hypothesis, the conjugate radius of M^n is $\geq 4r$, and by Lemma 1.8 it is therefore sufficient to show that the length $\ell_M(p)$ of the shortest nontrivial geodesic loop at p is bounded from below as follows:

$$\ell_M(p) \geq 2r \frac{\mathrm{vol}\, B(p, r)}{\mathrm{vol}\, B(p, r) + V_\lambda^n(2r)}. \tag{3.2}$$

The idea for proving this inequality is to compare the geometry of the ball $B(p, 4r) \subset M^n$ with the geometry of its local unwrapping $\tilde{B}_{4r}$, which is the ball $\tilde{B}(0, 4r) \subset T_p M$ equipped with the metric $\exp_p^* g$. The exponential map provides a length-preserving local diffeomorphism $\exp_p : \tilde{B}_{4r} \to B(p, 4r) \subset M^n$.

Let $\tilde{p}_1 = 0$ and let $\tilde{p}_2, \ldots, \tilde{p}_N$ be the various preimages of p in the domain $\tilde{B}_r \subset \tilde{B}_{4r}$. They correspond bijectively to the geodesic loops $\gamma_1, \ldots, \gamma_N$ of length $< r$ at p. Clearly, γ_1 is the trivial loop. Furthermore, for each point $\tilde{p}_i$ there exists precisely one isometric immersion $\varphi_i : \tilde{B}_r \to \tilde{B}_{4r}$ mapping 0 to $\tilde{p}_i$ and such that $\exp_p \circ \varphi_i = \exp_p$. Without loss of generality we may assume that $L(\gamma_2) = \ell_M(p)$ is the minimal length of a nontrivial loop at p.

Analyzing short homotopies, one can show that the maps φ_i constitute a pseudogroup of local covering transformations, hence $\varphi_i(\tilde{q}) \neq \varphi_j(\tilde{q})$ for $1 \leq i < j \leq N$ and for all $\tilde{q} \in \tilde{B}_r$. This fact has two implications:

First, $N \geq 2m + 1$, where $m := [r/\ell_M(p)]$. More precisely, we claim that the points $\varphi_2^\mu(\tilde{p}_1)$, for $-m \leq \mu \leq m$, are distinct preimages of p in $\tilde{B}_r$. For otherwise φ_2 would act as a permutation on the set $\{\varphi_2^\mu(\tilde{p}_1) \mid -m \leq \mu \leq m\}$. But this set has a unique center of mass $\tilde{q} \in \tilde{B}_{2r}$. Actually, $\tilde{q}$ lies in $\tilde{B}_r$, so $\tilde{q} = \varphi_2(\tilde{q})$, in contradiction with the fact that φ_2 is a local covering transformation.

Secondly, each point in $B(p, r)$ has at least N preimages in $\bigcup_{i=1}^N B(\tilde{p}_i, r) \subset \tilde{B}_{2r}$. Hence $N \, \mathrm{vol}\, B(p, r) \leq \mathrm{vol}\, \tilde{B}_{2r} \leq V_\lambda^n(2r)$, and inequality (3.2) follows upon combining this estimate with the fact that $N \geq 2 \, [r/\ell_M(p)] + 1$. $\square$

ON THE PROOF OF THEOREM 3.1. The basic idea is to pursue an indirect approach. If the theorem were false, there would exist a sequence $(M_j^n, g_j)_{j=1}^\infty$ of complete, simply connected Riemannian manifolds with $\frac{1}{4} \frac{j}{j+1} \leq K_{M_j} \leq 1$ such that none of these manifolds is diffeomorphic to a sphere Σ^n, possibly with an exotic differentiable structure, or to one of the projective spaces $\mathbb{CP}^{n/2}$, $\mathbb{HP}^{n/4}$, or $\mathrm{Ca}\mathbb{P}^2$.

By Myers' theorem diam $M_j^n \leq 2\pi\sqrt{1 + 1/j} < 3\pi$. Since we are in the even-dimensional case, we can apply Theorem 1.5 to conclude that inj $M_j^n \geq \pi$. Therefore the volume of each M_j^n is bounded from below by the volume of the standard sphere $\mathbb{S}^n$. (See Proposition 3.5 and Theorem 3.7 for more information about the equivalence of a lower volume bound and a lower injectivity radius bound.)

Hence Theorem 3.2 asserts that the manifolds M_j^n in our sequence belong to finitely many diffeomorphism types. One of these types must appear infinitely often; restricting ourselves to the corresponding subsequence, we are dealing in fact with a sequence of Riemannian metrics $\phi_j^* g_j$ on a fixed compact manifold M^n. Furthermore, the compactness theorem asserts that there exists a subsequence $(\phi_{j_\nu}^* g_{j_\nu})_{\nu=1}^\infty$ that converges in the $C^{1,\alpha}$-topology. As a limit space we thus obtain a compact, simply connected $C^{3,\alpha}$-manifold M^n that is neither homeomorphic to a sphere nor diffeomorphic to one of the projective spaces $\mathbb{CP}^{n/2}$, $\mathbb{HP}^{n/4}$, or $\mathrm{Ca}\mathbb{P}^2$, and that carries a Riemannian metric g of class $C^{1,\alpha}$ with $\frac{1}{4} \leq K \leq 1$ in distance comparison sense.

Thus it remains to extend Berger's rigidity theorem so it holds under the weak regularity properties of the limit spaces that appear in Gromov's compactness theorem. It is by no means a priori clear whether or not such an extension of Theorem 2.1 exists, since the smoothness of the Riemannian metric has been used in the arguments in [Berger 1960a] in a significant way. For instance, the regularity properties of the metric really matter in the theory of Riemannian manifolds with $-1 \leq K_M \leq 0$ and vol $M^n < \infty$ where the smooth category exhibits vastly more phenomena than the real analytic category. Concerning Berger's rigidity theorem, however, the alternate proof given by Cheeger and Ebin [1975] is much more robust than the original proof. Following their approach to some extent, Berger succeeded in eliminating all arguments that still required smoothness, replacing them by purely metric constructions [Berger 1983].

More precisely, he is able to recognize the cut locus C_p of an arbitrary point $p \in M^n$ as a k-dimensional, totally geodesic submanifold. At the same time the projective lines are recovered as totally geodesically embedded spheres of curvature 1 and of the complementary dimension $n - k$. Furthermore, the cut locus C_p is shown to have the property that any closed geodesic $c : \mathbb{R}/\mathbb{Z} \to C_p$ of length $L(c) = 2\pi$ spans a totally geodesic $\mathbb{RP}^2 \subset M^n$ of curvature $\frac{1}{4}$. Combining all this information, he can then construct the geodesic symmetries $\phi_p : M^n \to M^n$ directly, recognizing the limit space M^n as a symmetric space and thus a posteriori as a smooth Riemannian manifold. $\square$

4. An Improved Injectivity Radius Estimate

Already before Berger discussed the metrics g_ϵ of Remark 1.7(ii) on odd-dimensional spheres, it had been considered an interesting question whether or not the pinching constant in Klingenberg's injectivity radius estimate for simply connected, odd-dimensional manifolds could be improved. The extension of Theorem 1.6 to weakly quarter-pinched manifolds was summarized in our discussion of Berger's rigidity theorem in Remark 2.2.

With Berger's pinching below-$\frac{1}{4}$ theorem the problem became even more intriguing. Nevertheless, the first result in this direction was achieved only very recently:

THEOREM 4.1 (INJECTIVITY RADIUS ESTIMATE [Abresch and Meyer 1994]). *There exists a constant $\delta_{\mathrm{inj}} \in (0.117, 0.25)$ such that the injectivity radius $\mathrm{inj}\, M^n$ and the conjugate radius $\mathrm{conj}\, M^n$ of any compact, simply connected Riemannian manifold M^n with δ_{inj}-pinched sectional curvature coincide:*

$$\mathrm{inj}\, M^n = \mathrm{conj}\, M^n \geq \pi/\sqrt{\max K_M}\,.$$

The pinching constant δ_{inj} in this result is explicit and independent of the dimension. In fact, the theorem holds for $\delta_{\mathrm{inj}} = \frac{1}{4}(1 + \varepsilon_{\mathrm{inj}})^{-2}$, where $\varepsilon_{\mathrm{inj}} = 10^{-6}$. Its proof is based on direct comparison methods, not involving the concept of convergence of Riemannian manifolds. Yet the constant δ_{inj} obtained by this method is by no means optimal, since the argument involves several curvature-controlled estimates that are not simultaneously sharp. Currently there is not even a natural candidate for the optimal value of the pinching constant δ_{inj} in the preceding theorem. The Berger metrics described in Remark 1.7(ii) merely show that the number must be at least $0.117 > \frac{1}{9}$.

Notice that the conclusion $\mathrm{inj}\, M^n = \mathrm{conj}\, M^n$ is best possible, and in this respect the result can be considered as a natural generalization of Klingenberg's injectivity radius estimate in Theorem 1.6. In particular, the preceding estimate can be used not only to justify the extension of Berger's rigidity theorem to odd-dimensional manifolds, but to yield a corresponding extension of the pinching below-$\frac{1}{4}$ theorem:

THEOREM 4.2 (SPHERE THEOREM [Abresch and Meyer 1994]). *For any odd integer $n > 0$ there exists some constant $\delta_n \in (0, \frac{1}{4})$ such that any complete, simply connected Riemannian manifold M^n with δ_n-pinched sectional curvature K_M is homeomorphic to the sphere $\mathbb{S}^n$.*

Here, as in Berger's pinching below-$\frac{1}{4}$ theorem, the pinching constants $\delta_n \in (0, \frac{1}{4})$ are not explicit, and there is no reason why they should not approach $\frac{1}{4}$ as the dimension n gets large. The proof of Theorem 4.2 relies on the same convergence methods as the proof of Berger's pinching below-$\frac{1}{4}$ theorem, discussed in the previous section. The details for the odd-dimensional case have been worked out by Durumeric [1987]. His result requires some uniform lower bound for the injectivity radius as an additional hypothesis. Such a bound is now provided by Theorem 4.1.

In the remainder of this section we *explain the proofs* of the injectivity radius estimates for odd-dimensional, simply connected manifolds with $\delta \leq K_M \leq 1$, as stated in Theorems 1.6 and 4.1. By Myers' theorem one has a diameter bound in terms of the positive lower bound on the sectional curvature. Nevertheless, in the absence of a lower volume bound, the arguments for establishing Theorems 1.6 and 4.1 must be very different from the proofs of Cheeger's propeller lemma (Proposition 3.5) and Theorem 3.7. As in the even-dimensional case, the hypothesis that the manifold M^n be simply connected must be used in a significant way, and, as in the proof of Theorem 1.5, the starting point is to conclude from Lemma 1.8 that it is sufficient to rule out the existence of a closed geodesic $c_0 : \mathbb{R}/\mathbb{Z} \to M^n$ of length $L(c_0) < 2 \operatorname{conj} M^n$.

However, it is not possible to use Synge's lemma. Much more sophisticated global arguments, based on a combination of lifting constructions and Morse theory, are needed. The following result essentially goes back to [Klingenberg 1962, p. 50].

LEMMA 4.3 (LONG HOMOTOPY LEMMA). *Let (M^n, g) be a compact Riemannian manifold, and let $c_t : \mathbb{R}/\mathbb{Z} \to M^n$, for $0 \leq t \leq 1$, be a continuous family of rectifiable, closed curves such that*

(i) *c_0 is a nontrivial geodesic digon of length $L(c_0) < 2 \operatorname{conj} M^n$, and*
(ii) *$c_1 : \mathbb{R}/\mathbb{Z} \to \{c_1(0)\} \subset M^n$ is a constant curve.*

Then this family contains a curve c_τ of length $L(c_\tau) \geq 2 \operatorname{conj} M^n$.

Here, in contrast to Klingenberg's original version of the lemma, the family $(c_t)_{t \in [0,1]}$ can be any free null homotopy of c_0.

PROOF. The idea is to proceed indirectly and assume that $L(c_t) < 2 \operatorname{conj} M^n$ for all $t \in [0, 1]$. Without loss of generality we may suppose that the curves c_t are parametrized proportional to arclength.

We consider the family of curves as a continuous map $c : \mathbb{R} \times [0, 1] \to M^n$ such that $c(s + 1, t) = c(s, t) = c_t(s)$ and such that $c(0, 0)$ is a vertex of the geodesic digon. Then the segments $c_t|_{[-\frac{1}{2}, 0]}$ and $c_t|_{[0, \frac{1}{2}]}$ are strictly shorter than $\operatorname{conj} M^n$, and hence they can be lifted under $\exp_{c_t(0)}$. In this way one obtains a continuous map $\tilde{c} : [-\frac{1}{2}, \frac{1}{2}] \times [0, 1] \to TM$ such that $\exp \circ \tilde{c}(s, t) = c(s, t)$ and $\tilde{c}(0, t) = 0 \in T_{c_t(0)} M$. Since $\tilde{c}(s, 1) = 0$, it follows in particular that $\tilde{c}(-\frac{1}{2}, t) = \tilde{c}(\frac{1}{2}, t)$ for all $t \in [0, 1]$.

Since c_0 is a geodesic digon, it is clear that one of its arcs $c_0|_{[-\frac{1}{2}, 0]}$ or $c_0|_{[0, \frac{1}{2}]}$ lifts to a radial straight line segment in $T_{c_0(0)}$ of length $\frac{1}{2} L(c_0)$. The other arc lifts to a curve of equal length with the same end points, and hence their images coincide, contradicting the hypothesis that c_0 is a nontrivial digon. $\square$

PROOF OF THEOREM 1.6. Proceeding indirectly, we assume that $\operatorname{inj} M^n < \operatorname{conj} M^n$. By Lemma 1.8 one finds a closed geodesic $c_0 : \mathbb{R}/\mathbb{Z} \to M^n$ with $L(c_0) < 2 \operatorname{conj} M^n$. Since M^n is simply connected, there exist piecewise smooth free homotopies $c = (c_t)_{0 \leq t \leq 1}$ beginning at c_0 and ending at some constant curve c_1.

Recall that the energy functional and the length functional are related by the inequality $E(c_t) = \frac{1}{2} \int_{\mathbb{R}/\mathbb{Z}} |c_t'(s)|^2 \, ds \geq \frac{1}{2} L(c_t)^2$. Thus it follows from Lemma 4.3 that for any free homotopy c from c_0 to a constant curve the map $t \mapsto E(c_t)$ achieves its maximum at some $t_0 \in (0,1)$ and that, moreover,

$$E_{\min}(c_0) := \inf_c \max_{0 \leq t \leq 1} E(c_t) \geq 2 \big(\operatorname{conj} M^n\big)^2.$$

Let $(c^j)_{j=1}^\infty$ be a minimizing sequence of such homotopies, and let $t_{0,j}$ be parameters such that $E\big(c_{t_{0,j}}^j\big) = \max_{0 \leq t \leq 1} E(c_t^j)$. Clearly, $E\big(c_{t_{0,j}}^j\big) \to E_{\min}(c_0)$ as $j \to \infty$. Since the energy functional E on the free loop space satisfies Condition C of Palais and Smale, a subsequence of the curves $c_{t_{0,j}}^j$ converges towards a closed geodesic $\bar{c}_0 : \mathbb{R}/\mathbb{Z} \to M^n$ of length $L(\bar{c}_0) = \sqrt{2E_{\min}(c_0)} \geq 2 \operatorname{conj} M^n$ and Morse index $\operatorname{ind}_E(\bar{c}_0) \leq 1$. In this generality, the assertion about the Morse indices of the limiting geodesics obtained by the preceding minimax construction requires the degenerate Morse lemma from [Gromoll and Meyer 1969], since we have not made any attempt to perturb the Riemannian metric and to ensure that the energy functional E is a nondegenerate Morse function.

The contradiction appears when we look at the index form of the closed geodesic $\bar{c}_0$. The idea is to evaluate

$$I(Y,Y) = \int_{\mathbb{R}/\mathbb{Z}} \left| \frac{\nabla}{ds} Y \right|^2 - \langle R(Y, \bar{c}_0') \bar{c}_0', Y \rangle \, ds$$

on closed unit vector fields $v_1, \ldots, v_n$ that rotate with constant angular velocity and that are pairwise orthogonal. Since $K_M > \frac{1}{4}$ and $L(\bar{c}_0) \geq 2 \operatorname{conj} M^n \geq 2\pi$, it is indeed not hard to see that $\operatorname{ind}_E(\bar{c}_0) \geq n-1 \geq 2$. □

In the preceding proof we have used the degenerate Morse lemma, avoiding the bumpy metrics theorem [Abraham 1970], since we want to explain the additional difficulties that arise in the weakly quarter-pinched case. It should be pointed out that the standard minimax construction used in the proof yields a limiting geodesic $\bar{c}_0$ but nothing like a limiting homotopy $\bar{c}$.

REMARK 4.4. A similar argument can be given in the space $\Omega_{pp}M$ of loops with base point $p \in M^n$. In this case the minimax construction leads to a geodesic loop $\bar{c}_0$ of length $L(\bar{c}_0) \geq 2\pi$, and in order to estimate the index form one must use test fields Y that vanish at the initial and end points of $\bar{c}_0$. In this context it is customary to test the index form with vector fields obtained as the product of a parallel normal field and an appropriately transformed sine function. One concludes that the Morse index of $\bar{c}_0$ in $\Omega_{pp}M$ is $\geq n-1$, too.

This approach is actually pretty close to Klingenberg's original proof of Theorem 1.6. However, neither the degenerate Morse lemma nor the bumpy metrics theorem were known in 1961. Klingenberg's way out was to work in some path space $\Omega_{pq}M$, picking an end point $q \in M^n$ close to p such that in particular the energy functional E has only nondegenerate critical points on $\Omega_{pq}M$. This approach relies on the fact that the long homotopy lemma can be applied to nontrivial digons c_0 rather than merely to closed geodesics.

ON THE EXTENSION OF THE PROOF TO WEAKLY $\frac{1}{4}$-PINCHED MANIFOLDS. In the limiting case, where the simply connected, odd-dimensional manifold M^n has only weakly quarter-pinched sectional curvature, it has turned out to be necessary to work on the free loop space ΩM. Nevertheless, the index computation for the long geodesic $\bar{c}_0$ does not yield an immediate contradiction, but it leads to the following additional information:

(a) the long geodesic $\bar{c}_0$ has length $L(\bar{c}_0) = 2\pi$, and conj $M^n = \pi$;
(b) the holonomy action on the normal bundle of $\bar{c}_0$ is the map $-\mathrm{id}$;
(c) $K_M(\sigma) = \frac{1}{4}$ for any tangent plane σ of M^n containing $\bar{c}_0'$.

Properties (b) and (c) mean that in some sense $\bar{c}_0$ looks like a primitive closed geodesic in the real projective space $\mathbb{RP}^n_{1/4}$ with constant sectional curvature $\frac{1}{4}$.

The argument given in [Cheeger and Gromoll 1980] is based on the observation that the lifting construction in the proof of the long homotopy lemma actually proves a little more. The authors conclude that the long geodesic $\bar{c}_0$ is *nonliftable* in a suitable sense [Cheeger and Gromoll 1980, Lemma 1]. This property, which is technically fairly delicate to deal with, turns out to be quite strong since $L(\bar{c}_0) \leq 2\pi$ in the weakly quarter-pinched case. One concludes that for any $s_0 \in \mathbb{R}$ the first conjugate points of $\bar{c}_0$ on either side of $s = s_0$ appear at $s = s_0 \pm \frac{1}{2}$, an assertion much stronger than the statement about the conjugate radius in (a). It implies that there exists a closed unit normal field v_0 along $\bar{c}_0$ such that

$$K_M\left(\mathrm{span}\{\bar{c}_0', v_0\}\right) \equiv 1,$$

contradicting property (c) and thus obstructing the special long geodesic $\bar{c}_0$ as required.   □

The properties of the parallel unit vector field v_0 in the preceding proof assert that the long geodesic $\bar{c}_0$ looks in some sense like the boundary of a totally geodesically immersed hemisphere Σ of constant curvature $K_\Sigma = 1$. Loosely speaking, the contradiction that concludes the proof in [Cheeger and Gromoll 1980] is due to the fact that this picture of $\bar{c}_0$ is very different from the appearance of a primitive closed geodesic in $\mathbb{RP}^n_{1/4}$.

Nevertheless, any straightforward attempt to extend the argument of Cheeger and Gromoll to a proof of Theorem 4.1 fails badly.

REMARKS 4.5. (i) If the pinching constant δ is $< \frac{1}{4}$, that is, if $\delta = 1/(4(1+\varepsilon)^2)$ for some $\varepsilon > 0$, it is still possible to follow the approach in the proof of Theorem 1.6 for a while. The assumption inj $M^n <$ conj M^n still implies the existence of a long geodesic $\bar{c}_0$, which in this case has the following properties:

$$L(\bar{c}_0) \geq 2 \operatorname{conj} M^n \geq 2\pi, \quad \mathrm{ind}_E(\bar{c}_0) \leq 1. \tag{4.1}$$

The preceding upper bound for the Morse index of $\bar{c}_0$ yields additional information about the length and the *first holonomy angle* of the long geodesic:

$$L(\bar{c}_0) \leq 2\pi(1+\varepsilon), \quad \psi_1(\bar{c}_0) \leq \frac{\pi}{1+\varepsilon}.$$

By definition the *first holonomy angle* $\psi_1(c)$ of a closed curve $c : \mathbb{R}/\mathbb{Z} \to M^n$ is the angle of smallest absolute value among the rotation angles corresponding to the 2×2 blocks D_i in the expression of the holonomy matrix $U_c = \mathrm{diag}(1, D_1, \ldots, D_{(n-1)/2}) \in \mathrm{SO}(n)$ in canonical form. The other holonomy angles $0 \leq \psi_1(c) \leq \cdots \leq \psi_{(n-1)/2}(c) \leq \pi$ are defined similarly.

Refining the minimax construction, it is furthermore possible to obtain a long geodesic $\bar{c}_0$ that is *shortly null-homotopic*. This means that there exists a free homotopy $\bar{c}$ from $\bar{c}_0$ to a constant curve $\bar{c}_1$ consisting of closed curves $\bar{c}_t$ such that $L(\bar{c}_t) < L(\bar{c}_0)$ for all $t \in (0, 1]$.

(ii) By (i) the set of shortly null-homotopic closed geodesics $\bar{c}_0$ that obey (4.1) is nonempty and compact, and thus it contains an element of minimal length.

However, in contrast to the setup in the weakly quarter-pinched case, it is not possible to ensure that the curves $\bar{c}_t$ in a free null homotopy of the long geodesic $\bar{c}_0$ are strictly shorter than $2 \operatorname{conj} M^n$ for any $t \in (0, 1]$. Thus $\bar{c}_0$ cannot be recognized as the limit of a family of *liftable* curves. This is precisely the point where any direct attempt to generalize the argument from [Cheeger and Gromoll 1980] seems to fail.

The idea for the proof of Theorem 4.1 as given in [Abresch and Meyer 1994] is to consider a minimal closed geodesic $\bar{c}_0 : \mathbb{R}/\mathbb{Z} \to M^n$ as described in the preceding remark and to gain further geometric information by normalizing the short null homotopy $\bar{c}$ in three basic steps. Figure 1 is a Morse theory–type picture of the last two normalization steps.

On the tail of short null homotopies. We define the *tail* of a short null homotopy $\bar{c}$ as the family $(\bar{c}_t)_{t_0 < t \leq 1}$, where $t_0 := \inf\{t \mid L(\bar{c}_t) < 2\pi\}$. It will be convenient to introduce the space $\Omega M_{<2\pi}$ consisting of all closed curves in M^n of length $< 2\pi$ and the connected component $\Omega M_{<2\pi,0} \subset \Omega M_{<2\pi}$ containing the constant curves.

Since $2\pi \leq 2 \operatorname{conj} M^n \leq L(\bar{c}_0)$, it follows from the long homotopy lemma that a curve $\bar{c}_t$ in the short null homotopy of $\bar{c}_0$ is contained in $\Omega M_{<2\pi,0}$ if and only if $L(\bar{c}_t) < 2\pi$. The space $\Omega M_{<2\pi,0}$ in turn is sufficiently special to admit the following additional lifting construction:

LEMMA 4.6 [Abresch and Meyer 1994, Theorem 6.8]. *For any complete Riemannian manifold M^n with $K_M \leq 1$ there is a unique continuous map*

$$h : \Omega M_{<2\pi,0} \to M^n \times \Omega TM, \quad h(c_0) := (m_0, \tilde{c}_0),$$

such that the following conditions hold:

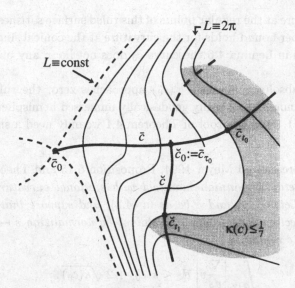

Figure 1. The initial segments $(\bar{c}_t)_{0 \leq t \leq t_0}$ and $(\check{c}_t)_{0 \leq t \leq t_1}$ of a normalized short null homotopy.

(i) $\tilde{c}_0$ is a lift of c_0 under the exponential map $\exp_{m_0} : B_{\pi/2}T_{m_0}M \to M^n$, where $B_{\pi/2}T_{m_0}M$ denotes the ball of radius $\frac{1}{2}\pi$ centered at $0 \in T_{m_0}M$, and
(ii) the origin in $T_{m_0}M$ is the center of the circumscribed ball around the image of $\tilde{c}_0$ in $B_{\pi/2}T_{m_0}M$.

It should be clear how to define the map h on a space of very short curves, say much shorter than the injectivity radius of M^n. The next step is to observe that by properties (i) and (ii) such partially defined maps h admit locally unique continuations. The existence of these continuations depends on the fact that in the standard sphere $\mathbb{S}^n$ of curvature 1 any curve of length $< 2\pi$ is contained in an open hemisphere, that is, in an open ball of radius $\frac{1}{2}\pi$. For the uniqueness part one should observe that the condition on the center of the circumscribed ball determines the tangent space containing the image of $\tilde{c}_0$. Finally, the long homotopy lemma asserts that there are no problems with monodromy.

Clearly, the map h extends continuously to the closure $\overline{\Omega M_{<2\pi,0}} \subset \Omega M$. Applying the preceding lemma with $c_0 = \bar{c}_{t_0}$, we can replace the tail of the short null homotopy $\bar{c}$ up to some reparametrization in t by the map

$$\hat{c} : \mathbb{R}/\mathbb{Z} \times [0,1] \to M^n, \quad \hat{c}(s,t) := \exp_{m_0} t\tilde{c}_0(s). \tag{4.2}$$

The image of this map has remarkable geometric properties if the total absolute curvature

$$\kappa(c_0) = \int_{\mathbb{R}/\mathbb{Z}} \left| \frac{\nabla}{ds} \frac{c_0'(s)}{|c_0'(s)|} \right| ds$$

is sufficiently small. For $\kappa(c_0) < \frac{1}{2}\pi$ it can be shown that $\hat{c}$ describes an immersed ruled surface $\Sigma \subset M^n$ with a conical singularity at $m_0 = \hat{c}(s,0)$. Clearly, the sectional curvature at the regular points of this ruled surface satisfies $K_\Sigma \leq K_M \leq 1$. The same upper bound holds for the curvature at the conical singularity, since by condition (ii) in Lemma 4.6 the curve $\tilde{c}_0$ does not lie in any open half-space in $T_{m_0}M$.

If the total absolute curvature $\kappa(c_0)$ approaches zero, the ruled surface Σ looks more and more like a totally geodesically immersed hemisphere of constant curvature $K_\Sigma = 1$. For the proof of Theorem 4.1 we only need a slightly weaker result:

LEMMA 4.7 [Abresch and Meyer 1994, Proposition 6.15 and Theorem 6.1]. *Let* M^n *be a complete Riemannian manifold with sectional curvature bounded by* $0 < K_M \leq 1$. *Let* c_0, $\tilde{c}_0$, *and* $\hat{c}$ *be as in* (4.2), *and suppose that* $\kappa(c_0) \leq \frac{1}{6}\pi$. *Then the unit vector field* v *along* c_0 *obtained by normalizing* $s \mapsto (\partial\hat{c}/\partial t)|_{(s,1)}$ *has total absolute rotation*

$$\int_{\mathbb{R}/\mathbb{Z}} \left| \frac{\nabla}{ds} v \right| ds \leq \kappa(c_0) + 2\sqrt{\kappa(c_0)}.$$

Furthermore, the first holonomy angle $\psi_1(c_0)$ *is bounded as follows*:

$$\psi_1(c_0) \leq \begin{cases} \kappa(c_0) & \text{if } n \text{ is even,} \\ \dfrac{\sqrt{2}\,\kappa(c_0) + 2\sqrt{\kappa(c_0)}}{\sqrt{1 - \sin\kappa(c_0)}} & \text{if } n \text{ is odd.} \end{cases}$$

This implies in particular that the initial curve $\bar{c}_{t_0}$ of the tail of the short null homotopy $\bar{c}$ has first holonomy angle $\psi_1(\bar{c}_{t_0}) < \frac{1}{3}\pi$ provided that the total absolute curvature of $\bar{c}_{t_0}$ satisfies $\kappa(\bar{c}_{t_0}) < \frac{1}{7}$.

A first normalization of the initial segment of the short null homotopy. The purpose of this normalization is to guarantee that the total absolute curvature of $\bar{c}_{t_0}$ is so much less than $\frac{1}{7}$ that the inequality $\kappa(\bar{c}_{t_0}) < \frac{1}{7}$ persists even after the second normalization step. This leads to the required contradiction, since the second normalization step will ensure that $\psi_1(\bar{c}_{t_0}) > \frac{1}{3}\pi$, provided that the pinching constant $\delta = 1/(4(1 + \varepsilon)^2)$ is sufficiently close to $\frac{1}{4}$.

Recall that $\bar{c}_0$ is a closed geodesic of length $\eta := L(\bar{c}_0) \in [2\pi, 2\pi(1 + \varepsilon)]$. Hence $\kappa(\bar{c}_0) = 0$ and $\bar{c}_t \in \Omega M_{<\eta,0}$ for $0 < t \leq 1$. But there is no bound for the length of the initial segment $(\bar{c}_t)_{0 \leq t \leq t_0}$ when considered as a curve in the free loop space ΩM. Nevertheless, it is possible to get away with some bounds for the differential $d\kappa$ on a suitable domain in $\Omega M_{<\eta,0}$. The key observation is that the total absolute curvature κ can be interpreted as the L^1-norm of the L^2-gradient $\mathrm{grad}_{L^2} L$ of the length functional L. Recall that the L^2-norm of a vector field X along $\bar{c}_t$ is defined by $\|X\|_{L^2}^2 = \int_{\mathbb{R}/\mathbb{Z}} |X|^2 |\bar{c}_t'| ds$. Since the L^2-Hessian of the length functional at some curve $\bar{c}_t$ is bounded from below by $-\max K_M \geq -1$, one obtains the following lemma:

LEMMA 4.8 (CURVE SHORTENING). *Let M^n, $\bar{c}_0$, and $\eta := L(\bar{c}_0)$ be as before. Suppose that the initial segment of the short null homotopy $\bar{c}$ beginning at $\bar{c}_0$ is a solution of the rescaled curve-shortening flow*

$$\frac{\partial \bar{c}_t}{\partial t} = -p(t) \operatorname{grad}_{L^2} L|_{\bar{c}_t},$$

except at finitely many $\theta_j \in [0, t_0]$ where the weight function $p : [0, t_0] \to (0, \infty]$ is unbounded. Then for any $t \in [0, t_0]$ the total absolute curvature of $\bar{c}_t$ satisfies

$$\kappa(\bar{c}_t)^2 \leq 2\left(L(\bar{c}_0) - L(\bar{c}_t)\right) \leq 4\pi\varepsilon.$$

REMARKS 4.9. (i) Clearly, $\theta_0 = 0$, and each $\theta_j > 0$ means that the trajectory approaches a closed geodesic $\bar{c}_{\theta_j}$. By our choice of the initial curve $\bar{c}_0$ such a geodesic must have Morse index $\operatorname{ind}_E(\bar{c}_{\theta_j}) \geq 2$, and thus it is possible to restart the curve-shortening flow below the critical level after performing some tiny, explicit deformation of $\bar{c}_{\theta_j}$ by means of the degenerate Morse lemma. By Condition C of Palais and Smale the closed interval $[2\pi^2, \frac{1}{2}\eta^2]$ contains only finitely many critical values of the energy functional E, and thus after at most finitely many restarts the flow passes through a curve $\bar{c}_{t_0}$ of length $L(\bar{c}_{t_0}) = 2\pi$.

(ii) In [Abresch and Meyer 1994, Proposition 5.1] we are actually working in some fixed, finite-dimensional subspace $\Omega_\ell^k M_{<\eta}$ of broken geodesics rather than on the Hilbert manifold ΩM, whose metric is induced by the H^1-inner products on the tangent spaces. This is to avoid the analytical difficulties concerning the long-time existence of the curve-shortening flow. A second benefit of working in a fixed finite-dimensional approximation space is the fact that two-sided bounds for the Hessians of the length and the energy functionals are available. These bounds will be used in a significant way in the next normalization step.

A second normalization of the initial segment of $\bar{c}$. From now on we suppose that the initial segment $(\bar{c}_t)_{0 \leq t \leq t_0}$ of the short null homotopy $\bar{c}$ has been normalized by means of the curve-shortening flow as discussed in the preceding step. Moreover, we assume that $\varepsilon \leq \frac{1}{64\,000}$. With this constraint on the pinching constant δ, Lemma 4.8 implies that $\kappa(\bar{c}_t) < \frac{1}{70}$ for all $t \in [0, t_0]$, and by Lemma 4.7 the first holonomy angle of the curve $\bar{c}_{t_0}$ is much smaller than $\frac{1}{3}\pi$. Since the map $t \mapsto \psi_1(\bar{c}_t)$ is continuous, we conclude that there exists some $\tau_0 \in (0, t_0)$ such that $\bar{c}_{\tau_0}$ has first holonomy angle equal to $\frac{1}{2}\pi$ and total absolute curvature $< \frac{1}{70}$. Our goal is to replace the segment $(\bar{c}_t)_{\tau_0 \leq t \leq t_0}$ of the short null homotopy $\bar{c}$ by some alternate arc $\check{c} : [0, t_1] \to \Omega M_{<\eta,0}$ from $\check{c}_0 = \bar{c}_{\tau_0}$ to some curve $\check{c}_{t_1}$ representing an element of the boundary of $\Omega M_{<2\pi,0}$. The construction of $\check{c}$ shall imply that the total absolute curvature of $\check{c}_{t_1}$ is still bounded by $\frac{1}{7}$ and that its first holonomy angle is still $> \frac{1}{3}\pi$, contradicting the assertion in Lemma 4.7.

The key observation for constructing $\check{c}$ is to realize that because of the equation $\psi_1(\check{c}_0) = \frac{1}{2}\pi$ there exists a subspace $W \subset T_{\check{c}_0}\Omega M$ of dimension ≥ 2 consisting of vector fields v along $\check{c}_0$ such that

$$\frac{\nabla}{\partial s} v \perp v \quad \text{and} \quad \left| \frac{\nabla}{\partial s} v \right| = \tfrac{1}{2} \pi \, |v| \, .$$

In other words, each of these fields v rotates with constant speed. On the one hand, the angular velocity is sufficiently large in comparison to the total absolute curvature $\kappa(\check{c}_0)$ to allow us to conclude that v and the tangent field $\check{c}_0'$ are almost perpendicular. On the other hand, the angular velocity of v is already sufficiently small for us to conclude that

$$\operatorname{hess}_{L^2}(L)_{|\check{c}_0}(v,v) \lesssim \left(\frac{\pi^2}{4L(\check{c}_0)^2} - \frac{1}{4(1+\varepsilon)^2} \right) \|v\|_{L^2}^2 \approx -\tfrac{3}{16} \|v\|_{L^2}^2 < 0. \quad (4.3)$$

It is customary to control the path dependence of parallel transport in terms of the norm of the Riemannian curvature tensor of M^n and the area of a spanning homotopy. A similar argument can be used to estimate the change in the first holonomy angle along $\check{c}$:

$$|\psi_1(\check{c}_{t_1}) - \psi_1(\check{c}_0)| \le \tfrac{4}{3} \operatorname{area}(\check{c}) \, . \quad (4.4)$$

The corresponding statements for the finite-dimensional approximation spaces $\Omega_\ell^k M_{<\eta}$ used in [Abresch and Meyer 1994] are somewhat more technical. Even if the number k of corners of the broken geodesics is chosen to be large, the constants appearing in Lemma 4.7 and in inequalities (4.3)–(4.4) differ significantly from the constants in the corresponding statements for $\Omega_\ell^k M_{<\eta}$ in [Abresch and Meyer 1994]. The reason is that the natural metrics g_k on the approximation spaces converge to the normalized L^2-inner product on ΩM rather than to the standard L^2-inner product used in the preceding discussion. Moreover, the estimates in [Abresch and Meyer 1994] are stated in terms of the energy functional E rather than the length functional L.

The advantage of working in the space $\Omega_\ell^k M_{<\eta}$ is that a two-sided bound for the Hessian of L is available, so there is a bound for the norm of the differential $d\kappa$. This means that the required bounds for the first holonomy angle and for the total absolute curvature of $\check{c}_{t_1} \in \partial \Omega_\ell^k M_{<2\pi,0}$ can be guaranteed if the length of $\check{c}$ with respect to the metric g_k on $\Omega_\ell^k M_{<\eta}$ is bounded by a small constant $t_{\max} > 0$ and if ε is sufficiently small.

In the finite-dimensional setup, we are dealing with a broken geodesic $\check{c}_0 = \bar{c}_{\tau_0} \in \Omega_\ell^k M_{<\eta,0}$. The tangent space $T_{\check{c}_0} \Omega_\ell^k M_{<\eta}$ also contains a subspace W_k of dimension at least 2 on which the Hessian of the length functional is bounded from above by an explicit negative constant. Now the idea is to pick the homotopy $\check{c}$ as a normal g_k-geodesic in $\Omega_\ell^k M_{<\eta}$ of length $\le t_{\max}$ starting at $\check{c}_0 = \bar{c}_{\tau_0}$ with initial vector $\frac{\partial}{\partial t} \check{c}_t|_{t=0} = v \in W_k$. Since $\dim W_k \ge 2$, we may assume that

$$dL\left(\frac{\partial}{\partial t} \check{c}_t \Big|_{t=0} \right) \le 0.$$

As we decrease $\varepsilon > 0$, the difference $L(\check{c}_0) - 2\pi$ becomes as small as we please, so the proof can be finished establishing a negative upper bound for

$$\frac{d^2}{dt^2} L(\check{c}_t) = \text{hess}_{L^2}(L) \left(\frac{\partial}{\partial t} \check{c}_t, \frac{\partial}{\partial t} \check{c}_t \right) \qquad (4.5)$$

for all $t \in [0, t_{\max}]$, and not just at $t = 0$ as in inequality (4.3). For this purpose we cannot refer to any modulus of continuity for the Hessian $\text{hess}_{L^2}(L)$ itself, since this would require a bound for the covariant derivative of the curvature tensor. However, the bounds for

$$\int_{\mathbb{R}/\mathbb{Z}} \left| \frac{\nabla}{\partial s} \frac{\partial}{\partial t} \check{c}_t \right|^2 \left| \frac{\partial}{\partial s} \check{c}_t \right|^{-1} ds \quad \text{and} \quad \left| \frac{\partial}{\partial s} \check{c}_t \wedge \frac{\partial}{\partial t} \check{c}_t \right|,$$

which have been used to prove inequality (4.3) for $\check{c}_0$, can be extended continuously along the g_k-geodesic $t \mapsto \check{c}_t$, and thus we obtain a uniform negative upper bound for the second derivative of the map $t \mapsto L(\check{c}_t)$ on $[0, t_{\max}]$, provided that $t_{\max}$ and ε are sufficiently small.

This concludes our sketch of the proof of Theorems 1.6 and 4.1.

5. A Sphere Theorem with a Universal Pinching Constant Below $\frac{1}{4}$

The injectivity radius estimate in Theorem 4.1 was the first result with a pinching constant below $\frac{1}{4}$ and independent of the dimension. In this context it is natural to ask whether the assertions in Berger's pinching below-$\frac{1}{4}$ theorem and the sphere theorem in the preceding section (Theorems 3.1 and 4.2) remain valid for some universal pinching constants. In the odd-dimensional case the answer is affirmative:

THEOREM 5.1 (SPHERE THEOREM [Abresch and Meyer a]). *There exists a constant* $\delta_{\text{odd}} \in (0, \frac{1}{4})$ *such that any odd-dimensional, compact, simply connected Riemannian manifold* M^n *with* δ_{odd}-*pinched sectional curvature is homeomorphic to the sphere* $\mathbb{S}^n$.

In fact, the constant δ_{odd} is explicit. Our proof works for $\delta_{\text{odd}} = \frac{1}{4}(1 + \varepsilon_{\text{odd}})^{-2}$, where $\varepsilon_{\text{odd}} = 10^{-6}$. In the even-dimensional case, however, our methods are not yet sufficient to generalize Berger's pinching below-$\frac{1}{4}$ theorem accordingly. So far, there is only the following partial result [Abresch and Meyer a]:

THEOREM 5.2 (COHOMOLOGICAL PINCHING BELOW-$\frac{1}{4}$ THEOREM). *There exists a constant* $\delta_{\text{ev}} \in (0, \frac{1}{4})$ *such that for any even-dimensional, compact, simply connected Riemannian manifold* M^n *with* δ_{ev}-*pinched sectional curvature the cohomology rings* $H^*(M^n; R)$ *with coefficients* $R \in \{\mathbb{Q}, \mathbb{Z}_2\}$ *are isomorphic to the corresponding cohomology rings of one of the compact, rank-one symmetric spaces* $\mathbb{S}^n$, $\mathbb{CP}^{n/2}$, $\mathbb{HP}^{n/4}$, *or* $\text{Ca}\mathbb{P}^2$, *or the rings* $H^*(M^n; R)$ *are truncated polynomial rings generated by an element of degree 8.*

Again, the constant δ_{ev} is explicit and independent of the dimension. The proof works for $\delta_{\mathrm{ev}} = \frac{1}{4}(1 + \varepsilon_{\mathrm{ev}})^{-2}$, where $\varepsilon_{\mathrm{ev}} = \frac{1}{27\,000}$.

Recall that $H^*(\mathrm{CaP}^2; R) = R[\xi_R]/(\xi_R^3)$, where $\deg \xi_R = 8$. However, we cannot exclude the possibility that $H^*(M^n; R) = R[\xi_R]/(\xi_R^{m+1})$, where $\deg \xi_R = 8$ and $m > 2$. For instance, we cannot apply J. Adem's result [1953, Theorem 2.2], which is based on Steenrod's reduced third power operations, since we do not have enough control on the cohomology ring of M^n with coefficients $\mathbb{Z}_3$.

Notice that in dimensions 2, 3, and 4 the more special results mentioned in the discussion of the topological sphere theorem in Section 1 are still much stronger than the assertions in Theorems 5.1 and 5.2. In fact, we even need to refer to Hamilton's result, since our proof for the sphere theorem depends on Smale's solution of the Poincaré conjecture in dimensions ≥ 5. We use Theorem 1.3 in order to handle the three-dimensional case.

The starting point for the proofs of both theorems is to establish the horseshoe conjecture of Berger [1962a], which had remained open until recently:

THEOREM 5.3 (HORSESHOE INEQUALITY [Abresch and Meyer a, Theorem 2.4]). *There exists a constant $\varepsilon_{\mathrm{hs}} > 0$ such that, for any complete Riemannian manifold M^n satisfying*

$$\delta_{\mathrm{hs}} := \tfrac{1}{4}(1 + \varepsilon_{\mathrm{hs}})^{-2} \leq K_M \leq 1$$

and

$$\pi \leq \operatorname{inj} M^n \leq \operatorname{diam} M^n \leq \pi(1 + \varepsilon_{\mathrm{hs}}),$$

the following statement holds: For any $p_0 \in M^n$ and any $v \in \mathbb{S}^{n-1} \subset T_{p_0}M$, the distance between the antipodal points $\exp_{p_0}(-\pi v)$ and $\exp_{p_0}(\pi v)$ is less than π.

If M^n is one of the projective spaces $\mathbb{CP}^m$, $\mathbb{HP}^m$, and CaP^2 with its Fubini–Study metric, the two points $\exp_{p_0}(-\pi v)$ and $\exp_{p_0}(\pi v)$ coincide. The horseshoe inequality asserts that their distance is less than the injectivity radius of M^n if the relevant geometric invariants of M^n do not deviate too much from the corresponding quantities of the projective spaces. Notice in particular that the pinching constant δ_{hs} is explicit and independent of the dimension. In fact, the assertion of Theorem 5.3 holds for $\varepsilon_{\mathrm{hs}} = \frac{1}{27\,000}$.

A horseshoe inequality for manifolds with nontrivial fundamental group had been established earlier by Durumeric [1984, Lemma 6]. His argument was based on a detailed investigation of the geometry of Dirichlet cells in the universal covering of M^n, and relied on the hypothesis $\pi_1(M^n) \neq 0$. The proof of Theorem 5.3 is very different. It requires some refined Jacobi field estimates, which might be useful in other contexts as well. Our plan is to explain these Jacobi field estimates in the next section and describe the proof of the horseshoe inequality in Section 7.

We conclude this section by explaining how to deduce Theorems 5.1 and 5.2 for manifolds M^n of dimension $n \geq 4$ by combining the injectivity radius estimate of Theorem 4.1, the diameter sphere theorem, and the horseshoe inequality of

Theorem 5.3 with some results from algebraic topology. This reduction had already been known to Berger [1962a]. Complete proofs are given in [Abresch and Meyer a]. Here we list only the basic steps.

In fact, Theorem 5.3 does not come into the proofs of Theorems 5.1 and 5.2 directly. Instead we need the following corollary:

COROLLARY 5.4 [Berger 1962a, Proposition 2]. *Let $\delta_{hs} \in (0, \frac{1}{4})$ be the constant of Theorem 5.3. Then any compact Riemannian manifold M^n satisfying $\delta_{hs} \leq K_M \leq 1$ and*

$$\pi \leq \operatorname{inj} M^n \leq \operatorname{diam} M^n \leq \frac{\pi}{2\sqrt{\delta_{hs}}}$$

admits a continuous, piecewise smooth map $f : \mathbb{RP}^n \to M^n$ of degree 1.

Here $\deg f$ denotes the standard integral mapping degree if M^n is odd-dimensional and orientable. Otherwise $\deg f$ has to be understood as the $\mathbb{Z}_2$-mapping degree.

REMARK 5.5. If $M^n = \mathbb{CP}^m$, with $n = 2m > 2$, the map $f : \mathbb{RP}^{2m} \to \mathbb{CP}^m$ obtained from the preceding corollary can be visualized in terms of the standard cell decompositions. Recall that $\mathbb{RP}^{2m} = \mathbb{RP}^{2m-1} \cup_\varphi e_{2m}$ and $\mathbb{CP}^m = \mathbb{CP}^{m-1} \cup_{\bar{\varphi}} e_{2m}$. Moreover, the fibers of the attaching map $\bar{\varphi} : \partial e_{2m} \to \mathbb{CP}^{m-1}$ are the Hopf circles in $\mathbb{S}^{2m-1} = \partial e_{2m}$. They are invariant under the antipodal map, and thus there is an induced map $\psi : \mathbb{RP}^{2m-1} \to \mathbb{CP}^{m-1}$ such that $\bar{\varphi}$ factors as $\psi \circ \varphi$. Hence the identity map on the $2m$-cell e_{2m} induces a map $f : \mathbb{RP}^{2m} \to \mathbb{CP}^m$ with $\deg_{\mathbb{Z}_2} f = 1$. This map coincides with the map constructed in the proof of the corollary.

It turns out that the mere existence of such a map $f : \mathbb{RP}^n \to M^n$ is a strong constraint for the topology of the manifold M^n:

THEOREM 5.6. *Let M^n be a compact, simply connected, odd-dimensional manifold. Suppose that there exists a continuous map $f : \mathbb{RP}^n \to M^n$ with $\deg_{\mathbb{Z}} f = 1$. Then M^n is a homology sphere.*

THEOREM 5.7. *Let M^n be a compact, simply connected, even-dimensional manifold. Suppose that there exists a continuous map $f : \mathbb{RP}^n \to M^n$ with $\deg_{\mathbb{Z}_2} f = 1$. Then the cohomology rings of M^n with coefficients $R \in \{\mathbb{Q}, \mathbb{Z}_2\}$ are isomorphic to truncated polynomial rings generated by an element ξ_R of degree 2, 4, 8, or n.*

Both theorems follow essentially from standard computations in algebraic topology, based on the Poincaré duality theorem, the Steenrod squares, and the universal coefficient theorem. The proof of Theorem 5.7 refers in addition to J. F. Adams's results on secondary cohomology operations and the Hopf invariant one problem [Adams 1960]. A summary of this work is given in [Milnor and Stasheff 1974, page 134]. Historically, Theorem 5.6 goes back to H. Samelson [1963], whereas Theorem 5.7 can be extracted from [Berger 1965, p. 135ff].

Theorem 5.7 is far from recognizing the manifold M^n up to homeomorphism. It determines the integral cohomology ring $H^*(M^n; \mathbb{Z})$ only up to the Serre class of torsion groups of odd order. This ambiguity reflects the fact that we can only work with the modulo-2 mapping degree of f. Furthermore, the *fake projective spaces* discovered by J. Eells and N. H. Kuiper [1962] show that it would not even be sufficient to recover the integral cohomology rings in order to recognize the manifold M^n up to homeomorphism.

ON THE PROOF OF THEOREM 5.1 IN DIMENSIONS $n \geq 5$. We set $\delta_{\text{odd}} := \max\{\delta_{\text{inj}}, \delta_{\text{hs}}\}$, where δ_{inj} and δ_{hs} are the constants from Theorems 4.1 and 5.3. It is convenient to scale the metric on M^n such that $\delta_{\text{odd}} \leq K_M \leq 1$. Because of the diameter sphere theorem of Grove and Shiohama it is sufficient to consider manifolds with diam $M^n \leq \pi/(2\sqrt{\delta_{\text{odd}}})$. By Theorem 4.1 we have inj $M^n \geq \pi$, so Corollary 5.4 yields a continuous, piecewise smooth map $f : \mathbb{RP}^n \to M^n$ of degree $\deg_{\mathbb{Z}} f = 1$. With Theorem 5.6 we conclude that the manifold M^n is a homology sphere. Since by hypothesis M^n is simply connected, Smale's solution of the Poincaré conjecture in dimensions $n \geq 5$ can be applied [Milnor 1965, p. 109; Smale 1961]. $\qquad\square$

ON THE PROOF OF THEOREM 5.2 IN DIMENSIONS $n \geq 4$. The argument is very similar. We estimate the injectivity radius of M^n by means of Theorem 1.5 rather than Theorem 4.1, so we may set $\delta_{\text{ev}} := \delta_{\text{hs}}$. Again we obtain a continuous, piecewise smooth map $f : \mathbb{RP}^n \to M^n$. We still have $\deg_{\mathbb{Z}_2} f = 1$, so we can apply Theorem 5.7. In order to conclude the proof, we observe that the truncated polynomial rings $R[\xi_R]/(\xi_R^{m+1})$ where $m = n/\deg \xi_R$ are precisely the cohomology rings of $\mathbb{S}^n$, $\mathbb{CP}^{n/2}$, or $\mathbb{HP}^{n/4}$, if the degree of the generator is n, 2, or 4, respectively. $\qquad\square$

6. New Jacobi Field Estimates

In this section the issue is to obtain precise control over normal Jacobi fields Y with $Y(0) = 0$ along any geodesic $\gamma : [0, r_2] \to M^n$. We are interested in mixed estimates for Y at some point $r_1 \in (0, r_2)$, which depend on information about the size of the initial derivative $\frac{\nabla}{dr} Y(0)$ and the boundary value $Y(r_2)$, and which refine the standard estimate provided by the Rauch comparison theorems. For this purpose it is essential to work with two-sided bounds for the sectional curvature K_M of the Riemannian manifold M^n. The basic estimates have been established in [Abresch and Meyer a, Theorem 5.6].

THEOREM 6.1 (MIXED JACOBI FIELD ESTIMATES). *Let $\lambda < \Lambda$ and let $0 < r_1 \leq r_2$. Suppose that $r_2 \leq \pi/\sqrt{\Lambda}$ if $\Lambda > 0$. Then there exists a continuous function $\Psi_{r_1 r_2} : [0, \infty) \times [0, \infty) \to [0, \infty)$ such that, for any Riemannian manifold M^n with sectional curvature bounded by $\lambda \leq K_M \leq \Lambda$ and for any geodesic $\gamma : [0, r_2] \to M^n$ with $|\gamma'| \equiv 1$, there is the following lower bound for a normal*

Jacobi field Y along γ with initial value $Y(0) = 0$:

$$|Y(r_1)| \geq \Psi_{r_1 r_2}\left(\left|\frac{\nabla}{dr}Y(0)\right|, |Y(r_2)|\right).$$

Furthermore, the function $\Psi_{r_1 r_2}$ has the following properties:

(i) *it is weakly convex and positively homogeneous of degree 1;*
(ii) *it is nondecreasing with respect to both variables;*
(iii) *it is locally of class $C^{1,1}$ except at $(\alpha, \eta) = (0,0)$;*
(iv) $\Psi_{r_1 r_2}(\alpha, \eta) \geq \max\{\alpha\operatorname{sn}_\Lambda(r_1), \eta\operatorname{sn}_\lambda(r_1)/\operatorname{sn}_\lambda(r_2)\}$ *for all $(\alpha, \eta) \in [0,\infty)^2$.*

REMARK 6.2. The lower bounds for $\Psi_{r_1 r_2}$ in (iv) reflect the standard Jacobi field estimates. The term $\alpha\operatorname{sn}_\Lambda(r_1)$ is due to the first Rauch comparison theorem, whereas the term $\eta\operatorname{sn}_\lambda(r_1)/\operatorname{sn}_\lambda(r_2)$ follows from the monotonicity of the map $s \mapsto \operatorname{sn}_\lambda(s)^{-1}|Y(s)|$ asserted by the infinitesimal Rauch comparison theorem. In fact,

$$\Psi_{r_1 r_2}(\alpha, \eta) = \begin{cases} \alpha\operatorname{sn}_\Lambda(r_1) & \text{if } \eta \leq \alpha\operatorname{sn}_\Lambda(r_2), \\ \eta\operatorname{sn}_\lambda(r_1)/\operatorname{sn}_\lambda(r_2) & \text{if } \alpha\operatorname{sn}_\lambda(r_2) \leq \eta. \end{cases} \tag{6.1}$$

The continuity of the first derivatives of $\Psi_{r_1 r_2}$ as asserted in (iii) implies that inequality (iv) is strict, provided that $\alpha\operatorname{sn}_\Lambda(r_2) < \eta < \alpha\operatorname{sn}_\lambda(r_2)$ and r_1 is sufficiently close to r_2. This is the range of parameters in which Theorem 6.1 improves the classical estimates.

The homogeneity property of $\Psi_{r_1 r_2}$ corresponds to the linearity of the Jacobi field equation. The convexity asserted in (i) is essential in order to apply Jensen's inequality when integrating the Jacobi field estimate from Theorem 6.1 over a family of geodesics. In particular, it enables us to deduce:

THEOREM 6.3 [Abresch and Meyer a, Theorem 5.4]. *Let $\lambda < \Lambda$ and let $0 < r_1 \leq r_2$. Suppose that $r_2 \leq \pi/\sqrt{\Lambda}$ if $\Lambda > 0$. Consider a Riemannian manifold M^n with sectional curvature bounded by $\lambda \leq K_M \leq \Lambda$ and a ruled surface $\gamma : [0, r_2] \times [0,1] \to M^n$ generated by normal geodesics $\gamma_\theta = \gamma(\cdot, \theta)$ emanating from a fixed point $p_0 \in M^n$. Then the lengths $\ell(r_i)$ of the circular arcs $\theta \mapsto \gamma(r_i, \theta)$ and the total angle*

$$\varphi_0 := \int_0^1 \left|\frac{\nabla}{\partial r}\frac{\partial\gamma}{\partial\theta}(0,\theta)\right| d\theta$$

satisfy the inequality

$$\ell(r_1) \geq \Psi_{r_1 r_2}(\varphi_0, \ell(r_2)),$$

where $\Psi_{r_1 r_2}$ is the comparison function introduced in Theorem 6.1.

For computational purposes it is necessary to have a more explicit description of the comparison functions $\Psi_{r_1 r_2}$ that appear in the preceding theorems. In fact, the proof of Theorem 6.1 provides the following information:

On the comparison functions $\Psi_{r_1 r_2}$. The values at all pairs (α, η) outside the cone $\alpha \operatorname{sn}_\Lambda(r_2) < \eta < \alpha \operatorname{sn}_\lambda(r_2)$ are given by formula (6.1). By homogeneity it is sufficient to define $\Psi_{r_1 r_2}(1, \eta)$ for $\operatorname{sn}_\Lambda(r_2) < \eta < \operatorname{sn}_\lambda(r_2)$. For this purpose we introduce the functions $\bar{y} : [0, r_2] \times [0, r_2] \to [0, \infty)$ by means of

$$
\bar{y}(r_0, r) := \begin{cases} \operatorname{sn}_\Lambda(r) & \text{if } r \le r_0, \\ \operatorname{sn}_\Lambda(r_0) \operatorname{cn}_\lambda(r - r_0) + \operatorname{cn}_\Lambda(r_0) \operatorname{sn}_\lambda(r - r_0) & \text{if } r_0 \le r \text{ and } n = 2, \\ \operatorname{sn}_\lambda(r) \bar{w}(r_0, r)^{1/2} & \text{if } r_0 \le r \text{ and } n > 2, \end{cases}
$$
(6.2)

where $\operatorname{sn}_\lambda$ and $\operatorname{cn}_\lambda$ are the generalized trigonometric functions defined on page 3, and where $\bar{w}(r_0, r)$ is the nonnegative number

$$
\bar{w}(r_0, r) := \frac{\operatorname{sn}_\Lambda^2(r_0)}{\operatorname{sn}_\lambda^2(r_0)} - 2 \det \begin{pmatrix} \operatorname{cn}_\lambda(r_0) & \operatorname{cn}_\Lambda(r_0) \\ \operatorname{sn}_\lambda(r_0) & \operatorname{sn}_\Lambda(r_0) \end{pmatrix} \int_{r_0}^r \frac{\operatorname{sn}_\Lambda(\varrho)}{\operatorname{sn}_\lambda^3(\varrho)} \, d\varrho.
$$

The graphs of the functions $\bar{y}_{r_0} : r \mapsto \bar{y}(r_0, r)$ with $0 \le r_0 < r_2$ foliate the domain $\{(r, y) \mid \operatorname{sn}_\Lambda(r) < y \le \operatorname{sn}_\lambda(r), \ 0 < r \le r_2\}$, as shown in Figure 2. In particular, for any $\eta \in [\operatorname{sn}_\Lambda(r_2), \operatorname{sn}_\lambda(r_2)]$ there is precisely one $r_0 \in [0, r_2]$ such that $\bar{y}(r_0, r_2) = \eta$, and $\Psi_{r_1 r_2}(1, \eta)$ is defined implicitly by the equation

$$
\Psi_{r_1 r_2}\big(1, \bar{y}(r_0, r_2)\big) = \bar{y}(r_0, r_1).
$$
(6.3)

REMARK 6.4. For $\lambda = \frac{1}{4}$ and $\Lambda = 1$ it is possible to evaluate the integral in the expression for $\bar{w}$ in terms of trigonometric functions. Moreover, the solution of the equation $\bar{y}(\hat{r}_0, \pi) = 1$, which is needed in order to compute $\Psi_{r_1 \pi}(1, 1) =$

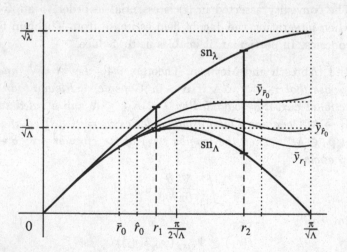

Figure 2. The functions $\bar{y}_{r_0}$ and the map $\eta \mapsto \Psi_{r_1 r_2}(1, \eta)$. The graphs of the functions $\bar{y}_{r_0}, \bar{y}_{\hat{r}_0},$ and $\bar{y}_{r_1}$ are only qualitative pictures with the correct number of local maxima, local minima, and saddle points, provided that $0 < \lambda < \frac{1}{4}\Lambda$. In an actual plot the qualitative properties of these functions would be almost invisible.

$\bar{y}(\hat{r}_0, r_1)$ for $r_1 \in (0, \pi)$, is given by

$$\hat{r}_0 = \begin{cases} 2 \arccos(2^{-1/3}) & \approx 0.416\,304\,\pi & \text{if } n = 2, \\ 2 \arcsin(\frac{1}{2} + \sin(\frac{1}{18}\pi)) & \approx 0.470\,548\,\pi & \text{if } n > 2. \end{cases}$$

In particular, if $r_1 = \frac{59}{120}\pi \in (\hat{r}_0, \frac{1}{2}\pi)$, it follows that $\Psi_{r_1\pi}(1,1) \geq (1+a_0)\sin(r_1)$ for $a_0 \approx 0.001\,663 > 0$. By continuity this inequality persists with a slightly smaller constant $a_\varepsilon > 0$ if $\lambda = \frac{1}{4}(1+\varepsilon)^{-2}$ and $\varepsilon > 0$ is sufficiently small. A numerical computation shows that $a_\varepsilon \approx 0.001\,661$ for $\varepsilon = \varepsilon_{\mathrm{hs}} = \frac{1}{27\,000}$.

With the preceding definition of the functions $\Psi_{r_1 r_2}$ in equations (6.1)–(6.3) it is straightforward, but tedious, to verify all the analytical properties listed as assertions (i)–(iv) in Theorem 6.1. Here we shall rather concentrate on the geometric ideas leading to the claimed lower bound for $|Y(r_1)|$.

ON THE PROOF OF THEOREM 6.1 IN DIMENSION $n = 2$. In this case the argument is quite easy. The normal Jacobi field Y can be written as a product yE of a nonnegative function $y : [0, r_2] \to [0, \infty)$ with a parallel unit normal field along the geodesic γ. The Jacobi field equation reduces to the scalar differential equation $y'' + K_{M|\gamma} y = 0$, and the Rauch comparison theorems assert that

$$y'(0)\,\mathrm{sn}_\Lambda(r_1) \leq y(r_1) \leq y'(0)\,\mathrm{sn}_\lambda(r_1)$$

for all $r_1 \in [0, r_2]$. The infinitesimal version of the Rauch comparison theorems provides the inequality $y(r_2)\,\mathrm{sn}_\lambda(r_1)/\mathrm{sn}_\lambda(r_2) \leq y(r_1)$. The latter inequality can be improved by applying the maximum principle to the differential inequality $y'' + \lambda y \leq 0$. We conclude that for any $r_0 \in [0, r_2)$ the restriction of y to the interval $[r_0, r_2]$ is bounded from below by any solution $\bar{y}_{r_0 r_2}$ of the differential equation $z'' + \lambda z = 0$ with boundary data $z(r_0) = y'(0)\,\mathrm{sn}_\Lambda(r_0)$ and $z(r_2) \leq y(r_2)$. Maximizing over r_0 leads to the functions $\bar{y}_{r_0} = \bar{y}(r_0, \cdot)$ introduced in formula (6.2). One finds that $y(r_1) \geq y'(0)\bar{y}_{r_0}(r_1)$ for any $r_1 \in [0, r_2]$ provided that $y'(0)\bar{y}_{r_0}(r_2) \leq y(r_2)$. $\square$

In dimensions $n > 2$ there is no way to apply the maximum principle directly. Nevertheless, it is still possible to write the normal Jacobi field Y as a product yE of a nonnegative function $y = |Y|$ and a unit normal field E along the geodesic γ. The difficulty is that the unit normal field E needs not be parallel.

Yet, the idea is to reduce to a two-dimensional situation by considering the ruled surface Σ defined by a variation of the geodesic γ which corresponds to the Jacobi field Y. The intrinsic sectional curvature K_Σ of this surface can be determined by means of the Gauss equations

$$K_\Sigma|_\gamma = K_M(T\Sigma|_\gamma) - \left| \frac{\nabla}{dr} E \right|^2,$$

so $\lambda - \left| \frac{\nabla}{dr} E \right|^2 \leq K_\Sigma \leq \Lambda$. We are able to proceed, since the angular velocity $\frac{\nabla}{dr} E$ of the Jacobi field Y can be bounded as follows:

LEMMA 6.5 [Abresch and Meyer a, Proposition 5.12]. *Let $\gamma : [0, r_2] \to M^n$ be a normal geodesic in a Riemannian manifold with sectional curvature bounded by $\lambda \leq K_M \leq \Lambda$. Suppose that $r_2 \leq \pi/\sqrt{\Lambda}$ if $\Lambda > 0$. Then on the interval $(0, r_2)$ the angular velocity $\frac{\nabla}{dr} E$ of a nontrivial normal Jacobi field $Y = yE$ (where $y = |Y|$), with initial value $Y(0) = 0$, can be estimated in terms of the function $u := y^{-1} y'$ as follows:*

$$\left| \frac{\nabla}{dr} E \right|^2 \leq \tfrac{1}{4} (\mathrm{ct}_\lambda - \mathrm{ct}_\Lambda)^2 - \left(u - \tfrac{1}{2}(\mathrm{ct}_\lambda + \mathrm{ct}_\Lambda) \right)^2 = -\mathrm{ct}_\lambda \, \mathrm{ct}_\Lambda + (\mathrm{ct}_\lambda + \mathrm{ct}_\Lambda) \, u - u^2.$$

PROOF. Since $Y(0) = 0$, the Jacobi field equation for Y can be expressed as a Riccati equation for the Hessian A of a local distance function along γ:

$$\frac{\nabla}{dr} Y = AY \qquad \text{and} \qquad \frac{\nabla}{dr} A + A^2 + R(\cdot, \gamma')\gamma' = 0.$$

Since E is a unit normal field, the standard comparison results for the Riccati equation assert that $\mathrm{ct}_\Lambda \, P_\gamma \leq A \leq \mathrm{ct}_\lambda \, P_\gamma$, where ct_λ and ct_Λ are the generalized cotangent functions introduced on page 3 and where $P_\gamma := \mathrm{id} - \langle \cdot, \gamma' \rangle \gamma'$. On the other hand it is easy to see that $\frac{\nabla}{dr} E + u E = AE$, or, equivalently:

$$\frac{\nabla}{dr} E + \left(u - \tfrac{1}{2}(\mathrm{ct}_\lambda + \mathrm{ct}_\Lambda) \right) E = A E - \tfrac{1}{2}(\mathrm{ct}_\lambda + \mathrm{ct}_\Lambda) E.$$

The lemma follows, since $\frac{\nabla}{dr} E$ is orthogonal to the unit vector field E and since $\tfrac{1}{2}(\mathrm{ct}_\lambda - \mathrm{ct}_\Lambda)$ is an upper bound for the norm of the right hand side. $\qquad\square$

The preceding lemma means that the lower bound for the curvature K_Σ of the ruled surface Σ is a function of the parameter r along the geodesic γ, and that this function depends on the logarithmic derivative u of $y = |Y|$. Expressing the Jacobi field equation in Σ in terms of u instead of y, we obtain the differential inequality

$$u' = -K_\Sigma - u^2 \leq -\lambda - \mathrm{ct}_\lambda \, \mathrm{ct}_\Lambda + (\mathrm{ct}_\lambda + \mathrm{ct}_\Lambda) u - 2u^2. \tag{6.4}$$

This differential inequality is still of Riccati type. However, because of the factor 2 in front of the quadratic term it corresponds to a linear differential inequality of second order for the function y^2, or more appropriately for $z := y^2 / \mathrm{sn}_\lambda$, rather than for y itself. A straightforward computation shows that z satisfies the inequality

$$z'' + (\mathrm{ct}_\lambda - \mathrm{ct}_\Lambda) z' + \left(\lambda - \mathrm{ct}_\lambda (\mathrm{ct}_\lambda - \mathrm{ct}_\Lambda) \right) z \leq 0.$$

By construction, sn_λ is a solution of the corresponding differential equation. Since $\mathrm{sn}_\lambda > 0$ on $(0, r_2]$, the maximum principle implies that for any interval $[r_0, r_2] \subset (0, r_2]$ the boundary value problem

$$z'' + (\mathrm{ct}_\lambda - \mathrm{ct}_\Lambda) z' + \left(\lambda - \mathrm{ct}_\lambda (\mathrm{ct}_\lambda - \mathrm{ct}_\Lambda) \right) z = 0$$

with

$$z(r_0) = \frac{\mathrm{sn}_\Lambda(r_0)^2}{\mathrm{sn}_\lambda(r_0)}\, y'(0)^2 \quad \text{and} \quad z(r_2) = \frac{\mathrm{sn}_\Lambda(r_2)^2}{\mathrm{sn}_\lambda(r_2)}$$

has a unique solution $\bar{z}_{r_0 r_2} \geq 0$, and that $y|_{[r_0,r_2]} \geq \sqrt{\mathrm{sn}_\lambda\, \bar{z}_{r_0 r_2}}$. By means of the Wronskian and the special solution sn_λ it is possible to compute the functions $\bar{z}_{r_0 r_2}$ explicitly. As in the two-dimensional case it remains to maximize over r_0, in order to arrive at the expression for $\bar{y}$ given in the third line in (6.2).

7. On the Proof of Berger's Horseshoe Conjecture

In this section we explain the proof of Theorem 5.3. Again we shall concentrate on the geometric ideas. For brevity we write ε rather than $\varepsilon_{\mathrm{hs}}$, and we always assume that ε is sufficiently small. Details can be found in [Abresch and Meyer a, § 4].

The hypothesis on the diameter of M^n implies that the distance between the points $p_1 := \exp_{p_0}(-\pi v)$ and $p_2 := \exp_{p_0}(\pi v)$ does not exceed $\pi(1+\varepsilon)$, and the geodesic $s \mapsto \exp_{p_0}(sv)$, for $s \in [-\pi,\pi]$, connecting p_1 to p_2 looks like a horseshoe. In order to use the upper bound on the diameter of M^n more efficiently, we consider the intermediate points $q_1^\varepsilon := \exp_{p_0}\left(-\frac{1}{2}(1+\varrho_\varepsilon)\pi v\right)$ and $q_2^\varepsilon := \exp_{p_0}\left(\frac{1}{2}(1+\varrho_\varepsilon)\pi v\right)$, where $\varrho_\varepsilon \approx \frac{4}{\pi}\varepsilon^{2/3}$ is defined as the solution of the equation

$$\sin\left(\tfrac{1}{2}\varrho_\varepsilon\pi\right) = \sin\left(\tfrac{1}{4}\varepsilon^{1/3}\pi\right)^{-1}\sin\left(\tfrac{1}{2}\varepsilon\pi\right).$$

Let $c^\varepsilon : [0,1] \to M^n$ be a minimizing geodesic from q_1^ε to q_2^ε. Since $\operatorname{diam} M^n \leq \pi(1+\varepsilon)$, it is clear that c^ε does not pass through p_0. The configuration described so far is depicted in Figure 3.

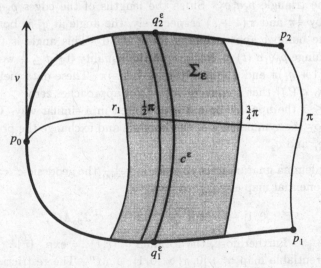

Figure 3. The horseshoe $p_1 p_0 p_2$, the geodesic c^ε, and the spherical ribbon Σ_ε.

We proceed indirectly and assume, contrary to the assertion of Theorem 5.3, that $d(p_1, p_2) \geq \pi$. This assumption enables us to control various properties of the geodesic c^ε, which finally lead to the contradiction $L(c^\varepsilon) > \pi(1 + \varepsilon) \geq \operatorname{diam} M^n$.

LEMMA 7.1. *If* $0 < \varepsilon \leq \frac{1}{64}$ *and* $d(p_1, p_2) \geq \pi$, *there are the following bounds for the distance between* p_0 *and the geodesic* c^ε *constructed above:*

$$\tfrac{7}{16}\pi \leq \tfrac{1}{2}\big(1 - \tfrac{1}{2}\varepsilon^{1/3}\big)\pi \leq d\big(p_0,\, c^\varepsilon(t)\big) < \tfrac{3}{4}\pi \qquad \text{for all } t \in [0,1].$$

Notice that the lower bound converges to $\frac{1}{2}\pi$ as ε approaches zero.

SKETCH OF PROOF. The idea for proving the lower bound is to consider the weak contraction $\Phi : M^n \to \mathbb{S}^n$ onto the sphere of constant curvature 1 which is induced by a linear isometry $T_{p_0} M^n \to T_{\bar{p}_0} \mathbb{S}^n$ and by the corresponding exponential maps. The map Φ collapses the entire complement of the ball $B(p_0, \pi)$ to the antipodal point of $\bar{p}_0$. Clearly,

$$d\big(p_0, c^\varepsilon(t)\big) \geq \inf\big\{ d(\bar{p}_0, \bar{q}) \mid \bar{q} \in \mathbb{S}^n,\, d(\bar{q}, \bar{q}_1^\varepsilon) + d(\bar{q}, \bar{q}_2^\varepsilon) \leq \pi(1 + \varepsilon) \big\},$$

where $\bar{q}_i^\varepsilon := \Phi(q_i^\varepsilon)$. In view of the fact that the infimum is achieved at some point $\bar{q}^\varepsilon \in \mathbb{S}^n$ with $d(\bar{q}^\varepsilon, \bar{q}_1^\varepsilon) = d(\bar{q}^\varepsilon, \bar{q}_2^\varepsilon) = \frac{1}{2}\pi(1 + \varepsilon)$, it is straightforward to compute the numerical value of this lower bound using the law of cosines.

For $0 \leq t \leq \frac{1}{2}$ the upper bound for the distance between p_0 and $c^\varepsilon(t)$ is obtained by applying Toponogov's triangle comparison theorem three times. Our model space is always the sphere $\mathbb{S}_\delta^2$ of constant curvature $\delta = \frac{1}{4}(1 + \varepsilon)^{-2}$. The first step is to consider the triangle $p_1 p_0 p_2$. By hypothesis $d(p_1, p_0) = d(p_0, p_2) = \pi$ and $d(p_1, p_2) \geq \pi$, so we get a lower bound for the length $d(p_1, q_2^\varepsilon)$ of the secant from p_1 to q_2^ε, which converges to π if $\varepsilon \to 0$. The next step is to consider the triangle $p_1 q_1^\varepsilon q_2^\varepsilon$. Since the lengths of the edges $p_1 q_1^\varepsilon$ and $q_1^\varepsilon q_2^\varepsilon$ are bounded by $\frac{1}{2}\pi$ and $\pi(1 + \varepsilon)$ respectively, the angle at q_1^ε is bounded from below by some number approaching $\frac{1}{2}\pi$ if $\varepsilon \to 0$. This angle is the exterior angle for the hinge $p_0 q_1^\varepsilon c^\varepsilon(t)$. Because of the inequality $0 \leq t \leq \frac{1}{2}$ we know that $d(p_0, q_1^\varepsilon) \leq \frac{1}{2}(1 + \varrho_\varepsilon)\pi$ and $d\big(q_1^\varepsilon, c^\varepsilon(t)\big) \leq \frac{1}{2}(1 + \varepsilon)\pi$. These data yield an upper bound for $d\big(p_0, c^\varepsilon(t)\big)$ that converges to $\frac{2}{3}\pi$ if ε approaches zero.

For $\frac{1}{2} \leq t \leq 1$ the upper bound is established in a similar way. One merely needs to employ the symmetry of the horseshoe and exchange the points p_1 and p_2 as well as q_1^ε and q_2^ε. $\qquad \square$

The preceding lemma guarantees that for $\varepsilon \leq \frac{1}{27\,000}$ the geodesic c^ε can be lifted under the exponential map $\exp_{p_0}$ to a curve

$$\tilde{c}^\varepsilon : [0,1] \to \overline{B(0, \tfrac{3}{4}\pi)} \smallsetminus B(0, r_1) \subset T_{p_0} M^n,$$

where $r_1 := \frac{59}{120}\pi$. Furthermore, the formula $\gamma^\varepsilon(r, t) := \exp_{p_0}\big(r\,|\tilde{c}^\varepsilon(t)|^{-1}\,\tilde{c}^\varepsilon(t)\big)$ defines a differentiable map $\gamma^\varepsilon : [0, \pi] \times [0,1] \to M^n$. The restriction of γ^ε to $[0, \pi) \times [0,1]$ describes an immersed ruled surface in M^n with a conical singularity at p_0. The geodesic c^ε lies in the image of this ruled surface. Its lift to the domain of γ^ε is the graph of the function $\tilde{r}^\varepsilon = |\tilde{c}^\varepsilon| : [0,1] \to [r_1, \frac{3}{4}\pi]$.

By the infinitesimal version of the Rauch comparison theorem the pullback metric $(\gamma^\varepsilon)^* g$ on the rectangular domain $[r_1, \frac{3}{4}\pi] \times [0,1]$ can be bounded from below in terms of the rescaled arclength function

$$\tilde{\varphi}^\varepsilon : t \mapsto \frac{1}{\sin(r_1)} \int_0^t \left| \frac{\partial}{\partial \theta} \gamma^\varepsilon(r_1, \theta) \right| d\theta$$

of the circular arc $t \mapsto \gamma^\varepsilon(r_1, t)$ as follows:

$$(\gamma^\varepsilon)^* g \geq dr^2 + \sin(r)^2 \, (d\tilde{\varphi}^\varepsilon)^2.$$

Hence the map $\mathrm{id} \times \tilde{\varphi}^\varepsilon$ yields a weak contraction from $\left([r_1, \frac{3}{4}\pi] \times [0,1], (\gamma^\varepsilon)^* g\right)$ to the spherical ribbon $\Sigma_\varepsilon = \left([r_1, \frac{3}{4}\pi] \times [0, \tilde{\varphi}^\varepsilon(1)], \bar{g}\right)$ where $\bar{g} := dr^2 + \sin(r)^2 \, d\varphi^2$ denotes the standard metric of constant curvature 1. In particular, the length of the geodesic c^ε in M^n is bounded from below by the distance of the points $\tilde{q}_1^\varepsilon := \left(\tilde{r}^\varepsilon(0), 0\right)$ and $\tilde{q}_2^\varepsilon := \left(\tilde{r}^\varepsilon(1), \tilde{\varphi}^\varepsilon(1)\right)$ in the inner metric space Σ_ε:

$$\mathrm{diam}\, M^n \geq L(c^\varepsilon) \geq \mathrm{dist}_{\Sigma_\varepsilon}\left(\tilde{q}_1^\varepsilon, \tilde{q}_2^\varepsilon\right). \tag{7.1}$$

By construction, $\frac{\partial}{\partial r} \gamma^\varepsilon(0,1) = -\frac{\partial}{\partial r}\gamma^\varepsilon(0,0) = v$, so the total angle

$$\varphi_0^\varepsilon := \int_0^1 \left| \frac{\nabla}{\partial r} \frac{\partial}{\partial t} \gamma^\varepsilon(0,t) \right| dt$$

at the conical singularity of the ruled surface described by γ^ε is bounded from below by π. The arc $t \mapsto \gamma^\varepsilon(\pi, t)$ connects the points $p_1 = \exp_{p_0}(-\pi v)$ and $p_2 = \exp_{p_0}(\pi v)$, and thus its length is $\geq d(p_1, p_2) \geq \pi$.

After these preparations we apply Theorem 6.3 with $\lambda = \delta_{\mathrm{hs}}$, $\Lambda = 1$, and $r_2 = \pi$, and use the monotonicity and homogeneity properties of $\Psi_{r_1\pi}$ asserted in Theorem 6.1(i) and (ii) to conclude that

$$\sin(r_1)\tilde{\varphi}^\varepsilon(1) \geq \Psi_{r_1\pi}(\varphi_0^\varepsilon, \pi) \geq \pi \Psi_{r_1\pi}(1,1).$$

As explained in Remark 6.4, it follows that $\tilde{\varphi}^\varepsilon(1) \geq (1+a_\varepsilon)\pi$ for $a_\varepsilon \approx 0.001\,661$. Notice that a_ε is a monotonically decreasing function of ε, and thus the preceding lower bound for $\tilde{\varphi}^\varepsilon(1)$ holds uniformly for all $\varepsilon \leq \varepsilon_{\mathrm{hs}}$.

Geometrically, this bound means that the length of the equatorial arc joining the two longitudinal segments in the boundary of the spherical ribbon Σ_ε exceeds π by a certain fixed amount. Since the points $\tilde{q}_1^\varepsilon$ and $\tilde{q}_2^\varepsilon$ on these longitudinal boundary arcs approach the equator if ε gets small, we conclude that their distance in the inner geometry of the ribbon is not only greater than π but even greater than $(1+\varepsilon)\pi$, provided ε is sufficiently small. But the inequality $\mathrm{dist}_{\Sigma_\varepsilon}\left(\tilde{q}_1^\varepsilon, \tilde{q}_2^\varepsilon\right) > (1+\varepsilon)\pi \geq \mathrm{diam}\, M^n$ contradicts (7.1).

8. Final Remarks

The injectivity radius estimate in Theorem 4.1, the sphere theorem and the cohomological pinching below-$\frac{1}{4}$ theorem stated in 5.1 and 5.2, and the horseshoe inequality in Theorem 5.3 have this feature in common: In each theorem the pinching constant is an explicit number $< \frac{1}{4}$, independent of the dimension. Nevertheless, we do not know the optimal value for any of the pinching constants δ_{inj}, δ_{odd}, δ_{ev}, or δ_{hs}. The current proofs combine several curvature controlled estimates in such a way that they do not become sharp simultaneously.

The values provided by these proofs are only slightly smaller than $\frac{1}{4}$, and thus they differ significantly from the values that are obstructed by the counterexamples known today. Yet, these lower bounds cannot be considered any less artificial than the numbers provided by the current proofs. For instance, because of the Berger spheres described in Remark 1.7(ii) it is necessary that $\delta_{\text{inj}} \geq 0.117 > \frac{1}{9}$. Furthermore, the examples in Table 1 show that $\delta_{\text{ev}} \geq \frac{1}{64}$ and $\delta_{\text{odd}} \geq \frac{1}{37}$. But the significance of the latter two numbers is impaired by the fact that only in low dimensions do we know any examples of compact, simply connected Riemannian manifolds with strictly positive sectional curvature that are not homeomorphic to spheres or projective spaces.

Appendix: Compact Manifolds of Positive Curvature

Table 1 lists all simply connected manifolds that are known to carry metrics with positive sectional curvature. To begin with, there are the symmetric spaces with $K_M > 0$. They must have rank one, and thus one is left with the spheres $\mathbb{S}^n$ and the projective spaces $\mathbb{CP}^n$, $\mathbb{HP}^n$, and $\text{Ca}\mathbb{P}^2$ equipped with the Fubini–Study metric.

Further examples of compact manifolds with $K_M > 0$ are already quite scarce. Currently, non-symmetric examples appear only in very few dimensions. They have been discovered as follows: First, Berger [1961] classified the simply connected, normal homogeneous spaces with $K_M > 0$. The only new examples that appeared in this classification are the two odd-dimensional spaces $\text{Sp}(2)/\text{SU}(2)$ and $M^{13} = \text{SU}(5)/(\text{Sp}(2) \times \text{S}^1)$. Their pinching constants $\delta_M := \min K_M / \max K_M$ were calculated later by H. Eliasson [1966] and Heintze [1971], respectively.

The next step was the classification by Wallach [1972] of simply connected, even-dimensional, homogeneous spaces with $K_M > 0$. Besides spheres and projective spaces there are only the three flag manifolds $M^6 = \text{SU}(3)/\text{T}^2$, $M^{12} = \text{Sp}(3)/(\text{SU}(2) \times \text{SU}(2) \times \text{SU}(2))$, and $M^{24} = \text{F}_4/\text{Spin}(8)$. They are, respectively, an $\mathbb{S}^2$-bundle over $\mathbb{CP}^2$, an $\mathbb{S}^4$-bundle over $\mathbb{HP}^2$, and an $\mathbb{S}^8$-bundle over $\text{Ca}\mathbb{P}^2$. As shown in [Valiev 1979], the pinching constants for the three flag manifolds are $\frac{1}{64}$; compare also [Grove 1989].

An infinite series of homogeneous, odd-dimensional examples has been found by S. Aloff and N. Wallach [1975]. They are quotient spaces $M_{k,l}^7 = \mathsf{SU}(3)/\mathsf{S}_{k,l}^1$, where the integers k and l label the various embeddings of S^1 into a maximal torus $\mathsf{T}^2 \subset \mathsf{SU}(3)$. H.-M. Huang [1981] computed that the pinching constants induced by a particular left-invariant metric defined in [Aloff and Wallach 1975] approach $\frac{16}{29\cdot37}$ as $\frac{k}{l} \to 1$. The manifolds $M_{k,l}^7$ are not only interesting for their geometric properties. M. Kreck and S. Stolz [1991] have found out that this sequence of examples contains seven-manifolds that are homeomorphic but not diffeomorphic.

M	dim	$\min K_M / \max K_M$	
symmetric spaces			
S^n	n	1	
$\mathbb{CP}^n$	$2n$	$\frac{1}{4}$	
$\mathbb{HP}^n$	$4n$	$\frac{1}{4}$	
$\mathrm{Ca}\mathbb{P}^2$	16	$\frac{1}{4}$	
normal homogeneous spaces [Berger 1961]			
$\mathsf{Sp}(2)/\mathsf{SU}(2)$	7	$\frac{1}{37}$	[Eliasson 1966]
$\mathsf{SU}(5)/(\mathsf{Sp}(2)\times\mathsf{S}^1)$	13	$\frac{16}{29\cdot37}$	[Heintze 1971]
ditto with a nonnormal metric		$\frac{1}{37}$	[Püttmann 1996]
flag manifolds [Wallach 1972]			
$\mathsf{SU}(3)/\mathsf{T}^2$	6	$\frac{1}{64}$	[Valiev 1979]
$\mathsf{Sp}(3)/(\mathsf{SU}(2)^3)$	12	$\frac{1}{64}$	[Valiev 1979]
$\mathsf{F}_4/\mathsf{Spin}(8)$	24	$\frac{1}{64}$	[Valiev 1979]
Aloff–Wallach examples [1975]			
$\mathsf{SU}(3)/\mathsf{S}_{k,l}^1$	7	$\to \frac{16}{29\cdot37}$	[Huang 1981]
ditto with certain Einstein metrics		$\to \frac{1}{37}$	[Püttmann 1996]
inhomogeneous orbit spaces [Eschenburg 1982; Bazaĭkin 1995]			
$\mathsf{SU}(3)/(\mathsf{T}^2\text{-action})$	6	$?$	
$\mathsf{SU}(3)/(\mathsf{S}_{klpq}^1\text{-action})$	7	$\to \frac{1}{37}$	[Püttmann 1996]
$\mathsf{S}_{p_1\ldots p_5}^1 \backslash \mathsf{U}(5)/(\mathsf{Sp}(2)\times\mathsf{S}^1)$	13	$\to \frac{1}{37}$	[Püttmann 1996]

Table 1. Compact, simply connected manifolds with $K_M > 0$. The arrows in front of some of the pinching constants indicate that the given values appear as limits for properly chosen subsequences.

L. Bérard-Bergery [1976] showed that there exist no other simply connected, odd-dimensional, homogeneous spaces of positive curvature. This result finishes the classification of the simply connected, homogeneous spaces with $K_M > 0$.

So far, nonhomogeneous examples have only been obtained as inhomogeneous orbit spaces, where a subgroup $H \subset G \times G$ acts on a simply connected Lie group G. Analyzing the two-sided T^2- and S^1-actions on $\mathsf{SU}(3)$, J. Eschenburg [1984; 1982; 1992] has found a six-dimensional inhomogeneous space of positive curvature and an infinite family of seven-dimensional inhomogeneous orbit spaces. These examples resemble the Aloff–Wallach examples in many respects.

Following Eschenburg's approach, A. Bazaĭkin [1995] has recently constructed an infinite sequence of 13-dimensional, simply connected, pairwise nonhomeomorphic orbit spaces with strictly positive sectional curvature. These examples are biquotients that are closely related to the second one of the normal homogeneous spaces discovered by Berger. A complete classification of all two-sided actions which lead to simply connected Riemannian manifolds with positive curvature has not been accomplished as yet.

Recently, I. Taimanov [1996] has discovered an isometric, totally geodesic embedding of the Aloff–Wallach space $M_{1,1}^7$, equipped with the metric considered by Huang, into the Berger space M^{13}, so explaining to some extent the curious coincidence of the pinching constants determined by Heintze and Huang, respectively. In Bazaĭkin's work the manifold $\mathsf{SU}(5)/(\mathsf{Sp}(2) \times \mathsf{S}^1)$ appears with a deformed metric that is homogeneous but not normal homogeneous. Under Taimanov's embedding $M_{1,1}^7 \hookrightarrow \mathsf{SU}(5)/(\mathsf{Sp}(2) \times \mathsf{S}^1)$ the corresponding one-parameter deformation induces the Aloff–Wallach metrics on $\mathsf{SU}(3)/\mathsf{S}_{1,1}^1$ with a slightly different parametrization. Th. Püttmann [1996] has computed the curvature tensors of these one-parameter deformations of metrics on $\mathsf{SU}(3)/\mathsf{S}_{1,1}^1$ and $\mathsf{SU}(5)/(\mathsf{Sp}(2) \times \mathsf{S}^1)$ in a systematic way, finding that $\frac{1}{37}$ is the optimal pinching constant in each case. The coincidence of the two constants is not a complete surprise, since Taimặnov's embedding stays totally geodesic for the entire deformation; however, we do not have any explanation why this value also coincides with the pinching constant of the seven-dimensional Berger space $\mathsf{Sp}(2)/\mathsf{SU}(2)$. As yet, the curvature computations cover all homogeneous metrics on $\mathsf{Sp}(2)/\mathsf{SU}(2)$ and $\mathsf{SU}(5)/(\mathsf{Sp}(2) \times \mathsf{S}^1)$, but it is still an open question whether or not $\frac{1}{37}$ remains optimal for the entire nine-parameter family of homogeneous metrics that exists on $\mathsf{SU}(3)/\mathsf{S}_{1,1}^1$.

As a consequence of Püttmann's improvments, any subsequence of the Aloff–Wallach examples $M_{k,l}^7$ where $\frac{k}{l} \to 1$ admits metrics whose pinching constants approach $\frac{1}{37}$. An analogous statement holds for the seven-dimensional inhomogeneous spaces of Eschenburg and the 13-dimensional spaces of Bazaĭkin.

It is a curious coincidence that the optimal, $\frac{1}{37}$-pinched metric on the Aloff–Wallach space $M_{1,1}^7$ is one of the Einstein metrics discovered by M. Wang [1982].

Furthermore, it can be shown that any $M_{k,l}^7$ with $\frac{k}{l}$ sufficiently close to 1 carries a homogeneous Einstein metric whose pinchimg constant is close to $\frac{1}{37}$. In contrast, the optimal metric on the 13-dimensional Berger space M^{13} is not Einstein.

References

[Abraham 1970] R. Abraham, "Bumpy metrics", pp. 1–3 in *Global analysis* (Berkeley, 1968) edited by S.-S. Chern and S. Smale, Proc. Symp. Pure Math. **14**, AMS, Providence, RI, 1970.

[Abresch 1985] U. Abresch, "Lower curvature bounds, Toponogov's theorem, and bounded topology I", *Ann. scient. Éc. Norm. Sup.* (4) **18** (1985), 651–670.

[Abresch 1987] U. Abresch, "Lower curvature bounds, Toponogov's theorem, and bounded topology II", *Ann. scient. Éc. norm. sup.* (4) **20** (1987), 475–502.

[Abresch and Meyer 1994] U. Abresch and W. T. Meyer, "Pinching below $\frac{1}{4}$, injectivity radius estimates, and sphere theorems", *J. Diff. Geom.* **40** (1994), 643–691.

[Abresch and Meyer a] U. Abresch and W. T. Meyer, "A sphere theorem with a pinching constant below $\frac{1}{4}$", to appear in *J. Diff. Geom.*

[Adams 1960] J. F. Adams, "On the non-existence of elements of Hopf invariant one", *Ann. of Math.* **72** (1960), 20–104.

[Adem 1953] J. Adem, "Relations on iterated reduced powers", *Proc. Nat. Acad. Sci. USA* **39** (1953), 636–638.

[Aloff and Wallach 1975] S. Aloff and N. R. Wallach, "An infinite family of distinct 7-manifolds admitting positively curved Riemannian structures", *Bull. Amer. Math. Soc.* **81** (1975), 93–97.

[Anderson 1990] M. Anderson, "Short geodesics and gravitational instantons", *J. Diff. Geom.* **31** (1990), 265–275.

[Bazaĭkin 1995] Я. В. Базайкин, "Об одном семействе 13-мерных замкнутых римановых многообразий положительной секционной кривизны" [On a family of 13-dimensional closed Riemannian manifolds of positive sectional curvature], Thesis, Univ. of Novosibirsk, 1995.

[Bérard-Bergery 1976] L. Bérard-Bergery, "Les variétés riemanniennes homogènes simplement connexes de dimension impaire à courbure strictement positive", *J. Math. pure et appl.* **55** (1976), 47–68.

[Berger 1960a] M. Berger, "Sur quelques variétés Riemanniennes suffisamment pincées", *Bull. Soc. math. France* **88** (1960), 57–71.

[Berger 1960b] M. Berger, "Les variétés riemanniennes $\frac{1}{4}$-pincées", *Ann. Scuola Norm. Sup. Pisa* **14** (1960), 161–170.

[Berger 1961] M. Berger, "Les variétés riemanniennes homogènes normales simplement connexes à courbure strictement positive", *Ann. Scuola Norm. Sup. Pisa* **15** (1961), 179–246.

[Berger 1962a] M. Berger, "On the diameter of some Riemannian manifolds", Technical report, Univ. of Califiornia, Berkeley, 1962.

[Berger 1962b] Berger, M.: "An extension of Rauch's metric comparison theorem and some applications", *Ill. Jour. of Math.* **6** (1962), 700–712.

[Berger 1965] M. Berger, *Lectures on geodesics in Riemannian geometry*, Tata Institute, Bombay, 1965.

[Berger 1983] M. Berger, "Sur les variétés riemanniennes pincées juste au-dessous de $\frac{1}{4}$", *Ann. Inst. Fourier* **33** (1983), 135–150.

[Cheeger and Ebin 1975] J. Cheeger and D. Ebin, *Comparison theorems in Riemannian geometry*, North-Holland, Amsterdam, and Elsevier, New York, 1975.

[Cheeger and Gromoll 1980] J. Cheeger and D. Gromoll, "On the lower bound for the injectivity radius of $\frac{1}{4}$-pinched Riemannian manifolds", *J. Diff. Geom.* **15** (1980), 437–442.

[Cheeger et al. 1982] J. Cheeger, M. Gromov, and M. Taylor, "Finite propagation speed, kernel estimates for functions of the Laplace operator, and the geometry of complete Riemannian manifolds", *J. Diff. Geom.* **17** (1982), 15–53.

[Durumeric 1984] O. C. Durumeric, "Manifolds with almost equal diameter and injectivity radius", *J. Diff. Geom.* **19** (1984), 453–474.

[Durumeric 1987] O. C. Durumeric, "A generalization of Berger's theorem on almost $\frac{1}{4}$-pinched manifolds, II", *J. Diff. Geom.* **26** (1987), 101–139

[Eells and Kuiper 1962] J. Eells and N. H. Kuiper, "Manifolds which are like projective spaces", *Publ. Math. IHES* **14** (1962), 181–222.

[Eliasson 1966] H. I. Eliasson, "Die Krümmung des Raumes $Sp(2)/SU(2)$ von Berger", *Math. Ann.* **164** (1966), 317–327.

[Eschenburg 1982] J.-H. Eschenburg, "New examples of manifolds with strictly positive curvature", *Invent. Math.* **66** (1982), 469–480.

[Eschenburg 1984] J.-H. Eschenburg, "Freie isometrische Aktionen auf kompakten Liegruppen mit positiv gekrümmten Orbiträumen", *Schriftenreihe Math. Inst. Univ. Münster* **32** (1984).

[Eschenburg 1992] J.-H. Eschenburg, "Inhomogeneous spaces of positive curvature", *Diff. Geom. Appl.* **2** (1992), 123–132.

[Freedman 1982] M. H. Freedman, "The topology of four-dimensional manifolds", *J. Diff. Geom.* **17** (1982), 357–453.

[Greub et al. 1973] W. Greub, S. Halperin, and R. Vanstone, *Connections, curvature, and cohomology*, vol. II, Academic Press, New York and London, 1973.

[Gromoll 1966] D. Gromoll, "Differenzierbare Strukturen und Metriken positiver Krümmung auf Sphären", *Math. Ann.* **164** (1966), 353–371.

[Gromoll and Grove 1987] D. Gromoll and K. Grove, "A generalization of Berger's rigidity theorem for positively curved manifolds", *Ann. scient. Éc. Norm. Sup.* (4) **20** (1987), 227–239.

[Gromoll and Grove 1988] D. Gromoll and K. Grove, "The low-dimensional metric foliations of Euclidean spheres", *J. Diff. Geom.* **28** (1988), 143–156.

[Gromoll and Meyer 1969] D. Gromoll and W. Meyer, "On differentiable functions with isolated critical points", *Topology* **8** (1969), 361–369.

[Gromoll and Meyer 1974] D. Gromoll and W. Meyer, "An exotic sphere with non-negative sectional curvature", *Ann. of Math.* **100** (1974), 401–408.

[Gromov 1981a] M. Gromov, *Structures métriques pour les variétés riemanniennes*, edited by J. Lafontaine and P. Pansu, CEDIC, Paris, 1981.

[Gromov 1981b] M. Gromov, "Curvature, diameter, and Betti numbers", *Comm. Math. Helv.* **56** (1981), 179–195.

[Grove 1987] K. Grove, "Metric differential geometry", pp. 171–227 in *Differential geometry* (Lyngby, 1985), edited by V. L. Hansen, Lect. Notes Math. **1263**, Springer, Berlin, 1987.

[Grove 1989] K. Grove, "The even-dimensional pinching problem and $SU(3)/T$", *Geom. Dedicata* **29** (1989), 327–334.

[Grove 1993] K. Grove, "Critical point theory for distance functions", pp. 357–385 in *Differential geometry: Riemannian geometry* (Los Angeles, 1990), edited by R. Greene and S. T. Yau, Proc. Symp. Pure Math. **54**, part 3, AMS, Providence, RI, 1993.

[Grove and Shiohama 1977] K. Grove and K. Shiohama, "A generalized sphere theorem", *Ann. of Math.* **106** (1977), 201–211.

[Grove et al. 1974a] K. Grove, H. Karcher, and E. Ruh, "Group actions and curvature", *Invent. Math.* **23** (1974), 31–48.

[Grove et al. 1974b] K. Grove, H. Karcher, and E. Ruh, "Jacobi fields and Finsler metrics with applications to the differentiable pinching problem", *Math. Annalen* **211** (1974), 7–21.

[Grove et al. 1990] K. Grove, P. Petersen, and J.-Y. Wu, "Geometric finiteness theorems via controlled topology", *Invent. Math.* **99** (1990), 205–213; erratum in **104** (1991), 221–222.

[Hamilton 1982] R. S. Hamilton, "Three-manifolds with positive Ricci curvature", *J. Diff. Geom.* **17** (1982), 255–306.

[Heintze 1971] E. Heintze, "The curvature of $SU(5)/(Sp(2) \times S^1)$", *Invent. Math.* **13** (1971), 205–212.

[Heintze and Karcher 1978] E. Heintze and H. Karcher, : "A general comparison theorem with applications to volume estimates for submanifolds", *Ann. scient. Éc. Norm. Sup.* (4) **11** (1978), 451–470.

[Huang 1981] H.-M. Huang, "Some remarks on the pinching problems", *Bull. Inst. Math. Acad. Sin.* **9** (1981), 321–340.

[Im Hof and Ruh 1975] H.-C. Im Hof and E. A. Ruh, "An equivariant pinching theorem", *Comment. Math. Helv.* **50** (1975), 389–401.

[Karcher 1989] H. Karcher, *Riemannian comparison constructions*, MAA Stud. Math. **27**, Math. Assoc. of America, Washington, DC, 1989.

[Klingenberg 1959] W. Klingenberg, "Contributions to Riemannian geometry in the large", *Ann. of Math.* **69** (1959), 654–666.

[Klingenberg 1961] W. Klingenberg, "Über Mannigfaltigkeiten mit positiver Krümmung", *Comm. Math. Helv.* **35** (1961), 47–54.

[Klingenberg 1962] W. Klingenberg, "Über Mannigfaltigkeiten mit nach oben beschränkter Krümmung", *Ann. Mat. Pura Appl.* **60** (1962), 49–59.

[Klingenberg and Sakai 1980] W. Klingenberg and T. Sakai, "Injectivity radius estimate for $\frac{1}{4}$-pinched manifolds", *Arch. Math.* **34** (1980), 371–376.

[Kreck and Stolz 1991] M. Kreck and S. Stolz, "Some nondiffeomorphic homeomorphic homogeneous 7-manifolds with positive sectional curvature", *J. Diff. Geom.* **33** (1991), 465–486.

[Meyer 1989] W. T. Meyer, "Toponogov's theorem and its applications", Lecture Notes, College on Differential Geometry, Trieste, 1989.

[Micallef and Moore 1988] M. Micallef and J. D. Moore, "Minimal two-spheres and the topology of manifolds with positive curvature on totally isotropic two-planes", *Ann. of Math.* **127** (1988), 199–227.

[Milnor 1965] J. Milnor, "Lectures on the h-cobordism theorem", Princeton Mathematical Notes, Princeton University Press, Princeton, NJ, 1965.

[Milnor and Stasheff 1974] J. Milnor and J. Stasheff, *Characteristic classes*, Ann. of Math. Studies **76**, Princeton University Press, Princeton, NJ, 1974.

[Myers 1935] S. Myers, "Riemannian manifolds in the large", *Duke Math. J.* **1** (1935), 39–49.

[Myers 1941] S. Myers, "Riemannian manifolds with positive mean curvature", *Duke Math. J.* **8** (1941), 401–404.

[Nikolaev 1983] I. G. Nikolaev, "Smoothness of the metric of spaces with two-sided bounded Aleksandrov curvature", *Siberian Math. J.* **24** (1983), 247–263; Russian original in *Sibirsk. Mat. Zh.* **24** (1983), 114–132.

[Peters 1987] S. Peters, "Convergence of Riemannian manifolds", *Comp. Math.* **62** (1987), 3–16.

[Püttmann 1996] Th. Püttmann, "Improved pinching constants of odd-dimensional homogeneous spaces", preprint, Bochum, 1996.

[Rauch 1951] H. E. Rauch, "A contribution to differential geometry in the large", *Ann. Math.* **54** (1951), 38–55.

[Ruh 1982] E. A. Ruh, "Riemannian manifolds with bounded curvature ratios", *J. Diff. Geom.* **17** (1982), 643–653.

[Samelson 1963] H. Samelson, "On manifolds with many closed geodesics", *Portugaliae Math.* **22** (1963), 193–196.

[Seaman 1989] W. Seaman, "A pinching theorem for four manifolds", *Geom. Dedicata* **31** (1989), 37–40.

[Shikata 1966] Y. Shikata, "On a distance function on the set of differentiable structures", *Osaka J. Math.* **3** (1966), 65–79.

[Smale 1961] S. Smale, "Generalized Poincaré's conjecture in dimensions greater than four", *Ann. of Math.* **74** (1961), 391–466.

[Synge 1936] J. Synge, "On the connectivity of spaces of positive curvature", *Quart. J. Math.* (Oxford Ser.) **7** (1936), 316–320.

[Taimanov 1996] I. Taimanov, "A remark on positively curved manifolds of dimensions 7 and 13", preprint, Novosibirsk, 1996.

[Valiev 1979] F. M. Valiev, "Sharp estimates of sectional curvatures of homogeneous Riemannian metrics on Wallach spaces" (Russian), *Sibirsk. Mat. Zh.* **20**:2 (1979), 248–262, 457.

[Wallach 1972] N. R. Wallach, "Compact homogeneous Riemannian manifolds with strictly positive curvature", *Ann. of Math.* **96** (1972), 277–295.

[Wang 1982] M. Wang, "Some examples of homogeneous Einstein manifolds in dimension seven", *Duke Math. J.* **49** (1982), 23–28.

[Weinstein 1967] A. Weinstein, "On the homotopy type of positively pinched manifolds", *Arch. Math.* **18** (1967), 523–524.

[Weiss 1993] M. Weiss, "Pinching and concordance theory", *J. Diff. Geom.* **38** (1993), 387–416.

[Wilhelm 1995] F. Wilhelm, "The radius rigidity theorem for manifolds of positive curvature", preprint, SUNY at Stony Brook, 1995.

[Wolf 1977] J. A. Wolf, *Spaces of constant curvature*, 4th edition, Publish or Perish, Berkeley, 1977.

UWE ABRESCH
FAKULTÄT FÜR MATHEMATIK
RUHR-UNIVERSITÄT BOCHUM
UNIVERSITÄTSSTRASSE 150
D-44780 BOCHUM
GERMANY
    abresch@math.ruhr-uni-bochum.de

WOLFGANG T. MEYER
WESTFÄLISCHE WILHELMS–UNIVERSITÄT
EINSTEINSTRASSE 62
D–48149 MÜNSTER
GERMANY
    meyerw@math.uni-muenster.de

Comparison Geometry
MSRI Publications
Volume 30, 1997

Scalar Curvature and Geometrization Conjectures for 3-Manifolds

MICHAEL T. ANDERSON

ABSTRACT. We first summarize very briefly the topology of 3-manifolds and the approach of Thurston towards their geometrization. After discussing some general properties of curvature functionals on the space of metrics, we formulate and discuss three conjectures that imply Thurston's Geometrization Conjecture for closed oriented 3-manifolds. The final two sections present evidence for the validity of these conjectures and outline an approach toward their proof.

Introduction

In the late seventies and early eighties Thurston proved a number of very remarkable results on the existence of geometric structures on 3-manifolds. These results provide strong support for the profound conjecture, formulated by Thurston, that every compact 3-manifold admits a canonical decomposition into domains, each of which has a canonical geometric structure.

For simplicity, we state the conjecture only for closed, oriented 3-manifolds.

GEOMETRIZATION CONJECTURE [Thurston 1982]. *Let M be a closed, oriented, prime 3-manifold. Then there is a finite collection of disjoint, embedded tori T_i^2 in M, such that each component of the complement $M \setminus \bigcup T_i^2$ admits a geometric structure, i.e., a complete, locally homogeneous Riemannian metric.*

A more detailed description of the conjecture and the terminology will be given in Section 1. A complete Riemannian manifold N is *locally homogeneous* if the universal cover $\tilde{N}$ is a complete homogenous manifold, that is, if the isometry group Isom $\tilde{N}$ acts transitively on $\tilde{N}$. It follows that N is isometric to $\tilde{N}/\Gamma$, where Γ is a discrete subgroup of Isom $\tilde{N}$, which acts freely and properly discontinuously on $\tilde{N}$.

Partially supported by NSF Grants DMS-9022140 at MSRI and DMS-9204093.

Thurston showed that, in dimension three, there are eight possible geometries, all of which are realized. Namely, the universal covers are either the constant curvature spaces $\mathbb{H}^3$, $\mathbb{E}^3$, $\mathbb{S}^3$, or products $\mathbb{H}^2 \times \mathbb{R}$, $\mathbb{S}^2 \times \mathbb{R}$, or twisted products $\widetilde{\mathrm{SL}}(2, \mathbb{R})$, Nil, Sol (see Section 1B).

It is perhaps easiest to understand the context and depth of this conjecture by recalling the classical uniformization (or geometrization) theorem for surfaces, due to Poincaré and Koebe. If M is a closed, oriented surface, the uniformization theorem asserts that M carries a smooth Riemannian metric of constant curvature K equal to -1, 0 or $+1$. This means that it carries a geometric structure modeled on $\mathbb{H}^2$, $\mathbb{E}^2$ or $\mathbb{S}^2$, respectively. Further, knowledge of the sign of the curvature and the area of the surface gives a complete topological description of the surface, via the Gauss–Bonnet formula

$$2\pi\chi(M) = \int_M K \, dV.$$

The validity of the Geometrization Conjecture in dimension three would similarly provide a deep topological understanding of 3-manifolds, as well as a vast array of topological invariants, arising from the geometry of the canonical metrics.

There is a noteworthy difference between these pictures in dimensions two and three, however. In two dimensions, there is typically a nontrivial space of geometric structures, that is, of constant curvature metrics—the moduli space, or the related Teichmüller space. Only the case $K = +1$ is rigid, that is, there is a unique metric (up to isometry) of constant curvature $+1$ on $\mathbb{S}^2$. The moduli space of flat metrics on a torus is a two-dimensional variety, and that of hyperbolic metrics on surfaces of higher genus g is a variety of dimension $3g - 3$. As we will indicate briefly below, these moduli spaces also play a crucial role in Thurston's approach to and results on the Geometrization Conjecture.

In dimension three, the geometric structures are usually rigid. The moduli spaces of geometric structures, if not a point, tend to arise from the moduli of geometric structures on surfaces. In any case, the question of uniqueness or moduli of smooth geometric structures on a smooth 3-manifold is by and large understood; what remains is the question of existence.

The Geometrization Conjecture may be viewed as a question about the existence of canonical or distinguished Riemannian metrics on 3-manifolds that satisfy certain topological conditions. This type of question has long been of fundamental interest to workers in Riemannian geometry and analysis on manifolds. For instance, it is common folkore that Yamabe viewed his work on what is now known as the Yamabe problem [1960] as a step towards the resolution of the Poincaré conjecture. Further, it has been a longstanding open problem to understand the existence and moduli space of Einstein metrics (that is, metrics of constant Ricci curvature) on closed n-manifolds. Most optimistically, one would like to find necessary and sufficient topological conditions that guarantee the existence of such a metric. The Thurston conjecture, if true, provides the answer to this in dimension three. (Einstein metrics in dimension three are metrics of constant curvature).

One of the most natural means of producing canonical metrics on smooth manifolds is to look for metrics that are critical points of a natural functional on the space of all metrics on the manifold. In fact, the definition of Einstein metrics is best understood from this point of view.

Briefly, let $\mathcal{M}_1$ denote the space of all smooth Riemannian structures of total volume 1 on a closed n-manifold M. Two Riemannian metrics g_0 and g_1 are *equivalent* or *isometric* if there is a diffeomorphism f of M such that $f^* g_0 = g_1$; we also say that they *define the same structure* on M. Given a metric $g \in \mathcal{M}_1$, let $s_g : M \to \mathbb{R}$ be its scalar curvature (the average of all the curvatures in the two-dimensional subspaces of TM), and let dV_g be the volume form determined by the metric and orientation. The total scalar curvature functional $\mathcal{S}$ is defined by

$$\mathcal{S} : \mathcal{M}_1 \to \mathbb{R}, \qquad \mathcal{S}(g) = \int_M s_g \, dV_g.$$

Hilbert showed that the critical points of this functional are exactly the *Einstein metrics*, that is, metrics that satisfy the Euler–Lagrange equation

$$Z_g := \mathrm{Ric}_g - \frac{s_g}{n} g = 0,$$

where Ric_g is the Ricci curvature of g (Section 2) and n is the dimension of M. It is an elementary exercise to show that in dimension three (and only in dimension three) the solutions of this equation are exactly the metrics of constant curvature, that is, metrics having geometric structure $\mathbb{H}^3$, $\mathbb{E}^3$ or $\mathbb{S}^3$.

In fact, $\mathcal{S}$ is the only functional on $\mathcal{M}_1$ known to the author whose critical points are exactly the metrics of constant curvature in dimension three. The fact that $\mathcal{S}$ is also the simplest functional that one can form from the curvature invariants of the metric makes it especially appealing.

The three geometries $\mathbb{H}^3$, $\mathbb{E}^3$, $\mathbb{S}^3$ of constant curvature are by far the most important of the eight geometries in understanding the geometry and topology of 3-manifolds. $\mathbb{H}^3$ and $\mathbb{S}^3$, in particular, play the central roles.

In this article, we will outline an approach toward the Geometrization Conjecture, based on the study of the total scalar curvature on the space of metrics on M^3. We formulate and discuss (Section 4) three conjectures on the geometry and topology of the limiting behavior of metrics on a 3-manifold that attempt to realize a critical point of $\mathcal{S}$. This conjecture, if true, implies that the geometrization of a 3-manifold can be implemented or performed by studying the convergence and degeneration of such a sequence of metrics.

R. Hamilton has developed another program toward resolution of the Geometrization Conjecture, by studying the singularity formation and long-time existence and convergence behavior of the Ricci flow on $\mathcal{M}$. This has of course already been spectacularly successful in certain cases [Hamilton 1982].

This article is intended partly as a brief survey of ideas related to the Thurston conjecture and of the approach to this conjecture indicated above. A number of new results are included in Sections 4–6, in order to substantiate this approach. However, by and large, only statements of results are provided, with references to proofs elsewhere, mainly in [Anderson a; b; c]. The paper is an expanded, but basically unaltered, version of talks given at the September 1993 MSRI Workshop.

1. Review of 3-manifolds: Topology, Geometry and Thurston's Results

1A. Topology. Throughout the paper, M will denote a closed, oriented 3-manifold and N will denote a compact, oriented 3-manifold with (possibly empty) compact boundary. There are two basic topological decompositions of M, obtained by examining the structure of the simplest types of surfaces embedded in M, namely spheres and tori.

THEOREM 1.1 (SPHERE DECOMPOSITION [Kneser 1929; Milnor 1962]). *Let M be a closed, oriented 3-manifold. Then M has a finite decomposition as a connected sum*

$$M = M_1 \# M_2 \# \cdots \# M_k,$$

where each M_i is prime. The collection $\{M_i\}$ is unique, up to permutation of the factors. (A closed 3-manifold is *prime* if it is not the three-sphere and cannot be written as a nontrivial connected sum of closed 3-manifolds.)

This sphere decomposition (or *prime decomposition*) is obtained by taking a suitable maximal family of disjoint embedded two-spheres in M, none of which bounds a three-ball, and cutting M along those spheres. The summands M_i are formed by gluing in three-balls to the boundary spheres. (This implicitly uses the Alexander–Schoenflies theorem, which says that any two-sphere embedded in $\mathbb{S}^3$ bounds a 3-ball.)

The sphere decomposition is canonical in the sense that the summands are unique up to homeomorphism. However, the collection of spheres is not necessarily unique up to isotopy; it is unique up to diffeomorphism of M.

A 3-manifold M is *irreducible* if every smooth two-sphere embedded in M bounds a three-ball in M. Clearly an irreducible 3-manifold is prime. The converse is almost true: a prime orientable 3-manifold is either irreducible or is $\mathbb{S}^2 \times \mathbb{S}^1$ [Hempel 1983].

The topology of an irreducible 3-manifold M is coarsely determined by the cardinality of the fundamental group. For then the sphere theorem [Hempel 1983] implies that $\pi_2(M) = 0$. Let $\tilde{M}$ be the universal cover of M, so that $\pi_1(\tilde{M}) = \pi_2(\tilde{M}) = 0$. If $\pi_1(M)$ is finite, $\tilde{M}$ is closed, and thus a homotopy three-sphere (that is, a simply connected closed 3-manifold), by elementary algebraic topology. If $\pi_1(M)$ is infinite, $\tilde{M}$ is open, and thus contractible (by the Hurewicz theorem); therefore M is a $K(\pi, 1)$, that is, M is aspherical.

Thus, the prime decomposition of Theorem 1.1 can be rewritten as

$$M = (K_1 \# K_2 \# \cdots \# K_p) \# (L_1 \# L_2 \# \cdots \# L_q) \# \left(\overset{r}{\underset{1}{\#}} (\mathbb{S}^2 \times \mathbb{S}^1) \right),$$

where the factors K_i are closed, irreducible and aspherical, while the factors L_j are closed, irreducible and finitely covered by homotopy three-spheres. Thus, one needs to understand the topology of the factors K_i and L_j.

It is worth emphasizing that the sphere decomposition is perhaps the simplest topological procedure that is performed in understanding the topology of 3-manifolds. In contrast, in dealing with the geometry and analysis of metrics on 3-manifolds, we will see that this procedure is the most difficult to perform or understand.

From now on, we make the further assumption that M and N are irreducible. Before stating the torus decomposition theorem, we introduce several definitions. Let S be a compact, oriented surface embedded in N (and thus having trivial normal bundle), with $\partial S \subset \partial N$. The surface S is *incompressible* if, for every closed disc D embedded in N with $D \cap S = \partial D$, the curve ∂D is contractible in S. This happens if and only if the inclusion map induces an injection $\pi_1(S) \to \pi_1(N)$ of fundamental groups (see [Jaco 1980, Lemma III.8]; his definition of incompressibility disagrees with ours for $S = \mathbb{S}^2$). If S is not incompressible, it is *compressible*. A 3-manifold N is *Haken* if it contains an incompressible surface of genus $g \geq 1$.

Incompressible tori play the central role in the torus decomposition of a 3-manifold, just as spheres do in the prime decomposition. Note, however, that when one cuts a 3-manifold along an incompressible torus, there is no canonical way to cap off the boundary components thus created, as is the case for spheres. For any toral boundary component, there are many ways to glue in a solid torus, corresponding to the automorphisms of T^2; typically, the topological type of the resulting manifold depends on the choice. Thus, when a 3-manifold is split along incompressible tori, one leaves the compact manifolds with toral boundary fixed. This leads to another definition: a compact 3-manifold N is *torus-irreducible* if every incompressible torus in N is isotopic to a boundary component of N.

THEOREM 1.2 (TORUS DECOMPOSITION [Jaco and Shalen 1979; Johannson 1979]). *Let M be a closed, oriented, irreducible 3-manifold. Then there is a finite collection of disjoint incompressible tori $T_i^2 \subset M$ that separate M into a finite collection of compact 3-manifolds with toral boundary, each of which is either torus-irreducible or Seifert fibered. A minimal such collection (with respect to cardinality) is unique up to isotopy.*

A 3-manifold N is *Seifert fibered* if it admits a foliation by circles with the property that a foliated tubular neighborhood $D^2 \times \mathbb{S}^1$ of each leaf is either the trivial foliation of a solid torus $D^2 \times \mathbb{S}^1$ or its quotient by a standard action of a cyclic group. The quotient or leaf space of the foliation is a two-dimensional orbifold with a finite number of isolated cone singularities. The orbifold or cone points correspond to the exceptional fibers, that is, fibers whose foliated neighborhoods are nontrivial quotients of $D^2 \times \mathbb{S}^1$.

The tori appearing in the Geometrization Conjecture give a torus decomposition of M. Thus, the Geometrization Conjecture asserts that the torus-irreducible and Seifert fibered components of a closed, oriented, irreducible 3-manifold admit canonical geometric structures.

Of course, it is possible that the collection of incompressible tori is empty. In this case, M is itself a closed irreducible 3-manifold that is either Seifert fibered or torus-irreducible.

The Geometrization Conjecture thus includes the following important special cases (recall M is closed, oriented and irreducible):

HYPERBOLIZATION CONJECTURE. *If $\pi_1(M)$ is infinite and M is atoroidal, then M is hyperbolic, that is, admits a hyperbolic metric.* (M is atoroidal if $\pi_1(M)$ has no subgroup isomorphic to $\mathbb{Z} \oplus \mathbb{Z} = \pi_1(T^2)$.)

ELLIPTIZATION CONJECTURE. *If $\pi_1(M)$ is finite, then M is spherical, that is, admits a metric of constant positive curvature.*

In fact, these are the only remaining open cases of the Geometrization Conjecture. If M has a nontrivial torus decomposition (equivalently, if M contains an incompressible torus), then in particular M is Haken. Thurston [1982; 1986; 1988] has proved the conjecture for Haken manifolds; see Theorem 1.4 below, and also [Morgan 1984]. If M has no incompressible tori, recent work on the Seifert fibered space conjecture [Gabai 1992; Casson and Jungreis 1994] implies that M is either Seifert fibered or atoroidal. It is known that Seifert fibered spaces have geometric structures (Section 1B). In the remaining case, M is atoroidal, and so satisfies the hypotheses of either the elliptization or the hyperbolization conjectures. Note that the elliptization conjecture implies the Poincaré conjecture.

For later sections, we will require a generalization of Seifert fibered spaces. Let N be a compact manifold, possibly with boundary. Then N is a *graph manifold* if there is a finite collection of disjoint embedded tori $T_i \subset N$ such that each component N_j of $N \smallsetminus \bigcup T_i$ is an $\mathbb{S}^1$ bundle over a surface. To such a decomposition one assigns a graph G as follows: the vertices of G are the components of $N \smallsetminus \bigcup T_i$, and two vertices are joined by an edge if the associated components are separated by a torus $T \in \{T_i\}$. This description of the graph is somewhat of a simplification; consult [Waldhausen 1967] for full details.

Of course, Seifert fibered spaces are graph manifolds, as one sees by letting $\{T_i\}$ be the boundaries of tubular neighborhoods of the exceptional fibers. A graph manifold need not admit a globally defined free, or locally free, $\mathbb{S}^1$ action. However, by definition, there are always free $\mathbb{S}^1$ actions defined on the components N_j. These $\mathbb{S}^1$ actions commute on the intersections of their domains of definition (neighborhoods of $\{T_i\}$), and thus extend to give free T^2 actions in this region. These locally defined $\mathbb{S}^1$ and T^2 actions give a well-defined partition of N into orbits, called the *orbit structure* of the graph manifold. In most cases, although not always, this orbit structure is unique up to isotopy [Waldhausen 1967].

We note that, as a consequence of their structure, irreducible graph manifolds of infinite fundamental group necessarily have a $\mathbb{Z} \oplus \mathbb{Z}$ contained in the fundamental group; in fact, with few exceptions, they have incompressible tori. For further details, see [Waldhausen 1967; Cheeger and Gromov 1986; Rong 1990; 1993].

1B. Geometries of 3-Manifolds.
We summarize here the basic features of the eight 3-manifold geometries. For details, see [Scott 1983; Thurston 1996, Section 3.8]. A *geometric structure* on a simply connected space X is a homogenous space structure on X, that is, a transitive action of a Lie group G on X. Thus, X is given by G/H, for H a closed subgroup of G. In order to avoid redundancy, it is assumed that the identity component H_0 of the stabilizer H is a compact subgroup of G, and that G is maximal. Further, G is assumed to be unimodular; this is equivalent to the existence of compact quotients of X.

The possible geometric structures may be divided into three categories.

Constant curvature geometries. Here X is the simply connected space form $\mathbb{H}^3$ of constant curvature -1, or $\mathbb{E}^3$ of curvature 0, or $\mathbb{S}^3$ of curvature $+1$. The corresponding geometries are called hyperbolic, Euclidean, and spherical. The groups G are $\mathrm{PSL}(2, \mathbb{C})$, $\mathbb{R}^3 \times \mathrm{SO}(3)$, and $\mathrm{SO}(4)$. In all cases, $H_0 = \mathrm{SO}(3)$.

Product geometries. Here $X = \mathbb{H}^2 \times \mathbb{R}$ or $\mathbb{S}^2 \times \mathbb{R}$. The groups G are given by the orientation-preserving subgroups of $\mathrm{Isom}\,\mathbb{H}^2 \times \mathrm{Isom}\,\mathbb{E}^1$ and $\mathrm{SO}(3) \times \mathrm{Isom}\,\mathbb{E}^1$, with stabilizer $H_0 = \mathrm{SO}(2)$.

Twisted product geometries. Here there are three possibilities, called $\widetilde{\mathrm{SL}}(2, \mathbb{R})$, Nil and Sol. For $\widetilde{\mathrm{SL}}(2, \mathbb{R})$, the space X is the universal cover of the unit sphere bundle of $\mathbb{H}^2$, and $G = \widetilde{\mathrm{SL}}(2, \mathbb{R}) \times \mathbb{R}$, with $H_0 = \mathrm{SO}(2)$. For the Nil geometry, X is the three-dimensional nilpotent Heisenberg group (consisting of upper triangular 3×3 matrices with diagonal entries 1), and G is the semidirect product of X with $\mathbb{S}^1$, acting by rotations on the quotient of X by its center. Again $H_0 = \mathrm{SO}(2)$. For the Sol geometry, X is the three-dimensional solvable Lie group, $H_0 = \{e\}$, and G is an extension of X by an automorphism group of order eight.

A 3-manifold N is *geometric* if it admits one of these eight geometric structures. Geometric 3-manifolds modeled on six of these geometries, namely all but the hyperbolic and Sol geometries, are Seifert fibered. Thus, topologically, such manifolds are circle "bundles" over two-dimensional orbifolds, with isolated cone singularities. In particular, all such manifolds have finite covers that are $\mathbb{S}^1$ bundles over closed surfaces of genus $g \geq 0$.

These six geometries divide naturally into a pair of threes, corresponding to whether the $\mathbb{S}^1$ bundle is trivial or not. 3-manifolds with product geometries $\mathbb{H}^2 \times \mathbb{R}$, $\mathbb{E}^3$, and $\mathbb{S}^2 \times \mathbb{R}$ are, up to finite covers, trivial circle bundles over oriented surfaces of genus g, where $g \geq 2$, $g = 1$, and $g = 0$, respectively. 3-manifolds with the twisted product geometries $\widetilde{SL}(2, \mathbb{R})$, Nil, and $\mathbb{S}^3$ are, up to finite covers, nontrivial circle bundles over surfaces, again of genus $g \geq 2$, $g = 1$, and $g = 0$, respectively.

Conversely, it is not difficult to prove [Scott 1983] that any Seifert fibered space admits a geometric structure modeled on one of these six geometries.

Geometric 3-manifolds modeled on the Sol geometry all have finite covers that are torus bundles over $\mathbb{S}^1$, with holonomy given by a hyperbolic automorphism of T^2, that is, an element of $SL(2, \mathbb{Z})$ with distinct real eigenvalues. Again, conversely, all such T^2 bundles admit a Sol geometry. Note that Sol-manifolds are graph manifolds, that is, they may be split by incompresssible tori into a union of Seifert fibered spaces.

Thus, seven of the eight geometric 3-manifolds are topologically $\mathbb{S}^1$-fibered over surfaces or T^2-fibered over $\mathbb{S}^1$. Since, in any reasonable sense, most 3-manifolds do not admit such fibrations, the hyperbolic geometry is by far the most prevalent of the eight geometries (see Section 1C).

It is well known [Scott 1983] that the same 3-manifold cannot have geometric structures modeled on two distinct geometries.

Of course, it is not true that the geometric structure itself, that is, the homogeneous metric, is unique in general. In this respect, we recall:

THEOREM 1.3 (MOSTOW RIGIDITY [Mostow 1968; Prasad 1973]). *Let N be a 3-manifold carrying a complete hyperbolic metric of finite volume. Then the hyperbolic metric is unique, up to isometry. Further, if N and N' are 3-manifolds with isomorphic fundamental groups, and if N and N' carry complete hyperbolic metrics of finite volume, then N and N' are diffeomorphic.*

In particular, invariants of the hyperbolic metric such as the volume and the spectrum are topological invariants of the 3-manifold.

There is a similar rigidity for spherical 3-manifolds, in the sense that any metric of curvature $+1$ on the manifold is unique, up to isometry [Wolf 1977]. The fundamental group in this case does not determine the topological type of the manifold. There are further topological invariants, such as the Reidemeister torsion. The other six geometries are typically not rigid, but have moduli closely related to the moduli of constant curvature metrics on surfaces.

1C. Thurston's Results on the Geometrization Conjecture. As already mentioned, many cases of the Geometrization Conjecture have been proved by Thurston [1982; 1986; 1988] (see also [Morgan 1984] for a detailed survey). In particular:

THEOREM 1.4 (GEOMETRIZATION OF HAKEN MANIFOLDS). *A closed, oriented, irreducible Haken manifold that is atoroidal admits a hyperbolic structure. A compact, oriented, irreducible, and torus-irreducible 3-manifold whose boundary consists of a finite number of tori admits a complete hyperbolic metric of finite volume.*

We indicate in a few lines the approach to the proof of this result. It was shown by Haken [1961] and Waldhausen [1968] that, if the manifold M is Haken, one may successively split it along incompressible surfaces into a hierarchy, that is, a collection of (possibly disconnected) compact submanifolds with boundary:

$$M = M_k \supset M_{k-1} \supset \cdots \supset M_1 \supset M_0 = \text{union of balls,}$$

where each M_i has an incompressible surface S_i with $\partial S_i \subset \partial M_i$, and M_{i-1} is obtained from M_i by splitting along S_i. If M is atoroidal, so is each M_i. Thurston proves that, for an appropriate hierarchy, the manifolds M_i admit complete, geometrically finite hyperbolic metrics, typically of infinite volume. This is proved by induction on the length of the hierarchy. Thus, suppose that M_{i-1} admits a complete, geometrically finite hyperbolic metric. The manifold M_i is obtained by gluing together certain of the ends of M_{i-1}. The most difficult part of the proof is showing that the hyperbolic metric on M_{i-1} may be deformed appropriately so that the ends to be glued are isometric, so that M_i acquires a complete, geometrically finite hyperbolic metric. Thurston has developed a wealth of new geometric ideas and methods to carry this out.

McMullen [1989; 1990] has given a different proof of this gluing process, in the case where M does not fiber over $\mathbb{S}^1$.

Theorem 1.4 implies that the torus-irreducible pieces of a nonempty torus decomposition carry hyperbolic structures. As we saw in Section 1B, all the Seifert fibered pieces also carry geometric structures. Thus, the Geometrization Conjecture is proved in the case of manifolds that have a nonempty torus decomposition.

Nevertheless, many, perhaps most, 3-manifolds are not Haken. Thurston has established the Geometrization Conjecture for many further classes of non-Haken 3-manifolds. For instance, suppose M is a closed oriented 3-manifold, and N is the complement of a knot K in M. One may obtain new closed 3-manifolds by *Dehn surgery* on N, that is, by gluing in a solid torus to the boundary of N. The possible Dehn surgeries are classified by classes in $\mathrm{SL}(2, \mathbb{Z})$. Thurston [1979] showed that if N admits a complete hyperbolic metric g_∞ of finite volume, all but finitely many Dehn surgeries yield closed manifolds that admit hyperbolic structures. All of these hyperbolic manifolds obtained by closing the cusp of

(N, g_∞) have volume strictly less than that of (N, g_∞). If M itself is not Haken, then Dehn surgeries on N will often yield closed non-Haken 3-manifolds.

2. Preliminaries on the Space of Metrics

Let **M** denote the space of all smooth Riemannian metrics on the closed oriented 3-manifold M. Thus, **M** is an open convex cone in the space $S^2(M)$ of symmetric bilinear forms on M. The diffeomorphism group Diff M of M acts naturally on **M** by pullback, $(\psi, g) \mapsto \psi^* g$. The two metrics g and $\psi^* g$ are isometric, and ψ is an isometry between them. Since all intrinsic notions associated with the metric are invariant under isometries, it is natural to divide **M** by the action of Diff(M). We let $\mathcal{M} = \mathbf{M}/\text{Diff } M$ be the space of Riemannian structures on M, that is, isometry classes of Riemannian metrics. The space $\mathcal{M}$ is no longer an infinite-dimensional manifold, since the action of Diff M is not free; fixed points of the action correspond to metrics with nontrivial isometry group, that is, maps $\phi \in \text{Diff } M$ such that $\phi^* g = g$. This rarely presents a problem, however. We denote by $\mathcal{M}_1$ the subset of $\mathcal{M}$ consisting of metrics of volume 1 on M.

The tangent space $T_g \text{Diff } M$ to the orbit of Diff M in **M** is the image of the map δ^* that associates to a vector field X on M the element $\delta^*(X) = \mathcal{L}_X g \in S^2(M)$, where $\mathcal{L}$ denotes the Lie derivative. Since δ^* is (underdetermined) elliptic, there is a splitting

$$T_g \mathbf{M} = T_g \text{Diff } M \oplus N_g \text{Diff } M = \text{Im } \delta^* \oplus \text{Ker } \delta,$$

where δ is the divergence operator, the formal adjoint of δ^* on $S^2(M)$, given by $\delta(\alpha) = -(D_{e_i}\alpha)(e_i, \cdot)$, where D is the covariant derivative of the metric g and $\{e_i\}$ is an orthonormal basis. We note that the action of Diff M on **M** has a slice, that is, a locally defined submanifold of **M**, transverse to the orbits of Diff M in a neighborhood of any $g \in \mathbf{M}$ [Ebin 1970].

The space **M** will be endowed with a normed $L^{k,p}$ topology, given by

$$\|h\|_{k,p}^p = \int \left(|h|^p + |Dh|^p + \cdots + |D^k h|^p \right) dV, \tag{2.1}$$

for $h \in T_g \mathbf{M} = S^2(M)$. Here all norms, derivatives, and the volume form, are taken with respect to the metric g. This corresponds, locally, to the Sobolev topology on functions defined on domains of $\mathbb{R}^n$, namely the first k derivatives are in L^p. The exact values of k and p may depend on the problems at hand, but the minimal requirement is that

$$\alpha = k - \frac{n}{p} > 0, \qquad 1 < p < \infty, \tag{2.2}$$

corresponding to the Sobolev embedding $L^{k,p} \subset C^\alpha$. The completion of **M** with respect to this topology will also be denoted by **M**, and gives **M** the structure of a Banach manifold, or Hilbert manifold when $p = 2$. Further, these norms are invariant under the action of Diff M, and thus descend to define a topology on $\mathcal{M}$.

Suppose $\mathcal{F} : \mathcal{M} \to \mathbb{R}$ is a smooth function in the $L^{k,p}$ topology on $\mathcal{M}$ and suppose $\{g_i\} \in \mathcal{M}$ is a sequence that approaches a critical value of $\mathcal{F}$. By slightly perturbing g_i if necessary, we can assume that $\|d\mathcal{F}_{g_i}\|_{(L^{k,p})^*} \to 0$, that is,

$$\sup_{\|h\|_{k,p}=1} |d\mathcal{F}_{g_i}(h)| = \sup_{\|h\|_{k,p}=1} \left| \frac{d}{dt}(\mathcal{F}(g_i(t))) \right| \to 0, \qquad (2.3)$$

where $g_i(t) = g_i + th$ and the norm is taken with respect to the metric g_i. The dual space $(L^{k,p})^*$ is naturally identified, locally, with $L^{-k,q}$, where $p^{-1}+q^{-1} = 1$. One sees that (2.3) contains less information, that is, is a weaker condition, the larger k and p are. Thus, in general, one would like to choose values for k and p as small as possible, in order that (2.3) give as much information as possible. Of course, the pair (k, p) must satisfy (2.2), and also be chosen so that $\mathcal{F}$ is smooth in the $L^{k,p}$ topology.

Next, we fix some notation for later sections. Given a metric g, let R denote the Riemann curvature tensor, given by $R = \sum R_{ijkl}\, dx^i \otimes dx^j \otimes dx^k \otimes dx^l$ in local coordinates. The sectional curvature $K_P = K_{ij} = R_{ijji}$ in the direction of a two-plane P in $T_x M$ spanned by orthonormal vectors e_i and e_j may be defined as the Gauss curvature at x of the geodesic surface in M tangent to P at x. Knowledge of the sectional curvature K_P for all two-planes P determines the curvature tensor.

The Ricci curvature Ric is a symmetric bilinear form on TM, obtained by contracting R; more precisely, $\mathrm{Ric}(v, w) = \sum R(v, e_i, e_i, w)$, for an orthonormal basis e_i. The scalar curvature s is the contraction of the Ricci curvature, $s = \sum \mathrm{Ric}(e_i, e_i)$. In dimension two, these curvatures are all the same, up to multiplicative constants. In dimension three, but not in higher dimensions, the Ricci curvature determines the full curvature R. For instance, if λ_i are the eigenvalues of Ric, with eigenvectors e_i, then for distinct indices (i, j, k) we have

$$K_{ij} = \tfrac{1}{2}(\lambda_i + \lambda_j - \lambda_k).$$

The covariant derivative associated to g will be denoted by D. The Laplacian or Laplace–Beltrami operator Δ associated with g will be taken to have negative spectrum (so $\Delta f = f''$ on $\mathbb{R}$).

3. Functionals on the Space of Metrics

We consider functionals $\mathcal{F} : \mathcal{M} \to \mathbb{R}$ on the space of metrics on M. There are functionals that measure the global size or extent of the Riemannian manifold (M, g); for example, volume, diameter, radius, etc. We will only consider functionals that are Lagrangian, in the sense that

$$\mathcal{F}(g) = \int_M \mathcal{L}_{\mathcal{F}}(g)\, dV_g,$$

where $\mathcal{L}_{\mathcal{F}}(g)$ is the (scalar) Lagrangian density for $\mathcal{F}$ at g, and dV_g is the volume form of the metric g. The invariance of $\mathcal{F}$ under the action of Diff M requires that the Lagrangian satisfy the invariance property

$$\mathcal{L}_{\mathcal{F}}(\psi^* g) = \psi^* \mathcal{L}_{\mathcal{F}}(g) = \mathcal{L}_{\mathcal{F}}(g) \circ \psi, \qquad (3.1)$$

for all $\psi \in$ Diff M. We say that $\mathcal{L}_{\mathcal{F}}(g)$ is a k-th order Lagrangian if it is a smooth function of the k-jet of g, that is, depends only on g and its first k derivatives. Thus, one method to produce Lagrangians is to consider functions of g and its derivatives in some coordinate system that are in fact invariant, in the sense of (3.1), under changes of coordinates.

Recall that any Riemannian manifold admits (geodesic) normal coordinates x_i at any prescribed point p. In these coordinates, the metric satisfies $g_{ij} = \delta_{ij}$ and $\partial g_{ij}/\partial x_k = 0$ at p. In other words, at p the metric osculates, to first order, a flat Euclidean metric. An important and well-known consequence of this is that there are no nonconstant invariant Lagrangians of order ≤ 1 [Lovelock and Rund 1972; Palais 1968]. Thus, one is required to seek Lagrangians of order at least two. We note that most other problems in the calculus of variations can be expressed in terms of first-order Lagrangians: for example, harmonic functions or maps, geodesics, minimal surfaces, Yang–Mills fields, etc.

Consider the Taylor expansion of a metric $g = g_{ij}$ in a normal coordinate system about p. A fundamental fact, due to Cartan, is that the order-k Taylor coefficients can be (universally) expressed in terms of polynomials in the components of the curvature tensor R and its covariant derivatives $\nabla^m R$, for $m \leq k-2$. For example, in normal coordinates, one has, by Riemann,

$$g_{ij} = \delta_{ij} + \frac{2}{3} \sum_{k,l} R_{iklj} x_k \, x_l + O(|x|^3).$$

Thus, we seek Lagrangians whose expressions in normal coordinates are of the form $\mathcal{L}_{\mathcal{F}}(g) = \phi(R, \nabla R, \ldots, \nabla^{k-2} R)$, that is, ϕ is a function of the components of the curvature tensor and its covariant derivatives. The orthogonal group $O(n)$ acts freely and transitively on the possible normal coordinates (which are determined uniquely by an orthonormal frame at p); the action of $O(n)$ extends naturally to an action on the curvature tensor R and its derivatives. Thus, we seek functions ϕ that are $O(n)$-invariant. If one considers functions ϕ that are polynomials P in the components of the arguments, then one seeks to classify $O(n)$-invariant polynomials $P(T_1, T_2, \ldots, T_{k-2})$ on a sum of tensor spaces over R^n (the tensor spaces being the spaces of curvature tensors R, covariant derivatives ∇R, and so on). Now the fundamental theorem of invariance theory for $O(n)$ states that any such polynomial is a linear combination of terms, each obtained by fully contracting an even tensor product of the $\{T_i\}$ to a scalar [Atiyah et al. 1973; Palais 1968].

Thus, for $k = 2$, the simplest Lagrangian one can take is the full contraction of the curvature tensor R, that is, the scalar curvature. Thus, in a precise sense, the simplest metric functional on $\mathcal{M}$ is the total scalar curvature functional

$$\mathcal{S} : \mathcal{M} \to \mathbb{R}, \qquad \mathcal{S}(g) = \int_M s_g \, dV_g.$$

Next, again for $k = 2$, one could take contractions on $R \otimes R$. It turns out there are three possibilities, namely $|R|^2$, $|\mathrm{Ric}|^2$ and s^2. This gives rise to the functionals $\mathcal{R}^2$, $\mathcal{R}ic^2$, and $\mathcal{S}^2$, corresponding to the L^2 norms of the tensors R, Ric, and s.

One could also consider higher-order functionals of the curvature and its co-variant derivatives. Since they become rapidly more complicated, especially regarding the expressions of their Euler–Lagrange equations, we will not pursue their discussion here.

Consider the equation for a critical point of $\mathcal{F}$, that is, the Euler–Lagrange equation associated to the Lagrangian $\mathcal{L}_{\mathcal{F}}$. For a k-th order Lagrangian, this will have the form

$$\nabla \mathcal{F}(g) = A^{ij}(g, \partial g, \dots, \partial^m g) = 0,$$

where $m \leq 2k$ and, generically, $m = 2k$. The two-tensor $A = A^{ij}$ is symmetric and, as a consequence of the invariance of $\mathcal{L}_{\mathcal{F}}$, is divergence-free: $\delta A = 0$. Thus, for the second-order Lagrangians mentioned above, the Euler–Lagrange equations will typically be a fourth-order system of partial differential equations in the metric g.

The scalar curvature functional has the remarkable property that its Euler–Lagrange equation is of second order in g; in fact, when restricted to $\mathcal{M}_1$, the gradient $\nabla|_{\mathcal{M}_1} \mathcal{S}$ (with respect to the L^2 metric on $\mathcal{M}_1$, that is, the metric (2.1) with $k = 0$ and $p = 2$) is given by

$$\nabla|_{\mathcal{M}_1} \mathcal{S}(g) = \frac{s}{n} g - \mathrm{Ric} \equiv -Z.$$

The two-tensor Z is just the trace-free part of the Ricci curvature. Further, in dimensions three and four, it is known [Lovelock and Rund 1972] that $\mathcal{S}$ is the unique functional (expressed in terms of a second-order invariant Lagrangian) whose Euler–Lagrange operator is of second order in g. The Euler–Lagrange equations for $\mathcal{R}^2$, $\mathcal{R}ic^2$ and $\mathcal{S}^2$ are all of order four in g; see [Berger 1970; Besse 1987, p.133; Anderson 1993; a] for a discussion.

Finally, we briefly discuss the appropriate topologies on the space $\mathcal{M}_1$ for these functionals. All of the functionals discussed above are smooth in the $L^{2,2}$ topology on $\mathcal{M}_1$; compare (2.1). For the functionals $\mathcal{R}^2$, $\mathcal{R}ic^2$ and $\mathcal{S}^2$, this is the smallest topology in which they are smooth. The functional $\mathcal{S}$ is also smooth in the weaker topology $L^{1,q}$, for $q > 3$.

4. Conjectures on the Realization of the Sigma Constant

As indicated in the Introduction, researchers in Riemannian geometry and analysis on manifolds (and of course in mathematical physics and general relativity) have long been interested in the existence and moduli of Einstein metrics. In light of the discussion in Section 3, it is natural to seek such metrics variationally, as critical points of the total scalar curvature functional.

A number of immediate problems are encountered in the variational approach to existence. The functional S is bounded neither below nor above. Further, it is well-known that any critical point has infinite index and co-index, that is, there are infinite-dimensional subspaces of $T_g \mathcal{M}$ on which S can be infinitesimally, and thus locally, decreased or increased. Thus S is far from satisfying any of the usual compactness properties used in obtaining existence of critical points, such as the Palais–Smale condition, mountain-pass lemmas, etc.

There is, however, a well-known minimax procedure to obtain critical values of S. It goes as follows. Given a metric $g \in \mathcal{M}_1$, let $[g]$ denote the conformal class of g, that is, $[g] = \{g' \in \mathcal{M}_1 : g' = \psi^2 g\}$, for some smooth positive function ψ. The functional S is bounded below on $[g]$; define

$$\mu[g] = \inf_{g \in [g]} S(g).$$

The number $\mu[g]$ is called the *Yamabe constant* (or *Sobolev quotient*) of $[g]$. An elementary comparison argument [Aubin 1976] shows that

$$\mu[g] \leq \mu(S^n, g_{\text{can}})$$

for any conformal class $[g]$, where g_{can} is the canonical metric of constant positive sectional curvature and volume 1 on the n-sphere S^n. Thus, define the *Sigma constant* by

$$\sigma(M) = \sup_{[g] \in \mathcal{C}} \mu[g],$$

where $\mathcal{C}$ is the space of conformal structures on M. Thus, $\sigma(M)$ is a smooth invariant of the manifold M. (I don't know who first considered this minimax approach. One guesses that certainly Yamabe was aware of it, and it may well have been considered earlier. I have found no definite references, besides the relatively recent [Kobayashi 1987; Schoen 1989]).

It is reasonable to expect, and certainly conjectured (in [Besse 1987, p. 128], for example), that $\sigma(M)$ is a critical value of S, that is, any metric $g_0 \in \mathcal{M}_1$ such that $s_{g_0} \equiv \mu[g_0] = \sigma(M)$ is an Einstein metric. In full generality, this remains unknown, due partly to the possible lack of uniqueness of metrics realizing $\mu[g]$.

REMARKS 4.1. (i) Clearly, $\sigma(M) \leq \sigma(S^n) = n(n-1)(\text{vol}\,S^n)^{2/n}$, where the volume is that of the unit sphere. Further, $\sigma(S^n)$ is realized by the canonical metric on S^n of volume 1.

(ii) If $\sigma(M) \leq 0$, it is easy to prove that a metric $g \in \mathcal{M}_1$ realizing $\sigma(M)$ is Einstein; see [Besse 1987, p. 128], and also Section 5.

(iii) If $\dim M = 2$, the Gauss–Bonnet theorem gives

$$2\pi\chi(M) = \int_M s_g \, dV_g = \mathcal{S}(g),$$

for all g. Thus, $\mathcal{S}$ is a constant functional on $\mathcal{M}_1$, whose value is a topological invariant of the surface M. In some sense, one can think of $\sigma(M)$ as a generalization of the Euler characteristic to higher-dimensional manifolds, especially in dimension three.

Comparatively little is known regarding the Sigma constant in dimension three, and even less in higher dimensions. Two important and well-known open questions are:

QUESTION 4.2. *If M is a homotopy three-sphere, is $\sigma(M) > 0$?*

QUESTION 4.3. *If M is a hyperbolic 3-manifold, does the hyperbolic metric realize $\sigma(M)$, modulo renormalization to volume 1?*

In fact, not a single example is known of a 3-manifold with $\sigma(M) < 0$. There are, however, two important positive results on $\sigma(M)$, due to Gromov–Lawson [1983] and Schoen–Yau [Schoen 1984]. Namely, if a 3-manifold M is a $K(\pi, 1)$, or contains a $K(\pi, 1)$ factor in its prime decomposition (see Section 1), then $\sigma(M) \leq 0$. This is equivalent to saying that M admits no metric of positive scalar curvature.

On the other hand, if M has a "small" fundamental group, then $\sigma(M) > 0$, assuming the Poincaré conjecture is true. More precisely, if M is a connected sum of a finite number of manifolds, each of which is either $\mathbb{S}^2 \times \mathbb{S}^1$ or a quotient of $\mathbb{S}^3$ by a group of isometries, then $\sigma(M) > 0$ [Gromov and Lawson 1980; Schoen and Yau 1979a]. Here one explicitly constructs metrics of positive scalar curvature on such manifolds, starting from the canonical metrics on the component manifolds, which clearly have positive scalar curvature.

The minimax procedure to obtain $\sigma(M)$ has two parts: first minimize in a conformal class, then maximize over all conformal classes. Fortunately, the first step has been solved [Yamabe 1960; Trudinger 1968; Aubin 1976; Schoen 1984]:

THEOREM 4.4 (SOLUTION TO THE YAMABE PROBLEM). *For any conformal class $[g] \in \mathcal{C}$, the Yamabe constant $\mu[g]$ is realized by a smooth metric $g_\mu \in [g]$ whose scalar curvature s_μ is identically equal to $\mu[g]$.*

The metrics g_μ realizing $\mu[g]$ are called *Yamabe metrics*. The solution to the Yamabe Problem amounts to showing that the equation

$$4\,\frac{n-1}{n-2}\,\Delta u - s_g\,u = -\mu[g]\,u^{(n+2)/(n-2)}, \tag{4.1}$$

where $g \in [g]$ is a fixed background metric, has a smooth, positive solution on M. Equation (4.1) is the Euler–Lagrange equation of the variational problem $S|_{[g]}$. In fact, the metric $g_\mu := u^{4/(n-2)} g$ gives then the desired solution to the Yamabe problem. Equation (4.1) is a nonlinear elliptic (scalar) equation. In particular, this is a determined problem in the sense that there is one equation imposed on an unknown function u. The subtlety of the problem arises from the fact that the exponent $2n/(n-2)$ is borderline for the Sobolev embedding $L^{1,2} \to L^{2n/(n-2)}$.

The second part of the minimax procedure, maximizing over the conformal classes, is considerably more difficult. In fact, at least in dimensions three and four, it is known that there are topological obstructions to the existence of Einstein metrics. In dimension three, since Einstein metrics have constant curvature, no reducible 3-manifold admits an Einstein metric, that is, neither $S^2 \times S^1$ nor any 3-manifold that is a nontrivial connected sum admits an Einstein metric. More generally, among the 3-manifolds admitting Seifert geometries discussed in Section 1B, only those admitting S^3 or E^3 geometries admit Einstein metrics. Similarly, torus bundles over S^1, corresponding to Sol geometry, do not admit Einstein metrics.

Thus, if one tries to realize the value $\sigma(M)$ on $\mathcal{M}_1$ by taking an appropriate maximizing sequence $\{g_i\} \in \mathcal{M}_1$ for S, the sequence $\{g_i\}$ has in general no subsequence converging to a limit metric in $\mathcal{M}_1$. Whether such a sequence $\{g_i\}$ should have convergent subsequences or not depends on the topology of the underlying 3-manifold M. It is thus an interesting (and difficult) challenge to relate the possible degenerations of a sequence $\{g_i\}$ to the topology of the underlying manifold M.

The following three conjectures describe the geometry and topology of metrics that attempt to realize the Sigma constant on a 3-manifold M. For all three conjectures, the following assumption is made.

ASSUMPTION. M is a closed, *irreducible*, oriented 3-manifold.

CONJECTURE I (THE NEGATIVE CASE). *Suppose $\sigma(M) < 0$. Then there is a finite collection of disjoint, embedded incompressible tori T_i^2 in M such that the complement $M \setminus \bigcup T_i^2$ is a finite union of complete hyperbolic manifolds M_j of finite volume, together with a (possibly empty) finite union of graph manifolds S_k with toral boundaries.*

Further,

$$|\sigma(M)| = 6\left(\sum \mathrm{vol}_{-1} M_j\right)^{2/3}, \tag{4.2}$$

where vol_{-1} *denotes volume in the hyperbolic metric. In particular, if M is atoroidal, it admits a hyperbolic metric that realizes the Sigma constant (modulo renormalization):* $|\sigma(M)| = 6(\mathrm{vol}_{-1} M)^{2/3}$.

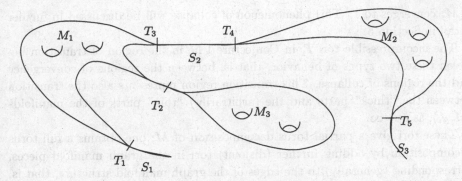

Figure 1. Conjecture I.

CONJECTURE II (THE ZERO CASE). *Suppose* $\sigma(M) = 0$. *Then M is a graph manifold, with infinite π_1. The Sigma constant $\sigma(M)$ is realized if and only if M is a flat 3-manifold; in particular, M must be finitely covered by a three-torus.*

Figure 2. Conjecture II.

CONJECTURE III (THE POSITIVE CASE). *Suppose* $\sigma(M) > 0$. *Then M is diffeomorphic to $\mathbb{S}^3/\Gamma$, for $\Gamma \subset \mathrm{SO}(4)$. The Sigma constant is realized by the standard metric of constant curvature on $\mathbb{S}^3/\Gamma$, that is, $\sigma(M) = 6(\mathrm{vol}_{+1} M)^{2/3}$, where vol_{+1} denotes volume in the metric of sectional curvature $+1$.*

It is worth discussing these conjectures in some further detail. Let $\{g_i\}$ be a maximizing sequence of Yamabe metrics for M, as described above, so that $\mu[g_i] \to \sigma(M)$. Figures 1 and 2 are schematic representations of the (near) limiting behavior of $\{g_i\}$ according to Conjectures I and II.

Conjecture I corresponds to the conjecture that, after possibly passing to a subsequence and altering g_i if necessary by a diffeomorphism ψ_i of M, the sequence $\{g_i\}$, when restricted to the domains $M_j \subset M$, can be chosen to converge (smoothly) to a complete hyperbolic metric (more precisely, a metric of constant negative curvature) of finite volume on $\bigcup M_j$. On the complementary part of M, namely the graph manifolds S_k, the sequence $\{g_i\}$ degenerates, so that there is no well-defined limiting Riemannian metric on $\bigcup S_k$. Briefly, the sequence $\{g_i\}$ collapses to a point each orbit $\mathcal{O}_x$ of the graph manifold structure (see Section 1), that is, $\mathrm{diam}_{g_i} \mathcal{O}_x \to 0$ as $i \to \infty$. In particular, the total volume of the graph manifold pieces $\bigcup S_k$ converges to 0, while the total volume of the complement

$\bigcup M_j$ converges to 1. This phenomenon of collapse will be discussed in further detail in Section 5.

The incompressible tori T_i in Conjecture I lie in the region of transition between these two types of behavior, that is, between the regions of convergence and the regions of collapse. This transition region represents also the transition between the "thick" parts and the (arbitrarily) "thin" parts of the manifolds (M, g_i), as $i \to \infty$.

These tori give a partial torus decomposition of M; one obtains a full torus decomposition by adding further (disjoint) tori in the graph manifold pieces, corresponding (generally) to the edges of the graph manifold structure, that is, to the decomposition of the graph manifolds into Seifert fibered pieces.

Conjecture I implies that the Sigma constant is realized by the union of the hyperbolic pieces; the graph manifold pieces play no role in the value of $\sigma(M)$. In particular, the set of hyperbolic pieces M_j is nonempty.

In the atoroidal case, there is no degeneration of $\{g_i\}$, that is, the sequence can be chosen to converge (smoothly) to a smooth metric of constant negative curvature on the closed manifold M.

The power $\frac{2}{3}$ in (4.2) is necessary, since the invariants $\sigma(M)$ and vol M of (4.2) behave differently under rescaling of the metric. Thus, (4.2) is a scale-invariant equality.

Conjecture II can now be understood by means of the discussion above—namely, there are no hyperbolic pieces. With the possible exception of the special case where M is a flat 3-manifold, the sequence $\{g_i\}$ fully collapses the graph manifold M along the orbits of the graph structure. The manifold M becomes arbitrarily long and thin in the metrics g_i, as $i \to \infty$, that is, $\text{diam}_{g_i} M \to \infty$ and $\text{inj}_x(g_i) \to 0$ for all $x \in M$, where inj_x denotes the injectivity radius at x (see Section 5).

It is necessary to explain how the condition that π_1 is infinite arises, since there are graph manifolds, such as $\mathbb{S}^3/\Gamma$, for which $\sigma(M) > 0$.

We claim that irreducible graph manifolds M of infinite π_1 satisfy $\sigma(M) = 0$. Indeed, it can be deduced from the results of Cheeger and Gromov (see Section 5) that an arbitrary graph manifold necessarily has $\sigma(M) \geq 0$. Further, as seen in Section 1, an irreducible graph manifold with infinite π_1 necessarily has a $\mathbb{Z} \oplus \mathbb{Z}$ contained in π_1. By a well-known result of Schoen and Yau [1979b], such manifolds do not admit metrics of positive scalar curvature; thus $\pi_1(M)$ infinite implies $\sigma(M) = 0$, as claimed.

On the other hand, irreducible graph manifolds of finite π_1 are known to be the spaces $\mathbb{S}^3/\Gamma$ and thus have $\sigma(M) > 0$.

Conjecture II thus amounts to the converse, that the only irreducible manifolds with $\sigma(M) = 0$ are graph manifolds; the discussion above then implies these manifolds have infinite π_1.

Conjecture III now needs no further explanation. Conjecturally, there should exist a sequence of Yamabe metrics g_i that converges smoothly to a metric of constant positive curvature on $\mathbb{S}^3/\Gamma$ and that realizes the Sigma constant.

We note that the decomposition of M via the collection $(\{M_j\}, \{T_i\}, \{S_k\})$ is unique, up to isotopy. This follows from the uniqueness of the torus decomposition (Theorem 1.2) and Mostow rigidity (Theorem 1.3). See also [Jaco and Shalen 1979].

These conjectures imply results about the geometry of 3-manifolds, namely about the structure of metrics realizing the Sigma constant, as described above, as well as the topology of 3-manifolds. We turn to a discussion of the topological consequences.

In fact Conjectures I–III imply the Geometrization Conjecture (page 49). For example, let us indicate how they imply the Poincaré conjecture, or more generally the Elliptization Conjecture (page 54). Thus, suppose M is a 3-manifold with finite fundamental group. Using the prime decomposition (Section 1), we may assume that M is irreducible. Then Conjecture I implies that $\sigma(M) \geq 0$, since Conjecture I implies that π_1 must be infinite if $\sigma(M) < 0$. For the same reason, Conjecture II implies $\sigma(M) > 0$. Thus, Conjecture III implies that M is S^3/Γ. Note that all three conjectures are needed to reach this conclusion; Conjecture III alone does not suffice, as Question 4.2 above indicates.

To see how Conjectures I–III imply the hyperbolization conjecture, suppose that M is irreducible, atoroidal, and has infinite fundamental group. Conjecture III implies that $\sigma(M) \leq 0$. By the discussion in Section 1B on graph manifolds, a graph manifold with infinite π_1 cannot be atoroidal; thus Conjecture II implies that $\sigma(M) < 0$. Finally, Conjecture I implies that M is hyperbolic.

In fact, it is not necessary to use Conjecture III in a proof of the hyperbolization conjecture. Namely, the assumptions of the hyperbolization conjecture imply that M is a $K(\pi, 1)$ (see Section 1A), so that by the above-mentioned results from [Gromov and Lawson 1983; Schoen 1984] one can conclude that $\sigma(M) \leq 0$. Thus, Conjectures I and II, together with known results, alternately imply the hyperbolization conjecture.

As pointed out in Section 1A, the remaining cases of the Geometrization Conjecture for closed, oriented, irreducible 3-manifolds have been proved by Thurston, Gabai, Casson, and Jungreis. We indicate briefly how these remaining cases also follow from Conjectures I–III. Conjecture III fully describes the topology of irreducible 3-manifolds with $\sigma(M) > 0$. If M is a 3-manifold as above with $\sigma(M) = 0$, Conjecture II implies that M is a graph manifold with infinite π_1. Such manifolds admit torus decompositions, all of whose components are Seifert fibered (including Sol manifolds). In fact the theory of collapse of 3-manifolds exhibits this splitting into Seifert fibered components, with the exception of Sol manifolds, which might not split by tori in the process of collapse. In other words, the torus decomposition of the graph manifolds M can be detected from the geometry of a collapsing sequence of metrics on M. We refer to [Cheeger

and Gromov 1986; Rong 1990] for further details. As indicated in Section 1,
Seifert fibered spaces are geometric. If $\sigma(M) < 0$, the tori of Conjecture I of the
conjecture decompose M into hyperbolic and graph manifold pieces, and by the
remarks above all graph manifold pieces are unions of geometric manifolds along
tori.

Conjectures I–III thus amount to the conjecture that an appropriate sequence
$g_i \in \mathcal{M}_1$ such that $\mu[g_i] \to \sigma(M)$ *implements* the Geometrization Conjecture,
provided M is irreducible. This will be discussed in some further detail in the
next sections.

Taken together, these conjectures imply that the Sigma constant $\sigma(M)$ of
an irreducible, oriented 3-manifold behaves in a remarkably similar way to the
Euler characteristic $\chi(M)$ of an oriented two-manifold (which is also just the total
scalar curvature). One sees immediately that Conjectures I, II, III, corresponding
to $\sigma(M)$ negative, zero, or positive, bear a strong resemblence to the classification
of surfaces with negative, zero, or positive Euler characteristic. Of course, the
Sigma constant alone cannot determine the topology of M, since for instance
all graph manifolds, in particular Seifert fibered spaces, with infinite π_1, have
$\sigma = 0$. With this "degeneracy" removed, one has quite a sharp correspondence
between the value of $\sigma(M)$ and the topology of M. For instance, the conjectures
imply there are only finitely many irreducible atoroidal 3-manifolds with a given
(necessarily nonzero) value of $\sigma(M)$.

For atoroidal manifolds with $\sigma(M) < 0$, the Sigma constant is related to the
hyperbolic volume by (4.2). Thurston [1979] has developed a beautiful theory
describing the basic structure of the values of the hyperbolic volume; see also
[Gromov 1981b].

5. Convergence and Degeneration of Riemannian Metrics

Let $\{g_i\}$ be a sequence of Yamabe metrics in $\mathcal{M}_1$ such that

$$\mu[g_i] \to \sigma(M), \tag{5.1}$$

so that $\{g_i\}$ attempts to realize the Sigma constant on M. If $\{g_i\}$ converges to
a metric $g \in \mathcal{M}_1$, then g is an Einstein metric on M, at least when $\sigma(M) \leq 0$,
and conjecturally in general.

Since an arbitrary 3-manifold does not admit an Einstein metric, $\{g_i\}$ cannot
converge in general to a metric $g \in \mathcal{M}_1$. Of course, when M admits an Einstein
metric g_0, since it is generally unique, $\{g_i\}$ should converge to g_0. In general,
however, there must exist subsets on M on which $\{g_i\}$ degenerates. How can
one relate the degeneration to the topology of M?

An examination of Conjectures I–III shows that they imply that the essential
two-spheres and tori in M are obstructions, and the only obstructions, to the
existence of Einstein metrics. For instance, the conjectures imply that if M is
irreducible and atoroidal, then $\{g_i\}$ should converge to an Einstein metric on

M; this would necessarily be of constant positive or negative curvature, since the atoroidal condition rules out flat metrics. Further, the conjectures implicitly describe the behavior of the degeneration of $\{g_i\}$ in a neighborhood of the essential tori in M: the metrics become very long and thin in this region. This will be described in more detail below. In the next section, we will describe the conjectural degeneration of $\{g_i\}$ in neighborhoods of essential two-spheres.

To summarize, the sequence $\{g_i\}$ should conjecturally degenerate along the two-spheres and tori corresponding to the sphere and torus decompositions of the 3-manifold, and should converge on the complement; this complement is also called the *characteristic variety* [Jaco and Shalen 1979].

To understand if a sequence of metrics converges, or understand how it degenerates, one needs to understand how to control the behavior of a sequence of metrics. First, we note that there is little or no reason to expect that one can control the convergence or degeneration of an arbitrary sequence $\{g_i\}$ satisfying (5.1). From (5.1), one controls the scalar curvature of the metric, as well as the gradient ∇S of $\{g_i\}$, restricted to the space of Yamabe metrics, in a weak topology (say $L^{-2,2}$, the dual of $L^{2,2}$). Controlling the scalar curvature of a metric gives good control of the metric in its conformal class, as indicated in Theorem 4.4; one seeks to control only the behavior of a function (the conformal factor), given that it satisfies an elliptic differential equation of the type (4.1). On the other hand, since the equivalence class of a metric (modulo the action of diffeomorphisms) depends locally on three unknown functions, one cannot expect to control the metric, or understand general degenerations, with the scalar curvature function alone. The condition $|\nabla S| \to 0$ in a topology such as the $L^{-2,2}$ topology is also too weak to lead to definite conclusions about the behavior of $\{g_i\}$.

An analogous visual picture can be obtained by considering minimizing sequences for the Plateau problem, that is, the problem of finding the disc of least area spanning a given smooth curve in $\mathbb{R}^3$. It is well-known that the Plateau problem has a smooth solution. However, the behavior of minimizing sequences, that is, the geometry or configuration of discs in $\mathbb{R}^3$ with given boundary whose area converges to the least area, can be quite bizarre; one may have very long, thin filaments whose contribution to the area is arbitrarily small. Thus, although the limit is well behaved (that is, smooth), the minimizing sequence may have very dense regions of bad behavior that are irrelevant to the geometry of the limit.

Under what circumstances can one control the convergence or degeneration of a sequence in $\mathcal{M}_1$? Arguing heuristically for the moment, the full curvature R of g involves all second derivatives of g (in local coordinates), so that one may expect that a bound on $|R|$ gives a bound on $|g|_{L^{2,\infty}}$. Assuming that one has a Sobolev inequality for g, it follows that g is bounded in $C^{1,\alpha}$ norm, for any $\alpha < 1$, again in local coordinates. If $g_i \in \mathcal{M}_1$ and there is a fixed coordinate system (atlas) in which $(g_i)_{kl}$ is bounded in $C^{1,\alpha}$, it follows from the Arzela–

Ascoli theorem that a subsequence converges, in the $C^{1,\alpha'}$ topology, for $\alpha' < \alpha$, to a limit metric g of class $C^{1,\alpha}$.

This heuristic reasoning can be made rigorous, and leads to the fundamental theory of Cheeger–Gromov on the behavior of metrics with uniform curvature bound. Although the result is valid in all dimensions, we will summarize it in dimension three only.

THEOREM 5.1 [Cheeger 1970; Gromov 1981a; Cheeger and Gromov 1986; 1990]. *Let $\{g_i\}$ be a sequence of metrics in $\mathcal{M}_1$. Suppose there is a uniform bound*

$$|R_{g_i}| \leq \Lambda. \tag{5.2}$$

Then there is a subsequence, also called $\{g_i\}$, and diffeomorphisms ψ_i of M such that exactly one of the following cases occurs:

I. (Convergence) *The metrics $\psi_i^* g_i$ converge in the $C^{1,\alpha'}$ topology, $\alpha' < \alpha$, to a $C^{1,\alpha}$ metric g_0 on M, for any $\alpha < 1$.*

II. (Collapse) *The metrics $\psi_i^* g_i$ collapse M along a graph manifold structure. Thus, M is necessarily a graph manifold. The metrics $\psi_i^* g_i$ collapse the orbits $\mathcal{O}_x$ (namely circles or tori) of a (sequence of) orbit structures to a point, as $i \to \infty$, that is, $\operatorname{diam}_{\psi_i^*(g_i)} \mathcal{O}_x \to 0$ for all $x \in M$.*

III. (Cusps) *There is a maximal domain $\Omega \subset M$ such that $\psi_i^* g_i|_\Omega$ converges, in the $C^{1,\alpha'}$ topology, $\alpha' < \alpha$, to a complete $C^{1,\alpha}$ metric g_0 of volume at most 1 on Ω. The complement $M \smallsetminus \Omega$ is collapsed along a sequence of orbit structures, as in case II. In particular, a neighborhood of $M \smallsetminus \Omega$ has the structure of a graph manifold.*

A sequence of metrics h_i defined on a domain V converges in the $C^{1,\alpha}$ topology to a limit metric h if there is a smooth coordinate atlas on V for which the component functions of h_i converge to the component functions of h; here the convergence is with respect to the usual $C^{1,\alpha}$ topology for functions defined on domains in $\mathbb{R}^3$. The convergence in cases I and III above is also in the weak $L^{2,p}$ topology, for any $p < \infty$. In the regions of collapse, the metrics g_i become long and thin: the injectivity radius converges to 0 in these regions, while the diameter of these regions diverges to infinity. The region Ω may not be connected; in fact, in general, it might have infinitely many components. We refer to [Cheeger and Gromov 1986; 1990] or to [Anderson 1993; a] for a more detailed discussion.

These results can be understood as a generalization of the coarse features of Teichmüller theory to higher dimensions and variable curvature. The three possibilities in Theorem 5.1 correspond to the three basic behaviors of sequences in the moduli of metrics of constant curvature on surfaces of genus 0, 1, and $g \geq 2$, respectively.

Teichmüller spaces play an important part in Thurston's results and work on the Geometrization Conjecture. Speaking very loosely, one is given hyperbolic structures on pieces of a 3-manifold, and studies the deformations and degenerations of hyperbolic structures on these pieces and their boundaries, in order to obtain hyperbolic structures on larger manifolds by a smooth gluing. Thus, the convergence and degeneration of hyperbolic metrics, on 3-manifolds and on surfaces, plays a central role.

In attempting to approach the Geometrization Conjecture by studying the space of all metrics on a 3-manifold, Theorem 5.1 plays an analogous central role.

Suppose, for instance, that there is a sequence $\{g_i\}$ of Yamabe metrics satisfying both (5.1) and (5.2), that is,

$$\mu[g_i] \to \sigma(M) \quad \text{and} \quad |R_{g_i}| \leq \Lambda, \tag{5.3}$$

for some Λ. As usual we are assuming that M is closed, oriented, and irreducible. We will outline how, in this case, one may approach and in fact come quite close to a proof of Conjectures I–III.

Let $\mathcal{C}$ be the space of conformal classes, represented by a choice of Yamabe metric (it is not known whether Yamabe metrics are unique in their conformal classes when $\mu > 0$). Although $\mathcal{C}$ is not an infinite-dimensional submanifold of $\mathcal{M}_1$, it does have a formal tangent space at every $g \in \mathcal{C}$, and we may write

$$T_g \mathcal{M}_1 = T_g \mathcal{C} \oplus N_g \mathcal{C}, \tag{5.4}$$

where $N_g \mathcal{C}$ is the normal space to $T_g \mathcal{C}$ in $T_g \mathcal{M}_1$, with respect to the L^2 metric on $\mathcal{M}_1$. (Note that $N_g \mathcal{C}$ is *not* tangent to the conformal class of metrics $[g] \subset \mathcal{M}_1$). Now $T_g \mathcal{C} = \{h \in S^2(M) : s(h) = \text{const}\} = \{h \in S^2(M) : \Delta(s'(h)) = 0\}$. One has the classical formula [Besse 1987, p. 63]

$$s'(h) = -\Delta \operatorname{tr}(h) + \delta\delta h - \langle \operatorname{Ric}, h \rangle. \tag{5.5}$$

Thus, $N_g \mathcal{C} = \operatorname{Im}(\Delta \circ s')^*$, which implies

$$N_g \mathcal{C} = \{\alpha \in S^2(M) : \alpha = D^2 u - (\Delta u)\, g - u \operatorname{Ric} \text{ for } u \text{ with } \int u \, dV_g = 0\}. \tag{5.6}$$

Applying this to the metrics g_i, we may then write

$$\nabla|_{\mathcal{M}_1} S_{g_i} = Z_{g_i} = D_i^2 u_i - (\Delta_i u_i)\, g_i - u_i \operatorname{Ric}_i + Z_i^T, \tag{5.7}$$

where $Z_i^T \in T_{g_i} \mathcal{C}$ is the tangential projection of Z. Since $\{\mu[g_i]\}$ approaches a critical (maximal) value of $S|_{\mathcal{C}}$, we may assume, as in (2.3),

$$Z_i^T \to 0 \quad \text{in } L^{-2,2}(T\mathcal{C}). \tag{5.8}$$

We apply Theorem 5.1 to $\{g_i\}$, and consider the three cases individually. It is implicitly assumed that appropriate subsequences and diffeomorphisms of M are taken where necessary.

Case I. Suppose the metrics g_i converge. Then the limit g is a $C^{1,\alpha} \cap L^{2,p}$ metric on M, which satisfies the equation

$$Z = D^2 u - (\triangle u)\, g - u \operatorname{Ric}, \qquad (5.9)$$

weakly. Taking the trace of (5.9) gives

$$\triangle u = -\tfrac{1}{2} su = -\tfrac{1}{2}\sigma(M)\, u, \qquad (5.10)$$

so that u is an eigenfunction of the Laplacian, with eigenvalue $-\tfrac{1}{2}\sigma(M)$. Elliptic regularity allows one to conclude that any weak $L^{2,p}$ solution g of (5.9) and (5.10) is smooth. Since the Laplacian has negative spectrum, we conclude that, if $\sigma(M) \leq 0$, then u is a constant. Further, since the integral of u is 0, by (5.6), it follows in this case that $u \equiv 0$. Thus, by (5.9), $Z = 0$, that is, g is an Einstein metric realizing $\sigma(M)$. It is conjectured that also in the positive case, $\sigma(M) > 0$ the only solution of (5.9) is again given by $Z = 0$.

Case II. If the metrics $\{g_i\}$ collapse, M is necessarily an (irreducible) graph manifold. As indicated in Section 4, M necessarily satisfies $\sigma(M) \geq 0$, and $\sigma(M) = 0$ if and only if $\pi_1(M)$ is infinite.

Case III. Suppose the metrics $\{g_i\}$ converge to a collection of complete, non-compact Riemannian manifolds (M_j, g_j) of finite volume, and collapse the complement. Then, arguing as in Case I, on each M_j the metric g_j satisfies

$$Z = D^2 u - (\triangle u)\, g - u \operatorname{Ric}. \qquad (5.11)$$

Taking the trace as in Case I implies that u is an eigenfunction of $\triangle$. Now we point out that in this case, we must have $\sigma(M) \leq 0$. Namely, if $\sigma(M) > 0$, so $\mu[g_i] > 0$ for i sufficiently large, it follows from (4.1) that g_i has a uniform bound on its Sobolev constant. This implies that the volume of unit balls in (M, g_i) is bounded below, that is, (M, g_i) does not become thin at any point. This is of course not the case due to the presence of cusps, that is, neighborhoods of infinity of M_j do not satisfy a (uniform) Sobolev inequality. Thus, $\sigma(M) \leq 0$ and the arguments above again imply that $u = 0$. It follows that g_j is a metric of constant negative curvature. With some further arguments, which are not difficult, it is possible to prove that the collection $\{M_i\}$ is finite and the metric $g = \{g_i\}$ on $\bigcup M_j$ realizes $\sigma(M)$. Since the complement of $\bigcup M_j$ is collapsed, it has the structure of a finite union of connected graph manifolds.

Thus, as indicated in outline form above, a significant portion of Conjectures I–III is resolved if there is a sequence of Yamabe metrics $\{g_i\}$ with $\mu[g_i] \to \sigma(M)$ having uniformly bounded curvature. More precisely, Conjecture III can be resolved modulo the conjecture that solutions of (5.9) are Einstein. Further, Conjecture II can be resolved.

Conjecture I requires however further consideration, for it remains to be proved that the tori in the hyperbolic cusps are incompressible in M. This is, of course, an important issue. For instance, Thurston has shown that every 3-manifold M has many hyperbolic knot or link complements, that is, there are knots or links L in M whose complements $\Omega = M \smallsetminus L$ admit complete hyperbolic metrics g_L of finite volume. In this case, the tori in the hyperbolic cusps are compressible in M; they just form the boundary components of a tubular neighborhood of L in M.

All of these metrics g_L can also be considered as critical points of $\{g_i\}$, in the sense discussed above. Namely, it is not hard to see that there are sequences $g_i \in \mathcal{M}_1$ such that $g_i|_\Omega \to g_L$ smoothly (and uniformly on compact sets), while $\|\nabla \mathcal{S}_{g_i}\|_{L^{-2,2}} \to 0$ as $i \to \infty$. These metrics g_L are not tied tightly to, and so do not reflect easily, the global topology of M; consider for instance that the three-sphere has many hyperbolic knot complements. The Thurston theory indicates that the hyperbolic knot or link complements tend to have large volume (when the curvature is -1); see Section 1C. This corresponds to large values of $|\mathcal{S}|$, that is of $|\mu|$, when the volume is normalized to 1. Since we are dealing with the case $\sigma(M) < 0$, the absolute value $|\sigma(M)|$ represents the *smallest* possible value of $|\mu[g]|$. Thus, it is not unreasonable to expect the critical metric corresponding to the value $\sigma(M)$ to have a special behavior.

In fact, we have the following result:

THEOREM 5.2. *Let (M_j, g_j) be a finite collection of hyperbolic manifolds realizing $\sigma(M)$, as discussed in Case III (page 72). Then each torus T_k in the cusp region of any hyperbolic component M_j is incompressible in M.*

The idea of the proof is that if one of the tori T in the hyperbolic cusps is compressible, T must bound a solid torus U in M. By explicitly constructing metrics on U that smoothly match the hyperbolic metric at T, one can prove that there is a smooth metric $g' \in \mathcal{M}_1$ with $\mu[g'] > \sigma(M)$. This contradicts the definition of $\sigma(M)$. Full details appear in [Anderson b]. This argument is similar in spirit, although quite different in proof, to Thurston's cusp closing theorem [Thurston 1979; Gromov 1981b].

In spite of the possible optimism implicit in the discussion above, Conjectures I–III remain very difficult to prove. The assumption that one can find sequences of Yamabe metrics $\{g_i\}$ satisfying (5.1) for which the curvature remains uniformly bounded is very strong, and it is not at all clear how to realize it. In fact, in general it is not true that a 3-manifold admits a sequence $\{g_i\}$ satisfying (5.1) with uniformly bounded curvature. Indeed, in all the arguments of this section, the (crucial) hypothesis of irreducibility has not been used. Thus, on manifolds of the form $M = N \# N$, where $N \neq \mathbb{S}^3$, or $M = \mathbb{S}^2 \times \mathbb{S}^1$, it is impossible to find metrics satisfying (5.1) with uniformly bounded curvature. Of course, when studying sequences in the space $\mathcal{M}_1$, it is not easy to distinguish the topology of the underlying manifold M.

Thus, there remains the fundamental problem of being able to understand if the sets in M where the curvature of $\{g_i\}$ diverges to infinity can be related with the essential two-spheres in M, that is, with the prime decomposition of M. This will be discussed further in the next section.

To summarize: assuming the conjecture that the only solutions of (5.9) are Einstein, we have outlined a proof that Conjectures I–III follow from the following conjecture:

CONJECTURE IV. *Let M be a closed, oriented, irreducible 3-manifold. Then there is a maximizing sequence of unit volume Yamabe metrics $\{g_i\}$ having uniformly bounded curvature.*

REMARK 5.3. Since the full curvature is needed to control the convergence or degeneration of metrics, the scalar curvature alone being too weak, it is perhaps natural to consider other functionals on $\mathcal{M}_1$, besides $\mathcal{S}$, that incorporate the full curvature. Thus, in dimension three, one may consider

$$\mathcal{R}^2 = \int_M |R|^2 \, dV,$$

the L^2 norm of the curvature tensor (see Section 3). This functional is clearly bounded below, so one can study the existence, regularity, and geometry of metrics that realize the infimum of $\mathcal{R}^2$. Such a program has been carried out in [Anderson 1993; a], where we obtain essentially the same results as in Theorem 5.1 (the L^∞ case). In addition, we show there that minimizing metrics are smooth, in contrast to the L^∞ case. However, the geometry of the minimizing metrics appears to be complicated: the Euler–Lagrange equations are of fourth order in g. Einstein metrics are solutions of the equations, but it would appear likely that there are other solutions as well on compact manifolds. In trying to implement the Geometrization Conjecture, as discussed above, by studying minimizing sequences of $\mathcal{R}^2$, one runs into the same difficulties as above, namely the behavior of the sequence, or the limit, near what one would conjecture to be essential two-spheres.

6. Essential Two-Spheres and "Black Holes": A Relation with General Relativity

We now discuss briefly some arguments supporting the validity of Conjecture IV. The scope of this paper will require this section to contain a number of oversimplificatons and to be even more terse than Section 5. Many important points will be neglected in order to keep the overall spirit of the argument simple.

Suppose, as above, that $g_i \in \mathcal{M}_1$ is a sequence of Yamabe metrics such that

$$\mu[g_i] \to \sigma(M). \tag{6.1}$$

Suppose further that $\sup |R_{g_i}| = R_i \to \infty$ as $i \to \infty$. Choose points $x_i \in M$ such that $|R_{g_i}|(x_i) = R_i$. In order to understand the degeneration of $\{g_i\}$ in a neighborhood of x_i, we rescale or "blow up" the metrics g_i by the factor R_i, that is, we consider the metrics

$$\tilde{g}_i = R_i g_i. \tag{6.2}$$

Then consider the behavior of the pointed Riemannian manifolds $(M, \tilde{g}_i, x_i)$ centered at x_i. Distances in $\tilde{g}_i$ are $\sqrt{R_i}$ times larger than distances in g_i, so that in effect, we are studying the degeneration behavior of $\{g_i\}$ in smaller and smaller neighborhoods of x_i, magnified to unit size. In particular, $\text{diam}_{\tilde{g}_i} M = \sqrt{R_i} \, \text{diam}_{g_i} M \to \infty$. The scaling properties of curvature imply that $|R_{\tilde{g}_i}| \leq 1$ and $|R_{\tilde{g}_i}|(x_i) = 1$.

Using an appropriate version of Theorem 5.1, one can conclude that, modulo diffeomorphisms, a subsequence of $\{\tilde{g}_i\}$ either converges uniformly on domains of bounded diameter to a limit metric $\tilde{g}$, or degenerates, that is, collapses (again uniformly on domains of bounded diameter), as described in Case II of Theorem 5.1. For simplicity, we will not deal with the latter case here. Thus, we assume that $\{\tilde{g}_i\}$ converges to a complete $C^{1,\alpha}$ metric $\tilde{g}$, defined on a 3-manifold X, with base point x. The triple $(X, \tilde{g}, x)$ is sometimes also called a *geometric limit* of $(M, \tilde{g}_i, x_i)$.

The gradient ∇S_{g_i} is given by (5.7), so that for the rescaled metrics $\tilde{g}_i$, one has

$$\tilde{Z}_i = \tilde{D}_i^2 u_i - (\tilde{\Delta}_i u_i) \, \tilde{g}_i - u_i \widetilde{\text{Ric}}_i + \tilde{Z}_i^T. \tag{6.3}$$

Note that u_i is scale-invariant. Now recall from (5.8) that $\|Z_i^T\|_{L^{-2,2}}(T\mathcal{C}) \to 0$ as $i \to \infty$. However, neither the $L^{-2,2}$ norm nor the functional S are scale-invariant. Taking the scaling behavior into account, one easily computes that

$$\|\tilde{Z}_i^T\|_{L^{-2,2}}(T\mathcal{C}) = R_i^{3/4} \|Z_i^T\|_{L^{-2,2}(T\mathcal{C})}. \tag{6.4}$$

Thus, it is no longer true, as in Section 5, that (6.1) implies automatically that $\|\tilde{Z}_i^T\|_{L^{-2,2}(T\mathcal{C})} \to 0$. The question of whether there exist sequences $\{g_i\}$ satisfying (6.1) and for which the tangential gradient $\tilde{Z}_i^T$ converges to 0 in $L^{-2,2}(\tilde{g}_i)$ norm is important. Because it involves deeper technicalities, it will not be discussed further here. We therefore suppose there exists a maximizing sequence $\{g_i\}$ of Yamabe metrics for which

$$\tilde{Z}_i^T \to 0 \quad \text{in } L^{-2,2}(\tilde{g}_i). \tag{6.5}$$

It follows that the limit metric $\tilde{g}$ satisfies the equation

$$Z = D^2 u - (\Delta u) \, g - u \, \text{Ric} \tag{6.6}$$

weakly, that is, in $L^{2,p}$. Here we have dropped the tildes. Note that, since the scalar curvature of $\{g_i\}$ is uniformly bounded, by scaling, the scalar curvature $\tilde{s}$ of the limit metric $\tilde{g}$ is identically 0 (in L^p) on X. In particular, for $\tilde{g}$, we have $Z = \text{Ric}$. Thus, setting $h = 1 + u$ in (6.6) gives

$$h \, \text{Ric} = D^2 h, \qquad \Delta h = 0. \tag{6.7}$$

It can be shown, using elliptic regularity, that weak $L^{2,p}$ solutions of the system (6.7) are in fact smooth. (Again, we have to assume that the function h thus obtained is not identically 0).

The equations (6.7) are classical equations arising in general relativity, called the *static vacuum Einstein equations*. Let $X_\pm = \{x \in X : \pm h(x) \geq 0\}$, and consider the product four-manifold $M_\pm^4 = X_\pm \times \mathbb{S}^1$, with warped product metric

$$g' = g_X + h^2 d\theta^2. \tag{6.8}$$

Then $(M_\pm^4, g')$ is a Ricci-flat four-manifold (its Ricci curvature vanishes identically), and is thus a vacuum solution of the Einstein equations. The length of the circle S_x^1 at $x \in M$ is given by $|h(x)|$; we note that the space M_+ or M_- may be singular on the locus where $h = 0$.

The equations (6.7) are defined on a space-like hypersurface of a Lorentzian four-manifold. In (6.8), we have changed the Lorentz signature $(-h^2 d\theta^2)$ to a Riemannian signature; this has no effect on computations of curvature and the like.

Summarizing, blow-ups of a sequence of metrics $\{g_i\}$ satisfying (6.5) and degenerating in a neighborhood of a point sequence $x_i \in M$ (of maximal curvature) have geometric limits that are solutions of the static vacuum Einstein equations.

The canonical solution of the static vacuum Einstein equations is the Schwarz-schild metric g_s, given by

$$g_s = \left(1 - \frac{2m}{r}\right)^{-1} dr^2 + r^2 ds_{\mathbb{S}^2}^2 + \left(1 - \frac{2m}{r}\right) d\theta^2, \tag{6.9}$$

with $h = (1 - 2m/r)^{1/2}$; here $m > 0$ is a free parameter, called the mass of the metric g_s. The metric on the space-like hypersurface

$$g_s = \left(1 - \frac{2m}{r}\right)^{-1} dr^2 + r^2 ds_{\mathbb{S}^2}^2 \tag{6.10}$$

is defined on $[2m, \infty) \times \mathbb{S}^2$, and is clearly spherically symmetric. Although the metric (6.9) or (6.10) may appear to be singular at $r = 2m$, this is only an apparent singularity, and can be removed by a change of coordinates. From (6.10), one sees that the set $B = h^{-1}(0)$ where $r = 2m$ is a totally geodesic two-sphere of constant curvature, while the metric (6.10) is asymptotically flat, that is, asymptotic to the flat metric on $\mathbb{R}^3$ for large r. The "horizon" $h^{-1}(0)$ is interpreted as the surface of an (isolated) black hole in general relativity. Asymptotically flat solutions to the equations (6.10) are often considered in the physics literature, since they serve as models of isolated black holes. The four-manifold Riemannian metric (6.9) is a smooth Ricci flat metric on $\mathbb{S}^2 \times \mathbb{R}^2$, asymptotic to the flat metric on $\mathbb{R}^3 \times \mathbb{S}^1$ at ∞.

This picture serves as a model for the general behavior of (appropriate) Yamabe sequences satisfying (6.1), in the neighborhoods of sets where the curvature goes to infinity. Namely, the two-sphere $B = h^{-1}(0)$ in the blow-up metric

$\tilde{g}$, when rescaled to the original sequence $\{g_i\}$, is being collapsed to a point. The fact that $\tilde{g}$ is asymptotically flat indicates, when blown down to $\{g_i\}$, that the curvature remains uniformly bounded in regions away from the central $\mathbb{S}^2$. Thus, $\{g_i\}$ is collapsing an $\mathbb{S}^2 \subset M$ to a point, giving rise to a limit metric g defined on the union of two manifolds (3-balls) M_1 and M_2 identified at one point. In other words, $\{g_i\}$ *performs* a surgery on the two-sphere $B \subset M$.

We illustrate how this behavior actually arises in a concrete example. It is known [Kobayashi 1987; Schoen 1989] that

$$\sigma(\mathbb{S}^2 \times \mathbb{S}^1) = \sigma(\mathbb{S}^3). \tag{6.11}$$

In fact, $\mathbb{S}^2 \times \mathbb{S}^1$ admits a sequence of conformally flat Yamabe metrics $g_i \in \mathcal{M}_1$ with $\mu[g_i] \to \sigma(\mathbb{S}^3)$. These metrics behave in the following way. View $\mathbb{S}^1$ as $I = [-1, 1]$ with endpoints identified. The metrics g_i have spherical $(\mathbb{S}^2)$ symmetry, and on domains of the form $\mathbb{S}^2 \times I_{\varepsilon_i}$, for $I_{\varepsilon_i} = [-1 + \varepsilon_i, 1 - \varepsilon_i]$, they converge smoothly to the canonical metric of volume 1 on $\mathbb{S}^3 \smallsetminus \{p \cup p'\}$, where p and p' are antipodal points on $\mathbb{S}^3$; here $\varepsilon_i \to 0$ as $i \to \infty$. In particular, $\{\mathbb{S}^2 \times \mathbb{S}^1, g_i\}$ converges to $\mathbb{S}^3/\{p \cup p'\}$, that is, $\mathbb{S}^3$ with two points identified, in the Gromov–Hausdorff topology [Gromov 1981a]. Note that the curvature is remaining uniformly bounded in the region $\mathbb{S}^2 \times I_{\varepsilon}$, for any fixed $\varepsilon > 0$. Let $J_{\varepsilon_i} = S^1 \smallsetminus I_{\varepsilon_i}$. On the complement $\mathbb{S}^2 \times J_{\varepsilon_i}$, whose diameter converges to 0, the curvature is blowing up, that is, diverging to $+\infty$. If one rescales, or blows up, the metrics $\{g_i\}$, as in the procedure described above, the blown-up metrics $\tilde{g}_i$ converge to the Schwarzschild metric. (Note that the Schwarzschild metric (6.10) is conformally flat.)

We see here, in this concrete example, and conjecturally in general, how $\{g_i\}$ is implementing the prime decomposition of the 3-manifold M. Of course, there remains the basic issue of proving that two-spheres that arise in this fashion are essential in M. We will not discuss this further here, beyond saying that it is natural to attempt to prove an analogue of Theorem 5.2 with spheres in place of tori.

There are however many other solutions to the equations (6.7) besides the Schwarzschild metric. Of course, there are the flat solutions on $\mathbb{R}^3$, where h is constant or linear. It is a classical result of Lichnerowicz [1955] that there are no complete, nonflat, asymptotically flat solutions of (6.7) with $h > 0$ everywhere. In fact, the assumption of asymptotic flatness can be dropped: There are no complete, nonflat solutions to (6.7) with $h > 0$ everywhere [Anderson c]. Thus, the cases of interest are where h vanishes somewhere.

In this regard there is a beautiful uniqueness theorem in the physics literature [Israel 1967; Robinson 1977; Bunting and Masood-ul-Alam 1987]:

THEOREM 6.1 (BLACK-HOLE UNIQUENESS THEOREM). *Let (X, g) be a smooth solution to the vacuum Einstein equations*

$$h \operatorname{Ric} = D^2 h, \qquad \triangle h = 0, \tag{6.12}$$

with $\partial X = h^{-1}(0)$ compact. If g is asymptotically flat, then (X, g) is the Schwarzschild solution.

This implies in particular that ∂X is connected: there are no smooth static solutions with multiple black holes. It is interesting to note that there are asymptotically flat solutions to (6.12) with multiple black holes, having only cone singularities along a line segment (geodesic) joining the black holes. In particular, these singularities are not curvature singularities: the curvature is uniformly bounded everywhere. The line singularity is interpreted as a "strut" keeping the black holes in equilibrium from their mutual gravitational attraction [Kramers et al. 1980].

Theorem 6.1 can be proved more generally under the single assumption that $h^{-1}(0)$ is compact, that is, it is not necessary to assume the metric asymptotically flat; the metric must be still be assumed to be smooth up to ∂X, see [Anderson c].

We note that there are examples of solutions to (6.12) with smooth non-connected, and in fact noncompact, boundary. The so-called B1 solution [Ehlers and Kundt 1962; Kramers et al. 1980] has 3-manifold metric given by

$$g_b = \left(1 - \frac{b}{r}\right)^{-1} dr^2 + \left(1 - \frac{b}{r}\right) d\phi^2 + r^2 \, d\theta^2. \tag{6.13}$$

Here the function h of (6.12) is given by $h = r\sin\theta$, for $r \in [b, \infty)$ and for $\phi \in [0, \pi]$, $\theta \in [0, 2\pi]$ the standard spherical coordinates on $\mathbb{S}^2$. This metric has the property that $h^{-1}(0)$ is two copies of $\mathbb{R}^2$, each asymptotic to a flat cylinder. This metric is not asymptotically flat in the usual sense, that is, not asymptotic to the flat metric on $\mathbb{R}^3$. However, it is asymptotic to the flat metric on $\mathbb{R}^2 \times \mathbb{S}^1$. Note that h is unbounded. Note further that when forming the four-manifold metric as in (6.8) with h as above, one obtains exactly the Schwarzschild metric on $\mathbb{S}^2 \times \mathbb{R}^2$. In fact, the metrics (6.10) and (6.13) are just different (orthogonal) three-dimensional slices to the four-dimensional Schwarzschild metric (6.9).

It seems that one should be able to classify completely the smooth solutions to the static vacuum Einstein equations with smooth boundary $B = h^{-1}(0)$, and that are complete away from B. We venture the following.

CONJECTURE 6.2. Let (X, g) be a smooth complete solution to the static Einstein vacuum equations (6.12), with $B = h^{-1}(0)$ smooth, that is, g smooth up to B. Then (X, g) is either flat, or the Schwarzschild solution, or the B1 solution.

There is a wealth of examples in the physics literature on singular solutions to the static vacuum equations (6.12); see [Ehlers and Kundt 1962; Kramers et al. 1980], for instance. These solutions are typically not complete, or have curvature singularities on B, that is, the curvature blows up on approach to some region in B. Since, in the context of our discussion, the spaces $(X, \tilde{g})$ arise as limits of spaces of bounded curvature, one might hope that these singular solutions could be ruled out.

In sum, the partly heuristic arguments presented above provide some evidence to support Conjecture IV, establishing a relation between the sets where the curvature of a maximizing sequence of Yamabe metrics diverges to infinity, and the sets of essential two-spheres in M.

Acknowledgements

I would like to express my thanks to F. Bonahon, J. Hass, S. Kerckhoff, G. Mess, C. McMullen, and Z. Sela for conversations and comments.

References

[Anderson 1993] M. Anderson, "Degenerations of metrics with bounded curvature and applications to critical metrics of Riemannian functionals", pp. 53–79 in *Differential geometry: Riemannian geometry* (Los Angeles, 1990), edited by R. Greene and S. T. Yau, Proc. Symp. Pure Math. **54**, AMS, Providence, RI, 1993.

[Anderson a] M. Anderson, "Extrema of curvature functionals on the space of metrics on 3-manifolds", preprint, SUNY at Stony Brook, 1996.

[Anderson b] M. Anderson, "Scalar curvature and geometrization of 3-manifolds I", in preparation.

[Anderson c] M. Anderson, "On the global structure of solutions to the static vacuum Einstein equations", in preparation.

[Atiyah et al. 1973] M. Atiyah, R. Bott, and V. Patodi, "On the heat equation and the Index theorem", *Invent. Math.* **19** (1973), 279–330.

[Aubin 1976] T. Aubin, "Équations différentielles non linéaires et problème de Yamabe concernant la courbure scalaire", *J. Math. Pures Appl.* **55** (1976), 269–296.

[Berger 1970] M. Berger, "Quelques formules de variation pour une structure riemannienne", *Ann. Sci. École Norm. Sup.* (4) **3** (1970), 285–294.

[Besse 1987] A. Besse, *Einstein manifolds*, Ergebnisse der Mathematik und ihrer Grenzgebiete (3) **10**, Springer, Berlin, 1987.

[Bunting and Masood-ul-Alam 1987] G. Bunting and A. Masood-ul-Alam, "Nonexistence of multiple black holes in asymptotically Euclidean static vacuum spacetimes", *Gen. Rel. and Grav.* **19** (1987), 147–154.

[Casson and Jungreis 1994] A. Casson and D. Jungreis, "Convergence groups and Seifert-fibered 3-manifolds", *Invent. Math.* **118** (1994), 441–456.

[Cheeger 1970] J. Cheeger, "Finiteness theorems for Riemannian manifolds", *Am. J. Math.* **92** (1970), 61–74.

[Cheeger and Gromov 1986] J. Cheeger and M. Gromov, "Collapsing Riemannian manifolds while keeping their curvature bounded I", *J. Diff. Geom.* **23** (1986), 309–346.

[Cheeger and Gromov 1990] J. Cheeger and M. Gromov, "Collapsing Riemannian manifolds while keeping their curvature bounded II", *J. Diff. Geom.* **32** (1990), 269–298.

[Ebin 1970] D. Ebin, "The manifold of Riemannian metrics", pp. 11–40 in *Global analysis* (Berkeley, 1968), edited by S.-S. Chern and S. Smale, Proc. Symp. Pure Math. **15**, AMS, Providence, 1970.

[Ehlers and Kundt 1962] J. Ehlers and W. Kundt, "Exact solutions of the gravitational field equations", pp. 49–101 in *Gravitation: An Introduction to Current Research*, edited by L. Witten, Wiley, NY, 1962.

[Gabai 1992] D. Gabai, "Convergence groups are Fuchsian groups", *Ann. of Math.* **136** (1992), 447–510.

[Gromov 1981a] M. Gromov, *Structures métriques pour les variétés riemanniennes*, edited by J. Lafontaine and P. Pansu, CEDIC, Paris, 1981.

[Gromov 1981b] M. Gromov, "Hyperbolic manifolds (according to Thurston and Jørgensen)", pp. 40–53 in *Bourbaki Seminar* 1979/80, Lecture Notes in Math. **842**, Springer, Berlin, 1981.

[Gromov and Lawson 1980] M. Gromov and B. Lawson, "The classification of simply connected manifolds of positive scalar curvature", *Ann. of Math.* **111** (1980), 423–434.

[Gromov and Lawson 1983] M. Gromov and B. Lawson, "Positive scalar curvature and the Dirac operator on complete Riemannian manifolds", *Publ. Math. IHES* **58** (1983), 83–196.

[Haken 1961] W. Haken, "Theorie der Normalflächen", *Acta Math.* **105** (1961), 245–375.

[Hamilton 1982] R. Hamilton, "Three manifolds with positive Ricci curvature", *J. Diff. Geom.* **17** (1982), 255–306.

[Hempel 1983] J. Hempel, *3-Manifolds*, Annals of Math. Studies **86**, Princeton Univ. Press, Princeton, 1976.

[Israel 1967] W. Israel, "Event horizons in static vacuum space-times", *Phys. Rev.* **164** (1967), 1776–1779.

[Jaco 1980] W. Jaco, *Lectures on three-manifold topology*, Regional conference series in mathematics **43**, AMS, Providence, 1980.

[Jaco and Shalen 1979] W. Jaco and P. Shalen, "Seifert fibered spaces in 3-manifolds", *Mem. Amer. Math. Soc.* **21** (1979), no. 220, 192 pp.

[Johannson 1979] K. Johannson, *Homotopy equivalence of 3-manifolds with boundaries*, Lecture Notes in Math, **761**, Springer, Berlin, (1979).

[Kneser 1929] H. Kneser, "Geschlossene Flächen in dreidimensionalen Mannifaltigkeiten", *Jahres. Deut. Math. Verein.* **38** (1929), 248–260.

[Kobayashi 1987] O. Kobayashi, "Scalar curvature of a metric with unit volume", *Math. Ann.* **279** (1987), 253–265.

[Kramers et al. 1980] D. Kramers, H. Stephani, M. MacCallum, and E. Herlt, *exact solutions of Einstein's field equations*, Cambridge University Press, Cambridge, 1980.

[Lichnerowicz 1955] A. Lichnerowicz, *Théories relativistes de la gravitation et de l'electromagnetisme*, Masson, Paris, 1955.

[Lovelock and Rund 1972] D. Lovelock and H. Rund, "Variational principles in the general theory of relativity", *Jahres. Deut. Math. Verein.* **74** (1972), 1–65.

[McMullen 1989]   C. McMullen, "Amenability, Poincaré series and quasiconformal maps", *Invent. Math.* **97** (1989), 95–127.

[McMullen 1990]   C. McMullen, "Iteration on Teichmüller space", *Invent. Math.* **99** (1990), 425–454.

[Milnor 1962]   J. Milnor, "A unique factorization theorem for 3-manifolds", *Amer. J. Math.* **84** (1962), 1–7.

[Morgan 1984]   J. Morgan, "On Thurston's uniformization theorem for three-dimensional manifolds", pp. 37–125 in *The Smith Conjecture*, edited by J. Morgan and H. Bass, Pure Appl. Math. **112**, Academic Press, Orlando, 1984.

[Mostow 1968]   G. Mostow, "Quasiconformal mappings in n-space and the strong rigidity of hyperbolic space forms", *Publ. Math. IHES* **34** (1968), 53–104.

[Palais 1968]   R. Palais, "Applications of the symmetric criticality principle in mathematical physics and differential geometry", pp. 247–301 in *Proceedings of the 1981 Shanghai symposium on differential geometry and differential equations*, Science Press, Beijing, 1984.

[Prasad 1973]   G. Prasad, "Strong rigidity of $\mathbb{Q}$-rank 1 lattices", *Invent. Math.* **21** (1973), 255–286.

[Robinson 1977]   D. Robinson, "A simple proof of the generalization of Israel's theorem", *Gen. Rel. and Grav.* **8** (1977), 695–698.

[Rong 1990]   X. Rong, Thesis, "Rationality of limiting 3-invariants of collapsed 3-manifolds", SUNY at Stony Brook, 1990.

[Rong 1993]   X. Rong, "The limiting eta invariant of collapsed 3-manifolds", *J. Diff. Geom.* **37** (1993), 535–568.

[Schoen 1984]   R. Schoen, "Conformal deformation of a Riemannian metric to constant scalar curvature", *J. Diff. Geom.* **20** (1984), 479–495.

[Schoen 1984]   R. Schoen, "Minimal surfaces and positive scalar curvature", pp. 575–578 in *Proceedings of the International Congress of Mathematicians* (Warsaw, 1983), PWN, Warsaw, 1984.

[Schoen 1989]   R. Schoen, "Variational theory for the total scalar curvature functional for Riemannian metrics and related topics", pp. 120–154 in *Topics in calculus of variations* (Montecatini Terme, 1987), edited by M. Giaquinta, Lect. Notes Math. **1365**, Springer, Berlin, 1989.

[Schoen and Yau 1979a]   R. Schoen and S. T. Yau, "Existence of incompressible minimal surfaces and the topology of 3-manifolds with non negative scalar curvature", *Ann. Math.* **110** (1979), 127–142.

[Schoen and Yau 1979b]   R. Schoen and S. T. Yau, "On the structure of manifolds with positive scalar curvature", *Manuscripta Math.* **29** (1979), 159–183.

[Scott 1983]   P. Scott, "The geometries of 3-manifolds", *Bull. London Math. Soc.* **15** (1983), 401–487.

[Thurston 1979]   W. Thurston, "The geometry and topology of three-manifolds", lecture notes, Princeton University, 1979.

[Thurston 1982]   W. Thurston, "Three-dimensional manifolds, Kleinian groups and hyperbolic geometry", *Bull. Amer. Math. Soc.* **6** (1982), 357–381.

[Thurston 1986] W. Thurston, "Hyperbolic structures on 3-manifolds I: Deformations of acylindrical manifolds", *Ann. of Math.* **124** (1986), 203–246.

[Thurston 1988] W. Thurston, "Hyperbolic structures on 3-manifolds II: Surface groups and 3-manifolds which fiber over the circle", preprint, 1988.

[Thurston 1996] W. Thurston, *Three-dimensional geometry and topology*, vol. 1, Princeton Univ. Press, Princeton, 1996.

[Trudinger 1968] N. Trudinger, "Remarks concerning the conformal deformation of Riemannian structures on compact manifolds", *Ann. Scuola Norm. Sup. Pisa* **22** (1968), 265–274.

[Waldhausen 1967] F. Waldhausen, "Eine Klasse von 3-dimensionalen Mannig-faltigkeiten I and II", *Invent. Math.* **3** (1967), 308–333 and **4** 1967, 87–117.

[Waldhausen 1968] F. Waldhausen, "On irreducible 3-manifolds which are sufficiently large", *Ann. of Math.* **87** (1968), 56–88.

[Wolf 1977] J. A. Wolf, *Spaces of constant curvature*, 4th edition, Publish or Perish, Berkeley, 1977.

[Yamabe 1960] H. Yamabe, "On a deformation of Riemannian structures on compact manifolds", *Osaka Math. J.* **12** (1960), 21–37.

MICHAEL T. ANDERSON
DEPARTMENT OF MATHEMATICS
SUNY AT STONY BROOK
STONY BROOK, NY 11794-3651
anderson@math.sunysb.edu

Comparison Geometry
MSRI Publications
Volume **30**, 1997

Aspects of Ricci Curvature

TOBIAS H. COLDING

This article is dedicated to my parents.

ABSTRACT. We describe some new ideas and techniques introduced to study
spaces with a given lower Ricci curvature bound, and discuss a number of
recent results about such spaces.

Introduction

In studying spaces with a given lower sectional curvature bound we have a
very powerful tool in the Toponogov triangle comparison theorem. This allows
us to study metric properties of such spaces (see for instance [Toponogov 1964;
Burago et al. 1992; Perelman 1995]), and topological properties (see for instance
[Cheeger 1991; Cheeger and Gromoll 1972; Grove and Shiohama 1977; Gromov
1981a; Grove and Petersen 1988; Perelman 1991]). Perhaps the most important
tool for studying topological properties of such manifolds is the notion of critical
points of distance functions in connection with the Toponogov triangle comparison theorem; see [Grove and Shiohama 1977] and compare with the remarks at
the end of Section 1.

When we only assume a lower Ricci curvature bound, no such estimate is
available. Classically, the only known general estimates of this type for Ricci
curvature are the volume comparison theorem [Bishop and Crittenden 1964;
Gromov 1981b] and the Abresch–Gromoll inequality [Abresch and Gromoll 1990].

In order to study manifolds with a given lower Ricci curvature bound there
are at least two obstacles to overcome. First, many results from the sectional
curvature case do not remain true for Ricci curvature. Second, due to the lack
of a good estimate on the distance function we do not have good control on the

Supported in part by NSF grant DMS-9504994 and an Alfred P. Sloan Research Fellowship.

local geometry in this case. We will see in this survey that in some sense the second obstacle is the most serious.

In Section 1 we will discuss a new estimate of the distance function. In later sections we will see that this type of estimate has a large number of consequences, some of which are given in Sections 2 and 3.

In Section 2 our main focus is the geometry and topology of manifolds with a lower Ricci curvature bound.

In Section 3 our focus is on regularity properties of general metric spaces that are (Gromov–Hausdorff) limits of n-dimensional manifolds with a given lower Ricci curvature bound. This is in part motivated by Gromov's compactness theorem [1981b], saying that the space of n-dimensional manifolds with a given lower Ricci curvature bound is precompact in the Gromov–Hausdorff topology.

In Section 4 we discuss some analytic properties of manifolds with a given lower Ricci curvature bound. The Harnack inequality and the gradient estimate of Yau and Cheng [Yau 1975; Cheng and Yau 1975] discussed in that section play a crucial role in the results of Section 1, 2 and 3.

The results described in this survey are to be found in the references listed under "Direct Sources" (page 94).

1. Integral Estimate of Angles and Distances Using the Hessian

Our main technical tool is a new estimate for distance functions. The first such estimate appeared in [Colding 1996a], and the reader should consult that reference for a more precise statement than we are about to give here (and in particular for the notion of "almost equal").

Suppose M is an n-dimensional closed manifold with $\mathrm{Ric}_M \geq n-1$. Consider the space of geodesics of some fixed length $l < \pi$, and identify each such geodesic with its initial velocity. Equip this space with the probability measure coming from the normalized Liouville measure.

Let $p, q \in M$ be points with $d(p, q)$ almost equal to π, where $d(p, q)$ denotes the distance between p and q. For any geodesic γ (not necessarily minimizing) of fixed length l, let Δ be the geodesic triangle of which γ is the side opposite the vertex p, and let $\underline{\Delta}$ be the comparison triangle on the unit sphere, in the sense of Toponogov (see Figure 1). For $0 \leq t \leq l$, let d_t be the distance between $\gamma(t)$ and p, and let $\underline{d}_t$ be the corresponding distance on the sphere.

(If the triangle in M does not satisfy the triangle inequality, so that no comparison triangle exists, we use an alternative analytical definition of $\underline{d}_t$ and $\underline{\angle}_t$ that still makes sense in this case.)

Then, in an L^2-sense (or equivalently for a set of geodesics of nearly full measure), d_t is almost equal to $\underline{d}_t$, and $\angle_t$ is almost equal to $\underline{\angle}_t$. More formally:

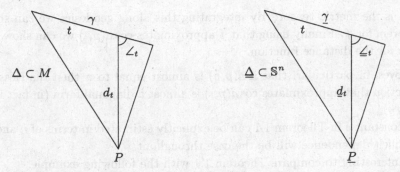

Figure 1. Estimate of distances and angles for positive Ricci curvature.

THEOREM 1.1. *Given $\varepsilon > 0$ and $0 < l < \pi$, there exists a $\delta = \delta(\varepsilon, n) > 0$ such that, for any M of dimension n with $\mathrm{Ric}_M \geq n - 1$, any $p, q \in M$ with $d(p, q) > \pi - \delta$, and any $0 \leq t \leq l$, we have $\|d_t - \underline{d}_t\|_2 < \varepsilon$ and $\|\angle_t - \underline{\angle}_t\|_2 < \varepsilon$.*

See [Colding 1996a] for details, and compare [Colding 1996b; Cheeger and Colding 1996] for later versions.

In the sequel we will let $\psi(\varepsilon \,|\, \ldots)$ and $\psi_i(\varepsilon \,|\, \ldots)$ denote nonnegative functions depending on ε and possibly on some additional parameters (written after the bar), satisfying the property that when the other parameters are fixed the function tends to 0 as ε approaches 0.

SKETCH OF THE PROOF OF THEOREM 1.1. First we use the fact that $d(p, q) > \pi - \delta$ and the bound $\mathrm{Ric}_M \geq n - 1$ to approximate $\cos d(p, \cdot)$ in the $(2, 1)$-Sobolev norm by a smooth function that satisfies $\|\Delta f + nf\|_2 < \psi_1(\delta \,|\, n)$; here the L^2-norm is normalized so that $\|1\|_2 = 1$. Next, from the Bochner formula for f and the fact that $\mathrm{Ric}_M \geq n - 1$, we get

$$\tfrac{1}{2}\Delta |\nabla f|^2 \geq |\mathrm{Hess}(f)|^2 + \langle \nabla \Delta f, \nabla f \rangle + (n - 1)|\nabla f|^2.$$

Integrating by parts over M gives

$$0 \geq \int_M |\mathrm{Hess}(f)|^2 - \int_M |\Delta f|^2 - (n - 1)\int_M f\Delta f.$$

Now since $\|\Delta f + nf\|_2 < \psi_1(\delta \,|\, n)$ we get from the Cauchy–Schwarz inequality

$$\psi_2(\delta \,|\, n) > \frac{1}{\mathrm{Vol}(M)}\int_M |\mathrm{Hess}(f)|^2 - \frac{n}{\mathrm{Vol}(M)}\int_M f^2.$$

From the Cauchy–Schwarz inequality we therefore get, again using the fact that $\|\Delta f + nf\|_2 < \psi_1(\delta \,|\, n)$,

$$\psi_3(\delta \,|\, n) > \frac{1}{\mathrm{Vol}(M)}\int_M |\mathrm{Hess}(f)|^2 + \frac{n}{\mathrm{Vol}(M)}\int_M f^2 + \frac{2}{\mathrm{Vol}(M)}\int_M f\Delta f$$

$$= \frac{1}{\mathrm{Vol}(M)}\int_M |\mathrm{Hess}(f) + fg|^2,$$

where g is the metric tensor. By integrating this along geodesics we can show the theorem for f. Finally using that f approximates $\cos d(p, \cdot)$ we can show the theorem for the distance function. $\qquad\square$

This proves, in particular, that if $d(p, q)$ is almost equal to π then the Hessian of a function that approximates $\cos d(p, \cdot)$ is almost a diagonal form (in fact it is almost fg).

The constant δ in Theorem 1.1 can be explicitly estimated in terms of n and ε. This explicit dependence will be the case throughout.

It is interesting to compare Theorem 1.1 with the following example:

EXAMPLE 1.2 [Anderson 1990b]. There exist metrics on $\mathbb{CP}^n\#\mathbb{CP}^n$, with $\mathrm{Ric} \geq 2n - 1$ and $\mathrm{Vol} \geq v > 0$, that are arbitrary close to a metric on $\mathbb{S}^{2n}$ with two conical singularities. Moreover, the diameter of $\mathbb{S}^{2n}$ with this metric is π.

The metric on $\mathbb{S}^4$ is constructed as follows. Let $\Pi : \mathbb{S}^3 \to \mathbb{S}^2$ be the Hopf fibration, and let $\sigma_X, \sigma_Y, \sigma_Z$ the standard left invariant coframing of $\mathbb{S}^3 = \mathrm{SU}(2)$, where σ_Z is tangent to the Hopf fibers. Define a metric on $\mathbb{S}^3$ by $C_1^2\sigma_Z^2 + C_2^2(\sigma_X^2 + \sigma_Y^2)$, where $C_1 \approx 0.08$ and $C_2 \approx 0.25$ are constants. Then $(\mathbb{S}^4, g)$ is the spherical suspension of $\mathbb{S}^3$ with this Berger metric.

In the metrics of [Anderson 1990b] the two embedded $\mathbb{CP}^1$'s in $\mathbb{CP}^2\#\mathbb{CP}^2$ are totally geodesic and round. Furthermore their curvature converges to infinity as $\mathbb{CP}^2\#\mathbb{CP}^2$ converges to $\mathbb{S}^4$. See [Anderson 1990b] for further details and examples of similar metrics in dimension > 4.

We will now consider a different measure on the set of all minimal geodesics with unit speed, one that behaves better when the volume of the manifold is small. For a manifold M, a point $p \in M$, and $r_0 > 0$, we identify the minimal geodesics contained in $B_{r_0}(p)$ with their endpoints: $\gamma \to (\gamma(0), \gamma(l))$, where $l = \mathrm{length}(\gamma)$. We equip this space of minimal geodesics with the natural measure coming from the product measure on $M \times M$; and we normalize this measure so that the space of minimal geodesics with endpoints in $B_{r_0}(p)$ has measure one.

The next theorem appeared in [Cheeger and Colding 1996]. It follows an earlier version given in [Colding a]. For $p, q, x \in M$, define the *excess* by $e_{p,q}(x) = d(p, x) + d(x, q) - d(p, q)$.

THEOREM 1.3. *Fix $r_0 > 0$. Given $\varepsilon > 0$, there exist $R = R(\varepsilon, r_0, n) > 0$, $\Lambda = \Lambda(\varepsilon, r_0, n) > 0$ and $\delta = \delta(\varepsilon, r_0, n) > 0$ such that, for any M of dimension n with $\mathrm{Ric}_M \geq -(n-1)R^{-2}\Lambda$ and any $p, q, x \in M$ with $e_{p,q}(x) < \delta$ and $d(p, x), d(q, x) > R$, we have $\|d_t - \underline{d}_t\|_2 < \varepsilon$ and $\|\angle_t - \underline{\angle}_t\|_2 < \varepsilon$.*

To prove Theorem 1.3, we show first that we can approximate the distance function in the $(2, 1)$-Sobolev norm by a harmonic function. We then show that the L^2-norm of the Hessian of this harmonic function is small. Finally we integrate this Hessian estimate along geodesics to get the theorem for the harmonic function, and we use the fact that the harmonic function approximates the distance function to get the theorem for the distance function.

One should compare these three steps with the corresponding three steps in the proof of Theorem 1.1.

It is important to note that in both Theorem 1.1 and Theorem 1.3 the difference in the angles is not necessarily uniformly small, but only small with respect to the L^2-norm. For examples where this difference is not uniformly small, see [Anderson 1992; Perelman 1997]. Such examples also show that we cannot control the critical points of the distance function for Ricci curvature (compare with the discussion in the Introduction).

For other estimates along these lines, see [Colding 1996a; Colding 1996b; Colding a; Cheeger and Colding 1996; Cheeger, Colding, and Tian b].

2. Almost Maximal Manifolds

Recall that the set of all metric spaces can be made into a metric space by means of the *Gromov–Hausdorff distance* d_{GH}. Denoting by $T_\varepsilon(X)$ the ε-neighborhood of a subset X of a metric space Y, this distance is defined as follows:

DEFINITION 2.1 [Gromov 1981b]. The Gromov–Hausdorff distance between two metric spaces (X_1, d_1) and (X_2, d_2) is the infimum of all $\varepsilon > 0$ such that there exist a metric space Y and isometric embeddings $j_1 : X_1 \to Y$ and $j_2 : X_2 \to Y$ with $j_1(X_1) \subset T_\varepsilon(j_2(X_2))$ and $j_2(X_2) \subset T_\varepsilon(j_1(X_1))$.

For noncompact metric spaces we say that a pointed sequence (X_i, x_i) converges to (X, x) in the pointed Gromov–Hausdorff topology if, for all $r > 0$, the sequence $X_i \cap B_r(x_i)$ converges to $X \cap B_r(x)$ in the Gromov–Hausdorff topology. This convergence should be thought of as convergence on compact subsets.

For our purposes the importance of this definition is that $d_{GH}(X_1, X_2) < \varepsilon$ if and only if there exist maps $f_1 : X_1 \to X_2$ and $f_2 : X_2 \to X_1$ such that, for $i = 1, 2$ and all $a_i, b_i \in X_i$,

$$|d(f_i(a_i), f_i(b_i)) - d(a_i, b_i)| < \psi(\varepsilon),$$

and for $i, j = 1, 2$ with $i \neq j$,

$$d(f_j \circ f_i(a_i), a_i) < \psi(\varepsilon).$$

One of the most useful properties of the Gromov–Hausdorff distance is that it is a good tool to measure the rough geometry of a metric space. In particular, we will see in Theorems 2.3 and 2.5 that if we have a space with a lower Ricci curvature bound then in many instances the rough (large-scale) geometry controls the small-scale geometry.

For the next few results the L^2-estimate on distance functions (and of the Hessian of a function that approximates the distance function) mentioned in Section 1 is crucial.

By $V_\Lambda^n(r)$ we will mean the volume of a ball of radius r in the n-dimensional simply connected space form of constant sectional curvature Λ.

THEOREM 2.2 [Colding 1996a]. *Given $\varepsilon > 0$, there exists $\delta = \delta(n, \varepsilon) > 0$ such that, if an n-dimensional manifold M has $\mathrm{Ric}_M \geq n-1$ and $\mathrm{Vol}(M) > V_1^n(\pi) - \delta$, then $d_{GH}(M, \mathbb{S}^n) < \varepsilon$.*

Theorem 2.2 was conjectured by Anderson–Cheeger and Perelman.

A key point in the proof of Theorem 2.2 is that the Bishop volume comparison theorem and the assumption on the volume imply that, for any $p \in M$, there exists $q \in M$ with $d(p, q) > \pi - \psi(\delta \,|\, n)$. We can therefore apply Theorem 1.1 for all $p \in M$.

Theorem 2.2 is the first result about Ricci curvature proved using "synthetic" techniques (see Section 3 for more on this).

The case of Alexandrov spaces, which are possibly singular spaces with a lower sectional curvature bound in the triangle comparison sense, was treated systematically from such a point of view in [Burago et al. 1992]; compare [Perelman 1995].

THEOREM 2.3 (VOLUME CONVERGENCE [Colding 1996b; a]). *For $r > 0$, consider all metric balls of radius r in all complete n-dimensional Riemannian manifolds with Ricci curvature greater or equal to $-(n-1)$. Equip this space with the Gromov–Hausdorff topology. Then the volume function is a continuous function.*

This result was conjectured by Anderson and Cheeger.

Using a covering argument and the volume comparison theorem we can reduce the proof of Theorem 2.3 to showing that, if a ball in an n-dimensional manifold for which the infimum of the Ricci curvature is almost zero is close to the corresponding ball in $\mathbb{R}^n$, the volumes are also close.

It is easy to see that a lower Ricci curvature bound is needed in Theorem 2.3; see for instance the examples in [Colding 1996b]. In the same reference we showed that Vol is continuous at the unit n-sphere, $\mathbb{S}^n$: more precisely, if M_i is a sequence of n-dimensional manifolds with $\mathrm{Ric}_{M_i} \geq n - 1$ and $M_i \xrightarrow{d_{GH}} \mathbb{S}^n$, then $\mathrm{Vol}(M_i) \to \mathrm{Vol}(\mathbb{S}^n)$.

THEOREM 2.4 [Colding a]. *There exists an $\varepsilon = \varepsilon(n) > 0$ such that if M is a closed n-dimensional manifold with $\mathrm{Ric}_M \, \mathrm{diam}_M^2 > -\varepsilon$ and $b_1(M) = n$ then M is homeomorphic to a torus if $n \neq 3$ and homotopically equivalent to a torus if $n = 3$.*

This result was conjectured by Gromov [1981b, p. 75].

To prove Theorem 2.4, we show first that a finite cover of M is close to a flat n-dimensional torus. This allows us to apply Theorem 2.3 to a finite cover of M to conclude that a finite cover is a homotopy torus. See also Theorem 2.6.

Theorem 2.4 should be compared with Bochner's theorem [Bochner 1946; Bochner and Yano 1953] and with Gromov's theorem [1981b]. Recall that

Bochner's theorem says that a closed n-manifold with nonnegative Ricci curvature has $b_1 \leq n$, and equality holds if and only if the manifold is isometric to a flat torus. Later Gromov showed (see also [Gallot 1983]) that there exists an $\varepsilon = \varepsilon(n) > 0$ such that any closed n-manifold M with $\mathrm{Ric}_M \, \mathrm{diam}_M^2 > -\varepsilon$ has $b_1 \leq n$.

Yamaguchi [1988] proved that, given $D > 0$ and $k \leq n$, there exists an $\varepsilon = \varepsilon(k, D, n) > 0$ such that if M is a closed n-manifold with $\mathrm{diam}_M \leq D$, $b_1(M) = k$, $\mathrm{Ric}_M \geq -\varepsilon$, and $|\mathrm{Sec}_M| \leq 1$, then a finite cover of M fibers over a k-torus. In [Yamaguchi 1991] he showed that if $\mathrm{Sec}_M \, \mathrm{diam}_M^2 > -\varepsilon$, then a finite cover of M fibers over a k-torus. Actually, in [Yamaguchi 1988] he made a stronger conjecture than the original one by Gromov, namely, that this should remain true for almost nonnegative Ricci curvature. This was disproved by Anderson [1992], who gave counterexamples for $k \leq n - 1$; Gromov's original conjecture, however, was left open.

Let M be an n-dimensional open Riemannian manifold with nonnegative Ricci curvature. By Gromov's compactness theorem [1981b], any sequence $r_i \to \infty$ has a subsequence $r_j \to \infty$ such that the rescaled manifolds $(M, p, r_j^{-2} g)$ converge in the pointed Gromov–Hausdorff topology to a length space M_∞.

Every such limit (an example of Perelman shows that M_∞ is not unique in general) is said to be a *tangent cone at infinity* of M. Even though uniqueness fails in general, one expects it to hold in the maximal case. In fact, we have the following result:

THEOREM 2.5 [Colding a]. *If an n-dimensional manifold M has nonnegative Ricci curvature and some M_∞ is isometric to $\mathbb{R}^n$ then M is isometric to $\mathbb{R}^n$.*

Theorem 2.5 was conjectured by Anderson and Cheeger [Anderson 1992].

SKETCH OF PROOF. Fix $p \in M$. By the assumption, if we rescale a large ball centered at p to unit size the rescaled ball is close to the unit ball in $\mathbb{R}^n$. Therefore, by Theorem 2.3, the volumes are close. Using the volume comparison theorem we can now conclude that a ball of a fixed size in M with center at p has the same volume as the corresponding ball in $\mathbb{R}^n$. From equality in the volume comparison theorem we conclude that M is isometric to $\mathbb{R}^n$. $\square$

As an immediate consequence of Theorem 2.3, small balls have almost maximal volume. The importance of this was pointed out to us by Anderson and Cheeger. In particular they observed that this, together with [Anderson 1990c] (see also [Anderson 1992]), implies Theorem 2.7 below. Moreover, this together with the result of [Perelman 1994] and methods from controlled topology [Ferry 1979; Ferry and Quinn 1991], as in [Grove et al. 1989; 1990; Petersen 1990], gives the next theorem:

THEOREM 2.6 (TOPOLOGICAL STABILITY [Colding a]). *If M is a closed n-manifold, there exists $\varepsilon(M) > 0$ such that, if N is a n-manifold with $\mathrm{Ric}_N \geq -(n-1)$ and $d_{GH}(M, N) < \varepsilon$, then M and N are homotopically equivalent (and even homeomorphic if $n \neq 3$).*

In the case of sectional curvature, Perelman [1991] proved that Theorem 2.6 remains true in the class of Alexandrov spaces. However, for Ricci curvature, if one allows M in Theorem 2.6 to have singularities, the conclusion does not hold. Indeed, Anderson [1990a] gave examples of metrics on $\mathbb{S}^2 \times \mathbb{S}^2$ with $|\mathrm{Ric}| \leq C$ that converge to a metric with two conical singularities on the suspension of $\mathbb{RP}^3$ (the diagonal and antidiagonal in $\mathbb{S}^2 \times \mathbb{S}^2$ are each collapsed to a point). Numerous further examples have since been considered. For instance, Example 1.2 shows that there exist metrics on $\mathbb{CP}^2 \# \mathbb{CP}^2$ arbitrary close to a metric suspension of a Berger metric on $\mathbb{S}^3$ and having $\mathrm{Ric} \geq 3$. Similarly, Otsu [1991] constructed metrics on $\mathbb{S}^3 \times \mathbb{S}^2$ with $\mathrm{Ric} \geq 4$ and arbitrarily close to a metric suspension of a metric on $\mathbb{S}^2 \times \mathbb{S}^2$. Even for Einstein metrics the topological stability of Theorem 2.6 fails if M has singularities; see for instance [Tian and Yau 1987].

Note that the ε in Theorem 2.6 must depend on M for the same reason that the conclusion fails if one allows M to have singularities. For instance, a trivial modification of Anderson's examples [1990b] gives a sequence where the manifolds are alternately homeomorphic to $\mathbb{S}^4$ and to $\mathbb{CP}^2 \# \mathbb{CP}^2$, and all metrics have $\mathrm{Ric} \geq 3$, but the limit is still the same spherical suspension of a Berger sphere.

THEOREM 2.7 (METRIC STABILITY [Colding a]). *Let M_i be a sequence of n-dimensional Einstein manifolds with $\mathrm{Ric}_{M_i} = c_i g_i$, for $|c_i| \leq n - 1$, converging to a closed n-manifold in the Gromov–Hausdorff topology. Then the M_i converge in the C^∞-topology.*

See also [Colding 1996a; 1996b; a] for further applications of these estimates.

The following theorem gives a generalization of Cheng's maximal diameter theorem [1975] to singular spaces that are limits.

THEOREM 2.8 (ALMOST MAXIMAL DIAMETER [Cheeger and Colding 1996]). *Given $\varepsilon > 0$, there exists $\delta = \delta(\varepsilon, n) > 0$ such that if $\mathrm{Ric}_M \geq n - 1$ and $\mathrm{diam}_M > \pi - \delta$ then for some metric space X we have $d_{GH}(M, S(X)) < \varepsilon$.*

Here $S(X) = (0, \pi) \times_{\sin r} X$ is the metric suspension of X.

THEOREM 2.9 (ALMOST SPLITTING [Cheeger and Colding 1996]). *Suppose that M_i is a sequence of n-dimensional manifolds with $\mathrm{Ric}_{M_i} \geq -(n - 1)\varepsilon_i$, where $\varepsilon_i \to 0$. If M_i converges in the pointed Gromov–Hausdorff topology to a metric space X that contains a line, then X splits isometrically, that is, $X = Y \times \mathbb{R}$ for some metric space Y.*

This was conjectured in [Fukaya and Yamaguchi 1992]. It generalizes the splitting theorem of [Cheeger and Gromoll 1971] to singular spaces that are limits.

THEOREM 2.10 (ALMOST VOLUME CONE IMPLIES ALMOST METRIC CONE [Cheeger and Colding 1996]). *If M has nonnegative Ricci curvature and Euclidean volume growth, every tangent cone at infinity is a Euclidean cone.*

Here a Euclidean cone is the metric completion of a space of the form $C(X) = (0, \infty) \times_r X$, for some metric space X.

The next theorem, which was conjectured by Gromov, was proved for manifolds of almost nonnegative sectional curvature by Fukaya and Yamaguchi [1992], who also observed that their proof would go through for almost nonnegative Ricci curvature provided that two conjectures could be established. One of these conjectures follows from Theorem 2.6; see [Colding a] for the exact statement. The other conjecture is Theorem 2.10. Therefore:

THEOREM 2.11 [Cheeger and Colding 1996]. *There exists an $\varepsilon = \varepsilon(n) > 0$ such that if M is a closed n-dimensional manifold with $\mathrm{Ric}_M \, \mathrm{diam}_M^2 > -\varepsilon$ then $\pi_1(M)$ is almost nilpotent* (that is, has a nilpotent subgroup of finite index).

3. The Structure of Spaces with Ricci Curvature Bounded Below

As mentioned in Section 2, the estimates in Section 1 and the way they occur give the possibility of treating Ricci curvature from a "synthetic" point of view (compare [Gromov 1980]). The first applications of such ideas were given in Section 2. In this section we will explore this further. Due to Gromov's compactness theorem we can think of the results of this section in two equivalent ways: as the study of smooth manifolds with a given lower Ricci curvature bound on a small but definite scale; or as the study of spaces that are Gromov–Hausdorff limits of such manifolds.

Throughout this section M_i will always be a sequence of n-manifolds with $\mathrm{Ric}_{M_i} \geq -(n-1)$, having M_∞ as a Gromov–Hausdorff limit. Unless otherwise stated, the examples, theorems, and definitions are to be found in [Cheeger and Colding a].

A tangent cone at $p_\infty \in M_\infty$ is a pointed metric space (X, x) that is a Gromov–Hausdorff limit of the rescaled metrics $(M_\infty, p_\infty, r_j d_\infty)$, where $r_j \to \infty$. Such limits exist by the Gromov compactness theorem.

We will sometimes also require that for all i and all $p_i \in M_i$ we have

$$\mathrm{Vol}(B_1(p_i)) \geq v > 0. \tag{3.1}$$

By the volume comparison theorem, this condition gives a uniform lower bound on the volume of all balls of a fixed radius. Condition (3.1) will often be referred to by saying that the sequence M_i does not collapse. Using in part Theorem 2.3, we can show that (3.1) is equivalent to requiring that M_∞ has Hausdorff dimension n.

The next example should be compared with the example of Perelman mentioned in Section 2.

EXAMPLE 3.2. There exists M_∞ where all M_i satisfy (3.1) and $p_\infty \in M_\infty$ such that the tangent cone at p_∞ is not unique.

An important feature in the noncollapsing case is that even though tangent cones are not unique, as seen in the preceding example, they are Euclidean cones. Indeed:

THEOREM 3.3. *Suppose that all M_i satisfy* (3.1). *Then, for all $p_\infty \in M_\infty$, every tangent cone at p_∞ is a Euclidean cone.*

The next example shows that, if the volume of balls of a fixed size is not uniformly bounded from below, tangent cones may not be Euclidean cones.

EXAMPLE 3.4. There exists M_∞ and $p_\infty \in M_\infty$ such that no tangent cone at p_∞ is a Euclidean cone.

We next define two notions of regular points.

DEFINITION 3.5. We say that $p \in M_\infty$ is a *weakly k-Euclidean point* if some tangent cone at p splits off a factor $\mathbb{R}^k$ isometrically.

DEFINITION 3.6. A point $p \in M_\infty$ is called *regular* if, for some k, every tangent cone at p is isometric to $\mathbb{R}^k$. In this case we write $p \in \mathcal{R}$.

DEFINITION 3.7. A point $p \in M_\infty$ is called *singular* if it is not regular. In this case we write $p \in \mathcal{S}$.

Note that, if the volume of balls of a fixed size is uniformly bounded from below (that is, if (3.1) is satisfied), weakly n-Euclidean implies regular by Theorem 2.5.

We next stratify the points of M_∞ according to how regular their tangent cones are.

DEFINITION 3.8. A point $p \in M_\infty$ is called k-*degenerate* if it is not $(k+1)$-weakly Euclidean. We let $\mathcal{D}_k$ denote the set of k-degenerate points.

Let dim denote the Hausdorff dimension.

THEOREM 3.9. *If all M_i satisfy* (3.1), $\dim(\mathcal{D}_k) \le k$.

If all M_i satisfy (3.1), then $\mathcal{S} = \mathcal{D}_{n-1}$. This follows from the fact that weakly n-Euclidean imply regular if the sequence M_i does not collapse.

THEOREM 3.10. *If all M_i satisfy* (3.1), *then $\mathcal{S} \subset \mathcal{D}_{n-2}$, so Theorem 3.9 implies* $\dim(\mathcal{S}) \le n - 2$.

In the next theorem the volume of subsets of M_∞ are measured with respect to the n-dimensional Hausdorff measure.

THEOREM 3.11 (VOLUME COMPARISON FOR LIMIT SPACES). *If all M_i satisfy* (3.1) *then M_∞ satisfies the relative volume comparison theorem.*

If we do not assume a lower bound on the volume, this is not always the case:

EXAMPLE 3.12. There exists M_∞ that does not satisfy the volume comparison theorem.

The next theorem was proved earlier by Fukaya and Yamaguchi [1994], under the additional assumption that all M_i have a uniform lower sectional curvature bound.

THEOREM 3.13 [Cheeger and Colding b]. *If all M_i satisfy (3.1), the isometry group of M_∞ is a Lie group.*

By assuming that all M_i are Einstein manifolds we get further regularity of the limit M_∞. This is the topic of the next two theorems.

THEOREM 3.14. *Suppose that all M_i satisfy (3.1) and are all Einstein with uniformly bounded Einstein constants. Then $\mathcal{S}$ is a closed subset and $\dim(\mathcal{S}) \leq n - 2$. Further, $\mathcal{R}$ is a smooth Einstein manifold and the convergence is in the C^∞-topology on compact subsets of $\mathcal{R}$.*

THEOREM 3.15 [Cheeger, Colding, and Tian b]. *Suppose that all (M, g_i) satisfy (3.1) and are Kähler–Einstein on M (where $\dim_\mathbb{C} M = n$), with uniformly bounded Einstein constants. Then $\dim(\mathcal{S}) \leq 2n - 4$. Further, there exists a subset $S \subset M_\infty$ with $H_{2n-4}(S) = 0$ (where H_{2n-4} is the Hausdorff measure) such that for all $p \in M_\infty \setminus S$ the tangent cone at p is unique and is equal to $\mathbb{C}^{n-2} \times \mathbb{C}^2/\Gamma$, where $\Gamma \subset \mathrm{SU}(2)$.*

Finally, in [Cheeger and Colding b] we give a generalization to the collapsed case of the volume convergence theorem of [Colding a].
See [Cheeger and Colding a; b; Cheeger, Colding, and Tian b] for more results in the spirit of those given in this section.

4. Function Theory on Spaces with a Lower Ricci Curvature Bound

In this section we will touch on some of the results on function theory on spaces with Ricci curvature bounded below (see also [Li 1993] for a discussion of this subject). For further details on the results of this section, see [Colding and Minicozzi 1996; a; Cheeger, Colding, and Minicozzi 1995].

DEFINITION 4.1. Let M be a open (complete, noncompact) manifold, and fix $p \in M$. Let r be the distance function from p. A harmonic function u on M has *polynomial growth of order at most d* if there exists some $C > 0$ so that $|u| \leq C(1 + r^d)$. We denote by $\mathcal{H}_d(M)$ the linear space of such functions.

Yau [1975] generalized the classical Liouville theorem of complex analysis to open manifolds with nonnegative Ricci curvature. Specifically, he proved that a positive harmonic function on such a manifold must be constant. This theorem was generalized in [Cheng and Yau 1975] by means of a gradient estimate that implies the Harnack inequality; in fact this gradient estimate played an important role in the proof of the results of Section 1. As a consequence, one sees that harmonic functions which grow less than linearly must be constant. In his study of these functions, Yau was motivated to conjecture that the space of harmonic functions of polynomial growth of a fixed rate is finite-dimensional on an open manifold with nonnegative Ricci curvature. On this conjecture of Yau we have the following result:

THEOREM 4.2 [Colding and Minicozzi a]. *For an open manifold with nonnegative Ricci curvature and Euclidean volume growth, the space of harmonic functions with polynomial growth of a fixed rate is finite-dimensional.*

Other important results on this conjecture of Yau can be found in [Christiansen and Zworski 1996; Donnelly and Fefferman 1992; Kasue a; Li 1995; Li and Tam 1989; 1991; Lin 1996; Wang 1995; Wu 1991].

For $d = 1$ we have the following theorem, which was proved earlier by Peter Li [1995] in the case where M is Kähler.

THEOREM 4.3 [Cheeger, Colding, and Minicozzi 1995]. *If* $\dim \mathcal{H}_1(M) = n + 1$ *for an open n-dimensional manifold M with nonnegative Ricci curvature, M is isometric to* $\mathbb{R}^n$.

To prove this, we show first that if M is an open manifold with nonnegative Ricci curvature and u is a nonconstant harmonic function with linear growth then any tangent cone at infinity M_∞ splits off a line. This implies in particular that if $\dim \mathcal{H}_1(M^n) = n + 1$ then $M_\infty = \mathbb{R}^n$; then Theorem 2.5 gives $M = \mathbb{R}^n$.

There exist manifolds with nonnegative (or even positive) Ricci curvature that do not split off a line, but admit a nonconstant linear growth harmonic function. In contrast, if M has nonnegative sectional curvature and M_∞ splits off a line, then M must split off a line.

As a final remark, I note that after the original version of this survey was written several related results were shown. In particular, jointly with Bill Minicozzi we settled the general case of Yau's conjecture mentioned above [Colding and Minicozzi c; d; e].

Acknowledgements

I am grateful to Mike Anderson, Jeff Cheeger, Chris Croke, Karsten Grove, Bill Minicozzi, Grisha Perelman, Peter Petersen, and Gang Tian for helpful discussions related to this survey.

Direct Sources

[Cheeger and Colding 1995] J. Cheeger and T. H. Colding, "Almost rigidity of warped products and the structure of spaces with Ricci curvature bounded below", *C. R. Acad. Sci. Paris* Sér. I **320** (1995), 353–357.

[Cheeger and Colding 1996] J. Cheeger and T. H. Colding, "Lower bounds on the Ricci curvature and the almost rigidity of warped products", *Ann. of Math.* **144** (1996), 189–237.

[Cheeger and Colding a] J. Cheeger and T. H. Colding, "On the structure of spaces with Ricci curvature bounded below, I", preprint, Courant Institute, 1996.

[Cheeger and Colding b] J. Cheeger and T. H. Colding, "On the structure of spaces with Ricci curvature bounded below, II", in preparation.

[Cheeger and Colding c] J. Cheeger and T. H. Colding, On the structure of spaces with Ricci curvature bounded below, III", in preparation.

[Cheeger, Colding, and Minicozzi 1995] J. Cheeger, T. H. Colding, and W. P. Minicozzi II, "Linear growth harmonic functions on complete manifolds with nonnegative Ricci curvature", *Geom. Funct. Anal.* **5** (1995), 948–954.

[Cheeger, Colding, and Tian a] J. Cheeger, T. H. Colding, and G. Tian, "Constraints on singularities under Ricci curvature bounds", to appear in *C. R. Acad. Sci. Paris.*

[Cheeger, Colding, and Tian b] J. Cheeger, T. H. Colding, and G. Tian, "On the singularities of spaces with bounded Ricci curvature", in preparation.

[Colding 1995] T. H. Colding, "Stability and Ricci curvature", *C. R. Acad. Sci. Paris* Sér. I **320** (1995), 1343–1347.

[Colding 1996a] T. H. Colding, "Shape of manifolds with positive Ricci curvature", *Invent. Math.* **124** (1996), 175–191.

[Colding 1996b] T. H. Colding, "Large manifolds with positive Ricci curvature", *Invent. Math.* **124** (1996), 193–214.

[Colding a] T. H. Colding, "Ricci Curvature and Volume Convergence", to appear in *Ann. of Math.*

[Colding and Minicozzi 1990] T. H. Colding and W. P. Minicozzi II, "On function theory on spaces with a lower Ricci curvature bound", to appear in *Math. Res. Lett.* **3** (1996), 241–246.

[Colding and Minicozzi a] T. H. Colding and W. P. Minicozzi II, "Harmonic functions with polynomial growth", to appear in *J. Diff. Geom.*

[Colding and Minicozzi b] T. H. Colding and W. P. Minicozzi II, "Large scale behavior of the kernel of Schrödinger operators", preprint, Courant Institute, 1996.

[Colding and Minicozzi c] T. H. Colding and W. P. Minicozzi II, "Generalized Liouville properties of manifolds", to appear in *Math. Res. Lett.*

[Colding and Minicozzi d] T. H. Colding and W. P. Minicozzi II, "Harmonic functions on manifolds", to appear in *Ann. of Math.*

[Colding and Minicozzi e] T. H. Colding and W. P. Minicozzi II, "Weyl-type bounds for harmonic functions", preprint, Courant Institute, 1996.

Other References

[Abresch and Gromoll 1990] U. Abresch and D. Gromoll, "On complete manifolds with nonnegative Ricci curvature", *J. Amer. Math. Soc.* **3** (1990), 355–374.

[Anderson 1990a] M. T. Anderson, "Short geodesics and gravitational instantons", *J. Diff. Geom.* **31** (1990), 265–275.

[Anderson 1990b] M. T. Anderson, "Metrics of positive Ricci curvature with large diameter", *Manuscripta Math.* **68** (1990), 405–415.

[Anderson 1990c] M. T. Anderson, "Convergence and rigidity of manifolds under Ricci curvature bounds", *Invent. Math.* **102** (1990), 429–445.

[Anderson 1992] M. T. Anderson, "Hausdorff perturbations of Ricci flat manifolds and the splitting theorem", *Duke Math. J.* **68** (1992), 67–82.

[Bochner 1946] S. Bochner, "Vector fields and Ricci curvature", *Bull. Amer. Math. Soc.* **52** (1946), 776–797.

[Bochner and Yano 1953] S. Bochner and K. Yano, *Curvature and Betti numbers*, Princeton University Press, Princeton, 1953.

[Bishop and Crittenden 1964] R. Bishop and R. Crittenden, *Geometry of manifolds*, Academic Press, New York, 1964.

[Burago et al. 1992] Y. Burago, M. Gromov, and G. Perelman, "A. D. Aleksandrov spaces with curvatures bounded below", *Uspehi Mat. Nauk* **47**:2 (1992), 3–51, 222; translation in *Russian Math. Surveys* **47**:2 (1992), 1–58.

[Cheeger 1991] J. Cheeger, "Critical points of distance functions and applications to geometry", pp. 1–38 in *Geometric topology: recent developments* (Montecatini Terme, 1990), edited by J. Cheeger et al., Lecture Notes in Math. **1504**, Springer, Berlin, 1991.

[Cheeger and Gromoll 1971] J. Cheeger and D. Gromoll, "The splitting theorem for manifolds of nonnegative Ricci curvature", *J. Diff. Geom.* **6** (1971), 119–128.

[Cheeger and Gromoll 1972] J. Cheeger and D. Gromoll, "On the structure of complete manifolds of nonnegative curvature", *Ann. of Math.* (2) 96 (1972) 413–443.

[Cheng 1975] S. Y. Cheng, "Eigenvalue comparison theorems and its geometric applications", *Math. Z.* **143** (1975), 289–297.

[Cheng and Yau 1975] S. Y. Cheng and S. T. Yau, "Differential equations on Riemannian manifolds and their geometric applications", *Comm. Pure Appl. Math.* **28** (1975), 333–354.

[Christiansen and Zworski 1996] T. Christiansen and M. Zworski, "Harmonic functions of polynomial growth on certain complete manifolds", *Geom. Funct. Anal.* **6** (1996), 619–627.

[Donnelly and Fefferman 1992] H. Donnelly and C. Fefferman, "Nodal domains and growth of harmonic functions on noncompact manifolds", *J. Geom. Anal.* **2** (1992), 79–93.

[Ferry 1979] S. Ferry, "Homotoping ε-maps to homeomorphisms", *Amer. J. Math.* **101** (1979), 567–582.

[Ferry and Quinn 1991] S. Ferry and F. Quinn, "Alexander duality and Hurewicz fibrations", *Trans. Amer. Math. Soc.* **327** (1991), 201–219.

[Fukaya and Yamaguchi 1992] K. Fukaya and T. Yamaguchi, "The fundamental groups of almost nonnegatively curved manifolds", *Ann. of Math.* (2) **136** (1992), 253–333.

[Fukaya and Yamaguchi 1994] K. Fukaya and T. Yamaguchi, "Isometry groups of singular spaces", *Math. Z.* **216** (1994), 31–44.

[Gallot 1983] S. Gallot, "A Sobolev inequality and some geometric applications", pp. 45–55 in *Spectra of Riemannian manifolds*, Kaigai, Tokyo, 1983.

[Gromov 1978] M. Gromov, "Almost flat manifolds", *J. Diff. Geom.* **13** (1978), 231–241.

[Gromov 1980] M. Gromov, "Synthetic geometry in Riemannian manifolds", pp. 415–419 in *Proceedings of the International Congress of Mathematicians* (Helsinki, 1978), Acad. Sci. Fennica, Helsinki, 1980.

[Gromov 1981a] M. Gromov, "Curvature, diameter, and Betti numbers", *Comment. Math. Helv.* **56** (1981), 53–78.

[Gromov 1981b] M. Gromov, *Structures métriques pour les variétés riemanniennes*, edited by J. Lafontaine and P. Pansu, CEDIC, Paris, 1981.

[Grove and Petersen 1988] K. Grove and P. Petersen, "Bounding homotopy types by geometry", *Ann. of Math.* (2) **128** (1988), 195–206.

[Grove and Shiohama 1977] K. Grove and K. Shiohama, "A generalized sphere theorem", *Ann. of Math.* (2) **106** (1977), 201–211.

[Grove et al. 1989] K. Grove, P. Petersen, and J. Y. Wu, "Controlled topology in geometry", *Bull. Amer. Math. Soc.* (N.S.) 20 (1989) 181–183.

[Grove et al. 1990] K. Grove, P. Petersen, and J. Y. Wu, "Geometric finiteness theorems via controlled topology", *Invent. Math.* **99** (1990), 205–213.

[Kasue a] A. Kasue, "Harmonic functions of polynomial growth on complete manifolds, II", to appear in *J. Math. Soc. Japan*.

[Li 1993] P. Li, "The theory of harmonic functions and its relation to Geometry", pp. 307–315 in *Differential geometry: partial differential equations on manifolds* (Los Angeles, 1990), edited by R. Greene and S. T. Yau, Proc. Symp. Pure Math. **54**, AMS, Providence, RI, 1993.

[Li 1995] P. Li, "Linear growth harmonic functions on Kähler manifolds with nonnegative Ricci curvature", *Math. Res. Lett.* **2** (1995), 79–94.

[Li and Tam 1989] P. Li and L.-F. Tam, "Linear growth harmonic functions on a complete manifold", *J. Diff. Geom.* **29** (1989), 421–425.

[Li and Tam 1991] P. Li and L.-F. Tam, "Complete surfaces with finite total curvature", *J. Diff. Geom.* **33** (1991), 139–168.

[Lin 1996] F. H. Lin, "Asymptotically conic elliptic operators and Liouville type theorems", preprint, Courant Institute, 1996.

[Myers 1941] S. Myers, "Riemannian manifolds with positive mean curvature", *Duke Math. J.* **8** (1941), 401–404.

[Otsu 1991] Y. Otsu, "On manifolds of positive Ricci curvature with large diameter", *Math. Z.* **206** (1991), 255–264.

[Otsu et al. 1989] Y. Otsu, K. Shiohama, and T. Yamaguchi, "A new version of differentiable sphere theorem", *Invent. Math.* **98** (1989) 219–228.

[Perelman 1991] G. Ya. Perelman, "Alexandrov's spaces with curvatures bounded below, II", preprint, 1991.

[Perelman 1994] G. Perelman, "Manifolds of positive Ricci curvature with almost maximal volume", *J. Amer. Math. Soc.* **7** (1994), 299–305.

[Perelman 1995] G. Perelman, "Spaces with curvature bounded below", pp. 517–525 *Proceedings of International Congress of Mathematicians* (Zürich, 1994), Birkhäuser, Basel, 1995.

[Perelman 1997] G. Perelman, "Construction of manifolds of positive Ricci curvature with big volume and large Betti numbers", pp. 157–163 in this volume.

[Petersen 1990] P. Petersen, "A finiteness theorem for metric spaces", *J. Diff. Geom.* **31** (1990), 387–395.

[Tian and Yau 1987] G. Tian and S. T. Yau, "Kähler-Einstein metrics on complex surfaces with $c_1(M)$ positive", *Comm. Math. Phys.* **112** (1987), 175–203.

[Toponogov 1964] V. Toponogov, "Spaces with straight lines", *Amer. Math. Soc. Transl.* **37** (1964), 287–290.

[Wang 1995] J. Wang, "Linear growth harmonic functions on complete manifolds", *Comm. Anal. Geom.* **3** (1995), 683–698.

[Wu 1991] H. Wu, "Polynomial functions on complete Kähler manifolds", pp. 601–610 in *Several complex variables and complex geometry*, Part 2 (Santa Cruz, CA, 1989), edited by E. Bedford et al., Proc. Sympos. Pure Math. **52**, Amer. Math. Soc., Providence, 1991.

[Yamaguchi 1988] T. Yamaguchi, "Manifolds of almost nonnegative Ricci curvature", *J. Diff. Geom.* **28** (1988), 157–167.

[Yamaguchi 1991] T. Yamaguchi, "Collapsing and pinching under a lower curvature bound", *Ann. of Math.* (2) **113** (1991), 317–357.

[Yau 1975] S. T. Yau, "Harmonic functions on complete Riemannian manifolds", *Comm. Pure and Appl. Math* **28** (1975), 201–228.

Tobias H. Colding
Courant Institute
251 Mercer Street
New York, NY 10012
colding@cims.nyu.edu

Comparison Geometry
MSRI Publications
Volume 30, 1997

A Genealogy of Noncompact Manifolds of Nonnegative Curvature: History and Logic

ROBERT E. GREENE

ABSTRACT. This article presents an approach to the theory of open manifolds of nonnegative sectional curvature via the calculus of nonsmooth functions. This analytical approach makes possible a very compact development of the by now classical theory. The article also gives a summary of the historical development of the subject of open manifolds of nonnegative sectional curvature and of related topics. Some very recent results are also discussed, including results of the author jointly with P. Petersen and S. H. Zhu on curvature decay.

Introduction

At first sight, the study of noncompact manifolds seems necessarily more complicated than that of compact ones: Removal of a single point from a compact manifold gives a noncompact one, but not all noncompact manifolds arise in this way—in general, the topological one-point compactification of a noncompact C^∞ manifold is not even a topological manifold. To look at the matter another way, more relevant to the subject of this article, a C^∞ compact manifold always admits a C^∞ function with only nondegenerate critical points, and compactness implies that there are only a finite number of such critical points; thus, a compact manifold has the homotopy type of a finite CW-complex, or what we shall call loosely finite topology. A C^∞ noncompact manifold (all manifolds will be C^∞ from now on) of course also always admits a proper C^∞ function with only nondegenerate critical points, but for some manifolds all such proper functions have infinitely many critical points. (Here *proper* means that the inverse of each set of the form $(-\infty, \alpha]$ is compact: this is the natural context for Morse theory.) Correspondingly, the manifold may fail to have finite topology, that is, it may not have the homotopy type of a finite complex.

Research supported by the National Science Foundation.

These elementary observations can be given more considerable depth. There is, for instance, a sense in which even just noncompact Riemann surfaces, let alone Riemannian manifolds of higher dimension, cannot be described by invariants, without special set-theoretic assumptions [Becker et al. 1980].

All this changes, however, if attention is restricted to manifolds admitting a complete Riemannian metric of nonnegative curvature.

Completeness itself imposes no restrictions: Every (paracompact) manifold admits a complete Riemannian metric, as can be seen by elementary arguments. Completeness alone does, however, yield a geometrically interesting object, as follows. If M is a complete, noncompact Riemannian manifold and $x_0 \in M$ is any point, we can take a sequence $\{x_i\}$ in M such that $\lim_{i \to +\infty} \mathrm{dist}(x_0, x_i) = +\infty$; if no such sequence existed, M would be bounded and hence compact. Now consider geodesics γ_i, parametrized by arclength, from x_0 to x_i, with $\gamma(0) = x_i$ and $\gamma(\mathrm{dist}(x_0, x_i)) = x_i$, and take a subsequence $\{\gamma_{i_j}\}$ such that $\{\gamma'_{i_j}(0)\}$ converges, say to V (such a subsequence must exist by the compactness of the unit ball in $T_{x_0} M$). Then the geodesic defined on $[0, +\infty)$ by

$$\gamma : t \mapsto \exp_{x_0}(tV)$$

is what is called a *ray*, that is, an arclength parameter geodesic defined on $[0, +\infty)$ that is a minimal-length connection between any two of its points. In Mark Twain's phrase, γ "lights out for the Territory" as fast as possible: intuitively, γ is a shortest connection to infinity.

In the absence of curvature hypotheses, the existence of a ray is no more than a geometric version of the essentially obvious fact that there is always a proper, injective map of $[0, +\infty)$ into a given noncompact manifold. But in the presence of everywhere nonnegative curvature, a new kind of meaning arises. Nonnegativity, and especially positivity, of curvature tends to make long geodesics nonminimizing. Thus some tension arises between the necessary existence of rays and the curvature's nonnegativity. This tension ultimately imposes strong restrictions on the topology of the manifolds that admit complete metrics of nonnegative curvature.

Completeness is essential here: every noncompact manifold admits a (generally noncomplete) metric of nonnegative sectional curvature [Gromov 1986].

Historically, the precise forms of these matters were discovered rather slowly and, as it were, almost indirectly in some cases. The historical development is shown on the genealogical chart on pages 132–134, albeit the lines of descent are from the imagination of the present author to some extent. The theme of the present article is, however, that, with the benefit of hindsight, and with the use of certain principles of analysis—which indeed could have been used from the start, very nearly—the whole subject can be developed very rapidly, essentially as a repeated application of the second variation formula. Thus, there is a certain contrast, though by no means a dichotomy, between the genealogy of the historical development shown in the chart, and the intrinsic logic, as the author perceives it at least, of the subject independent of history.

1. Of Second Variation and How Positive Curvature Can Be

Euclidean space with its standard flat metric is the separating case between spherical (constant positive sectional curvature) and hyperbolic (constant negative) geometries. This is a familiar classical concept, but it has certain subtle aspects. In particular, straight lines, which can be thought of as limits of longer and longer geodesics in larger and larger spheres, with curvature going to zero, are, from this viewpoint, just barely minimizing. One can attach precise meaning to this in the following way:

Consider a geodesic with arclength parameter, $\gamma : [0, L] \to M$, in some Riemannian manifold. And consider variations of γ with the variation vector field V along γ being everywhere perpendicular to γ (or equivalently to γ'). Suppose the variation has fixed endpoints so that $V(0) = 0$ and $V(L) = 0$. Finally, as a normalization, suppose that $\max_{t \in [0, L]} \|V(t)\| = 1$. Within this class of variations, the infimum of the second variation is in effect a measure of to what extent γ is minimizing.

Now what is interesting from the present viewpoint is that this infimum, for straight lines in Euclidean space, goes to 0 as L goes to infinity. (Its exact value is $4/L$.) The limit is nonzero (and positive) for a hyperbolic space, whereas for a sphere, negative second variation is of course possible for sufficiently long geodesics, that is sufficiently large values of L, implying, as is true, that such long geodesics are not of minimum length among nearby curves with the same endpoints.

The fact that straight lines in $\mathbb{R}^n$ are, in this sense, just barely minimizing suggests that there should be quantitative estimates on just how much positivity of curvature could be possible without forcing a geodesic to be nonminimal. The following lemma gives a specific version of this:

LEMMA 1.1. *Let $\gamma : [0, L] \to M$ be an arclength parameter geodesic and $N(t)$ be a parallel unit vector field along γ with $\langle N(t), \gamma'(t) \rangle \equiv 0$ for all $t \in [0, L]$. Let $k(t)$ be the sectional curvature of the two-plane spanned by $N(t)$ and $\gamma'(t)$. If $a, b > 0$ and $a + b = L$ and if γ is minimizing among nearby curves connecting its endpoints, then*

$$\frac{1}{a^2} \int_0^a t^2 k(t)\, dt + \frac{1}{b^2} \int_0^b t^2 k(L - t)\, dt \le \frac{1}{a} + \frac{1}{b}.$$

PROOF. Define a piecewise C^∞ vector field $V(t)$ along γ by

$$V(t) = \begin{cases} (t/a)N(t) & \text{if } 0 \le t \le a, \\ ((L - t)/b)N(t) & \text{if } a \le t \le L. \end{cases}$$

Then define a variation of γ by

$$(t, s) \mapsto \exp_{\gamma(t)}(sV(t)).$$

Since γ is locally minimizing, the associated second variation of arclength must be nonnegative. By the usual formula, this becomes

$$0 \leq \int_0^L \langle D_t V, D_t V \rangle \, dt - \int_0^L \|v\|^2 k(t) \, dt$$

$$= a\frac{1}{a^2} + b\frac{1}{b^2} - \frac{1}{a^2} \int_0^a t^2 k(t) \, dt - \frac{1}{b^2} \int_a^L (t-a)^2 k(t) \, dt$$

$$= \frac{1}{a} + \frac{1}{b} - \frac{1}{a^2} \int_0^a t^2 k(t) \, dt - \frac{1}{b^2} \int_0^b t^2 k(L-t) \, dt. \qquad \square$$

This lemma has two important corollaries. The first is an aspect of the Toponogov Splitting Theorem. The second, which is in effect a restriction on how much positivity of curvature there can be, is not so familiar, but forms an important complement to the results in Section 6 on how little positivity of curvature there can be.

COROLLARY 1.2 (TOPONOGOV). *Let $\gamma : (-\infty, +\infty) \to M$ be an arclength parameter geodesic with $\text{dist}(\gamma(t_1), \gamma(t_2)) = |t_1 - t_2|$ for all $t_1, t_2 \in (-\infty, +\infty)$. If, for some parallel unit vector field along γ, we have $k(t) \geq 0$ for all t, then $k(t) \equiv 0$ for all t.*

This follows immediately from the lemma by choosing $\gamma(a)$ of the lemma to correspond to $\gamma(t)$ of the corollary, for a given fixed t, and then letting a and b go to infinity.

By choosing $n - 1$ parallel, mutually perpendicular unit vector fields along γ, each perpendicular to γ', one easily obtains results corresponding to the lemma and corollary for Ricci curvature in place of sectional curvature.

For the second corollary, the following notation is needed: For a fixed $p \in M$, let $K_p(t)$ be the infimum of the sectional curvatures occurring at points q with $\text{dist}(q, p) = t$.

COROLLARY 1.3. *If M is a noncompact complete manifold of nonnegative sectional curvature and $p \in M$, then for each $r > 0$ we have*

$$\frac{1}{r} \int_0^r t^2 K_p(t) \, dt \leq 1.$$

This follows from Lemma 1.1 by choosing a ray $\gamma : [0, +\infty) \to M$ with $\gamma(0) = p$, setting $a = r$ and letting b go to infinity.

In terms of uniformly behaved $K_p(t)$, Corollary 1.3 makes clear that quadratic decay of $K_p(t)$ as a function of t is the critical case: $K_p(t)$ being of order $1/t^2$ gives the exact order of magnitude that the corollary allows.

This quadratic decay does in fact occur: For the paraboloid of revolution obtained by revolving the curve $y = \sqrt{x}$ around the x-axis, K_p equals $1/(2x + \frac{1}{2})^2$.

Here x and the arclength t from the origin $p = (0, 0)$ are uniformly comparable (since the derivative of $\sqrt{x}$ is bounded as $x \to +\infty$, and indeed goes to 0), so that $K(t)$ also decays quadratically as a function of t.

It is important to understand, however, that these observations about decay rates are really only relevant to cases where $K(t)$ has quite regular behavior. By way of example, note that even when M is a rotationally symmetric surface, with a metric in geodesic polar coordinates of the form $dr^2 + f^2(r)\, d\theta^2$, it is still possible for $\limsup_{r \to +\infty} K_p(r)$ to be $+\infty$ (and curvature to be everywhere > 0). An example can be constructed as a surface of revolution.

2. Of Analysis, Support Functions, and Comparisons

Curvature provides estimates on second variations of arclength, which one naturally thinks of as estimates (albeit in general one-sided only) on the second derivatives of distances. But this natural viewpoint does not apply literally, without some explication, since distances are not in general smooth functions on Riemannian manifolds. Fortunately, this lack of smoothness turns out not to matter, really. The essential reasoning needed to obviate the requirement of smoothness has deep roots from long ago in real and complex analysis, for instance, in the idea that subharmonicity needed to be defined and used for functions lacking any differentiability. Indeed, so widespread are this and other similar concepts in analysis that it is perhaps surprising that they began to appear explicitly in Riemannian geometry, it seems, only as recently as in [Calabi 1957]. Since then, however, they have been systematically studied and exploited in both Riemannian and Kähler contexts, in particular, in [Cheeger and Gromoll 1971; 1972; Greene and Wu 1976; 1978; Elencwajg 1975; Wu 1970; 1987].

The basic viewpoint needed is simply this: Define a function g to be *supporting* a function f at a point p if $f(p) = g(p)$ and $g(q) \leq f(q)$ for all q in a neighborhood of p. Then:

SUPPORT PRINCIPLE FOR CONVEXITY. *A continuous function f on a Riemannian manifold is convex (that is, convex along each geodesic) if it is supported at each point p by a convex function on some neighborhood of p.*

The proof is elementary, since in effect one can immediately reduce it to a standard and easy result about convex functions on the real line. In practice, a somewhat refined principle is more useful.

Define a C^∞ function f to be ε-*convex* if the eigenvalues of its second covariant differential are everywhere at least $-\varepsilon$. (Recall that the second covariant differential is the quadratic form $D_f^2(X, Y) = X(Yf) - (D_X Y)f$.) Then:

REFINED SUPPORT PRINCIPLE FOR CONVEXITY. *A continuous function f on a Riemannian manifold is convex if, for each point p and each $\varepsilon > 0$, there is an ε-convex function g defined on some neighborhood of p such that g supports f at p.*

The refined principle follows from the original principle easily using the fact that locally defined C^∞ strongly convex functions always exist on Riemannian manifolds; for example, $\text{dist}^2(x_0, \cdot)$ is C^∞ in a neighborhood of x_0 and the second covariant differential of this function has eigenvalues equal to 2 at x_0, and hence bounded away from 0 in a neighborhood U of x_0. The original principle can be applied to show that, for fixed but arbitrary x_0, and every $\delta > 0$, the functions $f(\cdot) + \delta \, \text{dist}^2(x_0, \cdot)$ are convex on U under the hypotheses of the refined principle. Since convexity is a local property, and since a limit of convex functions is convex, the conclusion of the refined principle follows by letting $\delta \to 0^+$.

The support principle can also be established for subharmonic functions, and also for plurisubharmonic functions on complex manifolds. Only the subharmonic case will be discussed here: for the complex case, see [Elencwajg 1975] and [Greene and Wu 1978].

By definition, a (real-valued) continuous function f on a Riemann manifold is *subharmonic* if it is a subsolution of Dirichlet problems, that is, if $f(x) \le h(x)$ for all $x \in K$, where K is any compact set with nonempty interior and $h : K \to \mathbb{R}$ is any continuous function harmonic in $\mathring{K}$ and satisfying $h(q) \ge f(q)$ for $q \in K - \mathring{K}$. We define a C^∞ function g to be ε-*subharmonic* if $\Delta g \ge -\varepsilon$ wherever g is defined (here $\Delta = \sum \partial^2 / \partial x_i^2$ at the center of a normal coordinate system $x_1, \dots, x_n$). Then:

REFINED SUPPORT PRINCIPLE FOR SUBHARMONICITY. *A continuous function f on a Riemannian manifold is subharmonic if, for each $p \in M$ and $\varepsilon > 0$, there is an ε-subharmonic function g defined on some neighborhood U of p such that g supports f at p.*

The proof runs almost parallel to that of the refined principle for convexity. On some (smaller) neighborhood V of p, the function $\text{dist}^2(p, \cdot)$ has positive Laplacian, bounded away from 0. If h is a (C^∞) function that is harmonic in a neighborhood of p, then for each $\delta > 0$, $f + \delta \, \text{dist}^2(p, \cdot) - h$ is supported at each point in a fixed neighborhood of p by the sum of a subharmonic function and a C^∞ function with positive Laplacian; this follows by choosing $\varepsilon > 0$ sufficiently small and using the function $g + \delta \, \text{dist}^2(p, \cdot) - h$, where g is as in the hypothesis. Thus, by calculus, $f + \delta \, \text{dist}^2(p, \cdot) - h$ cannot have a local maximum near p. Hence $f + \delta \, \text{dist}^2(p, \cdot)$ is subharmonic. Hence, since the limit of subharmonic functions is obviously subharmonic, f is subharmonic. There will be circumstances later when not only convexity or subharmonicity can be deduced, but also even smoothness.

SMOOTHNESS PRINCIPLE. *Let f be a function on a Riemannian manifold. If f and $-f$ are both convex, or both subharmonic, then f is C^∞ and harmonic.*

Actually, the convex situation is a special case of the subharmonic one, since convex functions are always subharmonic [Greene and Wu 1973]. But it is nice to notice that the convex case can be proved by completely elementary methods:

The function must be linear along all geodesics. To prove differentiability in a neighborhood of p, choose a geodesic γ_1 through p and a point x_1 near p but not on γ_1. Linearity shows that f is C^∞ on the local two-dimensional submanifold generated by geodesics from x_1 to (nearby) points of γ_1. Repeating this cone construction with x_2 near p but not on the two-dimensional submanifold gives a three dimensional local submanifold (containing p) on which f is differentiable. Further repetition eventually gives differentiability on a neighborhood of p. So f is differentiable and, since linear along geodesics, harmonic.

To treat the general subharmonic case, one proceeds as follows: For a given point p, choose a small closed ball B around p. By standard results in partial differential equations, there is a harmonic function h on $\mathring{B}$, with h continuous on B and $h \equiv f$ on $B - \mathring{B}$. By the subharmonicity of f and $-f$, $h \equiv f$ on $\mathring{B}$. Hence f is C^∞ (and harmonic) on $\mathring{B}$.

The support function method gives an almost immediate proof of the Toponogov Comparison Theorem, for the case of comparison with euclidean space, as pointed out originally by H. Karcher [1989]. (The negative and positive curvature comparisons are also obtained in the same work by similar methods, but some small technical subtleties are involved, and these cases, which are not directly relevant here, will be omitted.) The crucial observation is the following lemma, which follows from the support principle and an easy second variation argument.

LEMMA 2.1. *If M is a complete Riemannian manifold with nonnegative sectional curvature, with p a point of M, and if $\gamma(t)$ is an arclength parameter geodesic in M, then the function $F : t \mapsto t^2 - \mathrm{dist}^2(p, \gamma(t))$ is convex.*

PROOF. It suffices to construct a C^∞ support function g_{t_0} for F at $t = t_0$, fixed but arbitrary, with $d^2 g_{t_0}/dt^2 \geq 0$. For then, given $\varepsilon > 0$, there is a neighborhood of t_0 on which g_{t_0} is ε-convex, and the refined support principle applies. To construct such a g_{t_0}, choose a minimal geodesic $\gamma_1 : [0, L] \to M$ from p to $\gamma(t_0)$, where $L = \mathrm{dist}(p, \gamma(t))$. Define a vector field $W(s)$ along $\gamma_1(s)$ as the parallel translation of $\gamma'(t_0)$ along γ_1. Set $V = (s/L)W$ and define a variation of γ_1 by

$$(s, t) \mapsto \exp_{\gamma_1(s)}(tV(s)).$$

Let $L(t)$ be the length of the curve, as s varies over $[0, L]$ and t is fixed. Then a standard second variation calculation shows that $d^2 L(t)^2/ds^2 \leq 2$. On the other hand, clearly $\mathrm{dist}^2(p, \gamma(t)) \leq (L(t))^2$. Thus $g(t) = t^2 - L(t)^2$ satisfies the required conditions. □

Toponogov's Comparison Theorem can be easily deduced from this lemma in the form that estimates the third side of a geodesic triangle in terms of the other two sides and the included angle. Specifically, one wants to prove the following:

THEOREM 2.2 (TOPONOGOV COMPARISON FOR $k \geq 0$). *Suppose $\gamma_1 : [0, L] \to M$ is a minimal geodesic from a point $p = \gamma_1(0)$ to a point $q = \gamma_1(L)$ in a complete*

Riemannian manifold M of nonnegative curvature, and γ is an arclength geodesic through q with $\gamma(0) = q$ and $\langle \gamma'(0), -\gamma_1'(L) \rangle = \cos \alpha$ (so α is the angle between $\gamma'(0)$ and $-\gamma_1'(L)$). Then, for each positive t,

$$\mathrm{dist}^2(p, \gamma(t)) \leq t^2 + L^2 - 2Lt \cos \alpha.$$

Here the right-hand side of the inequality is of course the square of the length of the third side of a euclidean triangle whose other sides have lengths t and L and meet at an angle α.

PROOF. For each $\delta > 0$, the function

$$H_\delta(t) = t^2 + L^2 - 2Lt \cos \alpha + \delta t - \mathrm{dist}^2(p, \gamma(t))$$

is convex, by the lemma. Also $H(0) = 0$ and, by an easy first variation argument, $H(t)$ is positive for all small positive t. Thus $H_\delta(t)$ is positive for all positive t. Hence

$$H_0(t) = t^2 + L^2 - 2Lt \cos \alpha - \mathrm{dist}^2(p, \gamma(t)) \geq 0$$

for all positive t, which implies the conclusion of the theorem. $\square$

3. Of Noncriticality and Nonnegative Curvature

The idea of support functions discussed in the previous section makes it possible to deal with what amounts to the second derivative properties of functions that do not have second derivatives in the literal sense. It is natural and often very useful to have in addition a way to deal with first derivative properties.

The functions that arise in geometry are almost all locally Lipschitz continuous, that is, for each compact set K there is some constant B_K such that, for all $x, y \in K$,

$$|f(x) - f(y)| \leq B_K \, \mathrm{dist}_M(x, y).$$

Distance functions have this property by nature (and the triangle inequality), and geometrically constructed functions, which are derived in most cases from distance considerations, thus tend to have it, too. Lipschitz continuous functions on $\mathbb{R}$ are almost everywhere differentiable and equal the integral of their derivative, up to an integration constant. (These properties hold under the more general condition of absolute continuity, of course, but this additional generality is irrelevant for our present purposes.) Thus it is reasonable to try to relate general behavior of Lipschitz continuous functions to their first derivative properties.

For a function on a manifold, or what amounts to the same thing locally, a function of several variables, the most basic first derivative property is noncriticality, that is, nonvanishing of the gradient grad f, or, equivalently, of the differential df. This noncriticality has a natural possible generalization to the case of Lipschitz continuous functions. (This idea was introduced in [Greene and Shiohama 1981a]; see also [Greene and Wu 1974]. In the specific case of distance functions, related ideas occurred, for example, in [Grove and Shiohama 1977], but were tied to specifics of geodesic geometry.) In effect, one defines a point x to be a noncritical point of a function f if there is a continuously varying set of directions at and near x along which f decreases at a definite nonzero rate. (This set of directions would be $-\operatorname{grad} f$ in the case of C^1 functions f.)

To make this completely precise, define a vector $v \in T_x M$ to be a *subgradient* for a Lipschitz continuous function f if there is a continuous vector field V defined in a neighborhood of x with $V(x) = v$ and with the property that, for some $\varepsilon > 0$ and $\delta > 0$ and all y in a neighborhood of x,

$$f(\varphi_y(t)) - f(y) \le -\varepsilon t$$

for all $t \in (0, \delta)$, where φ_y is the geodesic emanating from y with $\varphi'_y(0) = V(y)$. This property is independent of the choice of Riemannian metric; indeed, the only function of the geodesics φ_y is to provide a suitable continuously varying family of curves with specified initial tangent. The definition could also be made with integral curves of a local C^∞ vector field, but it is sometimes convenient to be able to deal with vector fields that are just continuous. All such variants of the definition yield equivalent concepts.

A point x is a *noncritical point* for a Lipschitz continuous function f if there is a subgradient for f in $T_x M$ [Greene and Shiohama 1981b]. A point is *critical* for f if it is not noncritical. This definition of critical definitely does not coincide with the idea that f is "constant up to second order" at a critical point. For example, the function $x \mapsto \operatorname{dist}_M(x, p)$, for p fixed, always has p as a critical point, even though it increases at unit rate along arclength parameter geodesics emanating from p; similarly, if q is such that $\operatorname{dist}_M(p, q) = \sup_{x \in M} \operatorname{dist}_M(x, p)$, then q is a critical point even though there is a direction at q along which $\operatorname{dist}_M(\,\cdot\,, p)$ decreases at unit rate. The point is that such a direction of strict decrease (or increase) cannot be chosen to vary continuously in a neighborhood of q. Of course these definitions do coincide with the usual concept of being a noncritical point if f is C^1 differentiable.

Note that the property of being noncritical is open, almost by definition: If x is noncritical then so are all points y in a sufficiently small neighborhood of x, since the same V will work for such y as a subgradient. It is also easy to check, using the Lipschitz continuity of f, that if V is *any* continuous vector field defined in a neighborhood of x with $V(x)$ a subgradient at x for f, then V itself has the required property of the vector field in the definition of subgradient for some ε, δ and neighborhood of x. Finally, one can also check easily that a linear combination of subgradients with nonnegative coefficients and at least one positive coefficient is again a subgradient: this again uses the Lipschitz continuity of f.

The role of Lipschitz continuity in all this is primarily to insure that if two curves have nearly the same tangent vector at some common initial point, then the difference of f at parameter t along one curve from f along the other at parameter t is a small multiple of t. This provides the necessary control of one-sided (upper and lower) derivatives along curves.

Noncriticality in this extended sense is sufficient to make the main result of noncritical Morse theory work, to the extent that it could possibly work for nonsmooth functions:

LEMMA 3.1 (NONCRITICAL MORSE THEORY). *Suppose that $f : M \to \mathbb{R}$ is a Lipschitz continuous function.*

(1) *If, for some $a, b \in f(M)$, the set $f^{-1}([a,b])$ is compact and contains only noncritical points of f, then*

    (a) *$f^{-1}((-\infty, a])$ is homeomorphic to $f^{-1}((-\infty, b])$, and*

    (b) *$f^{-1}((-\infty, a])$ is homeomorphic to a C^∞ manifold-with-boundary $(N, \partial N)$; in particular $f^{-1}(\{a\})$ is homeomorphic to the C^∞ manifold ∂N.*

(2) *If, for some $a > 0$ in $f(M)$, the set $f^{-1}([a, +\infty))$ contains only noncritical points of f and $f^{-1}([a,b])$ is compact for all $b > a$, then M is diffeomorphic to $f^{-1}((-\infty, a))$.*

This lemma is established by smooth approximation techniques in [Greene and Shiohama 1981b]. The relevant smoothing ideas were introduced in [Greene and Wu 1973] and applied to noncritical point theory in [Greene and Shiohama 1981b]; compare [Greene and Wu 1974].

For literal distance functions $x \mapsto \mathrm{dist}_M(x, p)$ on a complete Riemannian manifold, the concept of noncriticality is equivalent to a condition on the geometry of geodesics: A point $x \neq p$ is noncritical if and only if there is a nonzero vector v in $T_x M$ such that, for every minimal geodesic $\gamma : [0,1] \to M$ with $\gamma(0) = p$ and $\gamma'(1) = x$, the angle between $\gamma(1)$ and v is $> \pi/2$. It is an easy exercise in the first variation formula to see that this geometric geodesic condition is equivalent to the definition already given. Indeed, the geodesic condition was used as a definition in some early works, such as [Grove and Shiohama 1977].

Although analytically the concept of noncriticality seems almost primordial, it took some time for its utility in geometry to become widely appreciated. Early investigations include [Greene and Wu 1974; Grove and Shiohama 1977; Greene and Shiohama 1981a; 1981b]. A history and some further details are given in [Greene 1989].

The relevance of these concepts to the theory of manifolds of nonnegative curvature is twofold: First, it applies to distance functions themselves, at large distances. And second, there are nontrivial convex functions on noncompact manifolds of nonnegative curvature; and convex functions have no critical points other than the points of the global minimum set.

To treat the first, one notes the following.

LEMMA 3.2. *If M is a complete manifold of nonnegative sectional curvature, if $p \in M$, and if $\delta > 0$, then there is a $B > 0$ with the following property: if*

$\mathrm{dist}_M(p, x) \geq B$ and if γ is a minimal geodesic from p to x, with $\gamma(0) = p$, then there is a ray $\gamma_1 : [0, +\infty) \to M$ with $\gamma_1(0) = p$ and such that the angle between $\gamma'(0)$ and $\gamma_1'(0)$ is less than δ.

The proof is an easy modification of the basic limiting argument establishing the existence of rays. From this lemma and Toponogov's theorem, one can reason about noncriticality of distance, as follows:

Suppose $\mathrm{dist}_M(p, x)$ is large enough to satisfy the lemma, for some fixed small δ. Choose γ fixed and then γ_1, in the notation of the lemma, and choose t very large. Then $\mathrm{dist}_M(x, \gamma_1(t))$ is estimated from above, compared to the euclidean triangle with sides $\mathrm{dist}_M(p, x)$ and $\mathrm{dist}_M(\gamma_1(0), \gamma_1(t)) = t$ and included angle δ. Choose a minimal geodesic γ_2 from x to $\gamma_1(t)$. Now consider the triangle formed by γ_2, $\gamma_1 \mid [0, t]$ and *any* minimal geodesic from p to x. Toponogov's theorem shows again that the angle at x must be larger than $\pi/2$; otherwise the third side could not have length as large as its actual length t. Thus the vector γ_2' at x satisfies the geodesic condition for noncriticality of $\mathrm{dist}_M(\cdot, p)$ at x. These arguments are given in [Gromov 1981a] and generalized to asymptotic nonnegativity of curvature in [Kasue 1988].

The relatively elementary argument just given already shows, when combined with Lemma 3.1 (noncritical Morse theory), the following striking fact:

THEOREM 3.3 (TOPOLOGICAL FINITENESS). *A complete noncompact manifold of nonnegative sectional curvature is diffeomorphic to the interior of a compact manifold-with-boundary.*

Historically, this result was discovered in the more involved context of the convex function considerations [Cheeger and Gromoll 1972]. These considerations are the subject of the next section.

4. Of the Distance from Infinity and Convexity

A more profound analysis of the structure of complete, noncompact, nonnegative curvature manifolds can be obtained by exploiting systematically a construction originally introduced by H. Busemann. This construction attaches to each ray $\gamma : [0, +\infty) \to M$ (parametrized by arclength) a function that we shall call B_γ, the Busemann function of γ. Intuitively, B_γ is a measure of how far a point is out toward infinity in the direction of γ. The precise definition is

$$B_\gamma(x) = \lim_{t \to +\infty} (t - \mathrm{dist}_M(x, \gamma(t))).$$

The function $t \mapsto (t - \mathrm{dist}_M(x, \gamma(t)))$ is monotone nondecreasing and bounded above by $\mathrm{dist}_M(x, \gamma(0))$ so that the limit exists and is finite-valued. Also, since each of the functions $t \mapsto t - \mathrm{dist}_M(x, \gamma(t))$ is globally Lipschitz continuous with Lipschitz constant 1, the function B_γ is also Lipschitz continuous with Lipschitz constant 1.

These considerations are independent of any curvature hypotheses. But in the presence of everywhere nonnegative curvature, the Busemann functions acquire a new virtue: they are necessarily convex. From the perspective of support functions and Toponogov's Comparison Theorem developed earlier, this convexity is almost immediate, as will now be shown.

Nonnegative curvature implies, either by Toponogov's Comparison Theorem or by a direct second variation argument, that the function $x \mapsto -\operatorname{dist}_M(x, \gamma(t))$ has second derivatives, along arclength geodesics in a neighborhood of a fixed x_0, whose value is at least $-2/\operatorname{dist}_M(x_0, \gamma(t))$. This is true in the sense of support functions even if the function $\operatorname{dist}(\cdot, \gamma(t))$ does not have two derivatives literally; that is, along such a geodesic $\theta(s)$, the function $\delta s^2 - \operatorname{dist}_M(\theta(s), \gamma(t))$ is convex for $\delta \geq \operatorname{dist}_M(x_0, \gamma(t))$. It follows that the same support function second derivative estimate is true of $t - \operatorname{dist}_M(x, \gamma(t))$. By taking the limit, one immediately derives the convexity of B_γ.

An elementary argument shows that, if $f : M \to \mathbb{R}$ is a (geodesically) convex function, then f is necessarily locally Lipschitz continuous. (This and subsequent remarks on general convex function theory are independent of curvature hypotheses.) Thus the whole machinery of noncriticality can be applied. And in this regard, convex functions have an extraordinary property:

THEOREM 4.1 (NONCRITICALITY OF CONVEX FUNCTIONS [Greene and Shiohama 1981a; 1981b]). *If $f : M \to \mathbb{R}$ is a convex function on a complete Riemannian manifold and if x is a point of M such that $f(x) > \inf_M f$, then x is a noncritical point for f.*

OUTLINE OF PROOF. By connecting x via a geodesic to a point y with $f(y) < f(x)$, one sees that $\operatorname{dist}_M(x, f^{-1}((-\infty, f(x) - \delta]))$ goes to 0 as $\delta \to 0^+$. The set $f^{-1}((-\infty, f(x) - \delta])$ is convex. From this convexity, one sees that there is a unique shortest connection from each point near x to this set when $\delta > 0$ is small enough. Uniqueness shows that this shortest connection varies continuously with the variation of x, near x. The tangent vectors to these geodesics satisfy the vector field condition for there to be a subgradient for f at x. (See [Greene and Shiohama 1981a; Greene and Shiohama 1981b] for this argument in detail.) □

It is actually most advantageous to apply these noncriticality considerations not to the individual Busemann functions but to the function $B(x) = \sup_\gamma B_\gamma(x)$, where the supremum is taken over all rays γ with fixed $\gamma(0) = p$. The supremum of convex functions is convex, so when M has everywhere nonnegative sectional curvature, the function B is again convex. Moreover, the function B has compact sublevel sets, that is, $B^{-1}((-\infty, \alpha])$ is compact for each $\alpha \in \mathbb{R}$.

The proof is again a variant of the basic ray construction: If $\{x_j\}$ is a sequence converging to infinity, that is, $\lim \operatorname{dist}(p, x_j) = +\infty$, if $B(x_j) \leq \alpha$ for all j and if γ_j is a sequence of minimal, arclength-parameter geodesics, from p to x_j for each j, then B is bounded above on γ_j, by $\max(B(p), B(x_j)) \leq \max(B(p), \alpha)$, by the convexity of B. Let γ be a ray which is a limit of a subsequence of the γ_j

(such a γ exists). Along γ, B is bounded by $\max(B(p), \alpha)$, by continuity. But B_γ is unbounded on γ, hence so is B. This contradiction completes the proof.

These observations give another proof for Theorem 3.3 on topological finiteness: M is diffeomorphic to the interior of a compact manifold-with-boundary homeomorphic to $B^{-1}((-\infty, \alpha])$, for any $\alpha > \inf_M B$. This follows by combining Lemma 3.1 (noncritical Morse theory), Theorem 4.1, and the convexity and properness of B.

But the function B enables one to refine this picture further. First, it is not hard to see that, if $\beta < \alpha$, then

$$B^{-1}((-\infty, \beta]) = \{x : B(x) \le \alpha, \text{dist}_M(x, B^{-1}([\alpha, +\infty))) = \alpha - \beta\}.$$

It follows that if $\beta_0 = \inf_M B = \min_M B$, then $B^{-1}(\{\beta_0\})$ is a compact convex subset of M with empty interior. This set thus lies in a totally geodesic submanifold of M, and indeed is the closure of (an open subset U of) such a submanifold. The proof of this is almost a copy of the corresponding structural result for compact, convex subsets of $\mathbb{R}^n$. The difference here is that the closure of the submanifold may equal the submanifold, that is, the submanifold may have no boundary. This can happen in $\mathbb{R}^n$ only if the submanifold is a single point.

Given such a compact convex subset C in M, equal to the closure of a totally geodesic submanifold U, one can consider the δ-push-in $C_\delta = \{x \in U : \text{dist}_M(x, C - U) \ge \delta\}$, provided that $C - U \ne \varnothing$.

It is not hard to see, using support functions as usual, that the function $x \mapsto -\text{dist}(x, C - U)$ is convex on U. In particular, the push-ins C_δ are convex, for each $\delta > 0$. There is a maximal δ for which C_δ is nonempty. And the same structural result for this C_δ, which is necessarily of lower dimension, enables one to do the push-in construction again in a lower-dimensional totally geodesic submanifold.

Since dimension drops at each stage, this process must eventually terminate with a maximal push-in that is a totally geodesic submanifold without boundary. Clearly, M has the homotopy type of this submanifold. But in fact, standard topological neighborhood constructions [Rushing 1973], together with the noncriticality of convex functions already discussed, show that M is diffeomorphic to a tubular neighborhood of the submanifold δ. Thus one obtains what is usually called the Soul Theorem:

THEOREM 4.2 (STRUCTURE OF NONNEGATIVELY CURVED MANIFOLDS [Cheeger and Gromoll 1972; Poor 1974]). *If M is a complete, noncompact manifold of everywhere nonnegative curvature, then there is a compact totally geodesic submanifold (without boundary) S of M such that M is diffeomorphic to a tubular neighborhood of S. In particular, M is diffeomorphic to the total space of a vector bundle over a compact manifold of nonnegative curvature.*

The submanifold S is usually called the *soul* of M, an unattractive but apparently permanent piece of terminology. The theorem was largely established in [Cheeger and Gromoll 1972], but the diffeomorphism statement was not obtained in full generality until [Poor 1974]. Of course, here and throughout the discussion, the fact has been used repeatedly that a totally geodesic submanifold of a manifold of nonnegative sectional curvature has itself nonnegative sectional curvature.

In the particular situation of complete noncompact manifolds of nonnegative curvature, it is not necessary to appeal to the topological generalities referred to before the statement of theorem. By using the noncritical Morse theory for nonsmooth functions that is a main theme in the present article, one can deduce more directly that such a manifold M is diffeomorphic to a tubular neighborhood of a soul S of M. For this, one needs only that the function $x \mapsto \text{dist}(x, S)$ is noncritical at every point $x \notin S$. For then, by the smooth approximation of subgradient flow as developed in [Greene and Shiohama 1981a; 1981b], for example, one obtains a diffeomorphism of M onto $\{x : \text{dist}(x, S) < \varepsilon\}$, for any $\varepsilon > 0$. For ε small enough, this latter set is a tubular neighborhood, and the proof is complete.

To see that $x \mapsto \text{dist}(x, S)$ is noncritical at each $x \notin S$, fix such an x and consider a minimal arclength parameter geodesic γ from x to S, so that $\text{length}(\gamma) = \text{dist}(x, S)$, while γ starts at x and terminates at a point in S. Suppose first that $B(x) > \min_M B$, where B is the supremum of the Busemann functions, as before. The convexity of B implies that the rate of decrease of B along γ is at least $(B(x) - \min B)/\text{length}\,\gamma$ at the point x. Moreover, the convexity of B also implies that the set of unit vectors in $T_x M$ along which B decreases at least at that rate is contained in a closed convex cone in $T_x M$ that lies in an open half-space. In particular, there is a unit vector $u \in T_x M$ and an $\alpha > 0$ such that the angle from u to any such geodesic γ is $\geq \pi/2 + \alpha$. Hence, by the logic already discussed in comparing the geometric noncriticality of [Grove and Shiohama 1977] with the analytic noncriticality idea used here, it follows that $\text{dist}(\,\cdot\,, S)$ is noncritical (in the analytic sense) at x.

If $B(x) = \min_M B$ but still $x \notin S$, then one can apply the same reasoning to deduce the noncriticality of $\text{dist}(\,\cdot\,, S)$ at x. In this case, one works not with B but with the relevant convex function associated to the push-in that takes x to the next-lower-dimensional totally convex set that is the next stage in the construction of the soul. By total convexity, all minimal connections γ from x to S lie in a totally convex set containing x and S on which the convex function for the push-in is defined. So the argument goes exactly as before.

Striking as this structural result is, it leaves a number of substantial questions unanswered. One of them is, which vector bundles over compact manifolds S of nonnegative sectional curvature actually do admit complete metrics of nonnegative sectional curvature on their total spaces? Trivial bundles obviously do: one just uses the product metric on $S \times \mathbb{R}^n$. In practice, it is not easy to find cases where one can see that a certain vector bundle does not occur. This makes the following results, proved in [Özaydin and Walschap 1994], particularly intriguing: The total space of a rank k vector bundle over a compact flat manifold M admits a complete metric of nonnegative sectional curvature if and only if E admits a complete flat metric, and also if and only if E is diffeomorphic to $\mathbb{R}^n \times_{\pi_1(M)} \mathbb{R}^k$, where $\pi_1(M)$ acts on $\mathbb{R}^n$ by covering transformations (of M) and on $\mathbb{R}^k$ by an orthogonal representation. This latter condition is in turn equivalent to E being diffeomorphic to the total space of a vector bundle of rank k over M that admits a flat Riemannian connection. From this, one deduces that the total space of a rank-2 oriented vector bundle over a compact flat manifold admits a complete metric of nonnegative sectional curvature if and only if its rational Euler class vanishes. This provides many examples of vector bundles over flat manifolds the total space of which does not admit a complete metric of nonnegative sectional curvature—for example, among oriented rank-2 bundles over the n-torus, only the trivial bundle has a complete metric of nonnegative sectional curvature.

So far in this section, attention has been concentrated on B, the supremum of the Busemann functions over all rays from a given point, for the logical reason that in general relatively little information can be derived by considering a single Busemann function. But there is one situation wherein much information can be derived about a single Busemann function: Suppose that a complete Riemannian manifold M contains a line $\gamma : (-\infty, +\infty) \to M$, that is a curve with

$$\operatorname{dist}(\gamma(t_1),\ \gamma(t_2)) = |t_1 - t_2|$$

for all $t_1, t_2 \in \mathbb{R} = (-\infty, +\infty)$. Associated to this situation are two Busemann functions: B_+, the Busemann function of the ray $\gamma|[0, +\infty)$, and B_-, that of the ray $t \mapsto \gamma(-t)$, for $t \in [0, +\infty)$.

It follows from the triangle inequality that $B_+ + B_- \leq 0$ everywhere on M and that $B_+ + B_- = 0$ at points of γ. This is not conspicuously instructive in complete generality. But if M is supposed to have nonnegative Ricci curvature, second variation arguments already discussed show that Busemann functions on M are subharmonic in the support sense (and even convex, under the more restrictive assumption of nonnegative sectional curvature).

Now the maximum principle holds for support-subharmonic functions, as is easy to see. Thus the fact that $B_+ + B_-$ is nonnegative on M and vanishes on γ implies that it vanishes in fact on all of M, if B_+ and B_- are subharmonic— e.g., when M has everywhere nonnegative Ricci curvature. Then B_+ is support-harmonic (since it is subharmonic and its negative B_- is also subharmonic). So, by the smoothness principle of Section 2, B_+ is a C^∞ harmonic function. (When M has nonpositive sectional curvature, the weaker and more elementary smoothness principle for functions linear along geodesics suffices here.)

Since B_+ is then C^∞, it has a gradient everywhere, of length 1. The gradient has length no more than 1 because B_+ is Lipschitz continuous with constant 1.

To see that the gradient has length at least 1 at a given $x \in M$, choose a sequence of minimal, arclength parameter geodesics γ_j from x to points $\gamma(t_j)$ with $t_j \to +\infty$ and unit vectors γ_j at x converging to a (unit) vector v. Then B_+ can easily be seen to increase at unit rate along $t \mapsto \exp_x(tv)$, $t \geq 0$.

Moreover, B_+ being harmonic implies that its level surfaces, which are smooth hypersurfaces, have mean curvature 0. Since they are "equidistant", it follows easily that they are totally geodesic hypersurfaces. One then sees immediately that M must be isometric to

$$\{x \in M : B_+(x) = 0\} \times \mathbb{R},$$

the isometry being the map that associates to each (x, t) the flow to time t along the integral curve of grad B_+ emanating from x at $t = 0$.

This gives the Splitting Theorem of [Cheeger and Gromoll 1971] for manifolds of nonnegative Ricci curvature (and, by the easier form of the arguments, Toponogov's Splitting Theorem [Toponogov 1964] for manifolds of nonnegative sectional curvature): If a complete n-dimensional manifold of nonnegative Ricci curvature contains a line, it is isometric to the product of an $(n-1)$-manifold and the real line $\mathbb{R}$.

5. Of Distance-Nonincreasing Retractions and Manifolds with Positive Curvature at One Point

If C is a closed subset of a complete Riemannian manifold M, then, for each $p \in M$, there is a point p' in C that is as close to p as possible, that is,

$$\text{dist}(p, p') = \text{dist}(p, C) := \inf_{q \in C} \text{dist}(p, q).$$

If C is a closed, convex set in $\mathbb{R}^n$, then, for each $p \in \mathbb{R}^n$, there is exactly one such closest point p' in C. In this case, one can define a retraction R of $\mathbb{R}^n$ onto C by setting $R(p) = p'$. This retraction R has the property of being distance-nonincreasing in the sense that, for all $p_1, p_2 \in \mathbb{R}^n$,

$$\text{dist}(R(p_1), R(p_2)) \leq \text{dist}(p_1, p_2).$$

For complete Riemannian manifolds, the situation is more complicated. The "closest point" p' may not be unique: for example, the north pole is equidistant from every point of the boundary of the southern hemisphere (which is convex). However, if C is a closed, convex set in a complete Riemannian manifold, there is a neighborhood U of C such that, for each $p \in U$, there is a unique closest point p' in C; a proof is given in [Greene and Shiohama 1981a], where the problem was investigated in detail.

Even if two points p_1, p_2 have unique closest points p_1' and p_2', and if C is convex, it may be that $\text{dist}(p_1', p_2') > \text{dist}(p_1, p_2)$. For instance, this happens when C is the southern hemisphere and p_1, p_2 are points of the same latitude in the northern hemisphere (excluding the equator and pole).

This latter example, if looked at in some detail, however, strongly suggests that there is some kind of control over closest-point retractions onto convex sets. If p_1, p_2 are on the parallel of latitude ε north, and are separated in longitude by θ, the law of spherical cosines gives

$$\cos \operatorname{dist}(p_1, p_2) = \sin^2 \varepsilon + \cos^2 \varepsilon \cos \theta,$$

where we have chosen units making the radius of the sphere equal to 1. On the other hand,

$$\cos \operatorname{dist}(p_1', p_2') = \cos \theta.$$

Since $\cos \varepsilon = 1 - \frac{1}{2}\varepsilon^2 + \ldots$ and $\sin \varepsilon = \varepsilon - \frac{1}{6}\varepsilon^3 + \ldots$, one sees that the retraction $R : p_i \mapsto p_i'$, as before, is *almost* distance-nonincreasing, where almost means up to an error that is second order in the distance ε of p_1, p_2 from C.

The historic concept of "infinitesimals of higher order", or Duhamel's Principle as it is often called, now suggests that, if some similar second-order statement holds for Riemannian manifolds in general, then it should be possible to construct distance-nonincreasing retractions onto (sub)level sets of convex functions. In outline, with p_1, p_2 in the same level, say α, one projects p_1, p_2 successively onto closely-spaced lower levels via closest-point mappings, until one reaches some fixed lower level β. These successive projections make at most second-order increases in distance. So by taking the limit with the number of intermediate levels going to infinity, one obtains a retraction that is distance-nonincreasing. The "errors" have vanished in the limit, having been infinitesimals of higher order.

This old-fashioned, almost eighteenth-century, way of thinking can in fact be made easily into a precise proof of the following result, first obtained by Sharafutdinov [1977] for retraction on the soul of a manifold of nonnegative curvature (compare [Greene and Shiohama 1981a]):

DISTANCE-NONINCREASING RETRACTION THEOREM. *If M is a complete Riemannian manifold, and if $f : M \to \mathbb{R}$ is a convex function, then, for each $\alpha \in f(M)$, there is a retraction*

$$R : M \to f^{-1}((-\infty, \alpha])$$

satisfying

$$\operatorname{dist}_M(R(x_1), R(x_2)) \le \operatorname{dist}_M(x_1, x_2)$$

for all $x_1, x_2 \in M$.

To make the proof of this result more explicit, note first that it is enough to construct, for a given fixed $\beta > \alpha$, a distance-nonincreasing retraction $R_{\beta,\alpha}$ from $f^{-1}((-\infty, \beta])$ onto $f^{-1}((-\infty, \alpha])$. Then the required R on M can be obtained as a composition of $R_{\alpha+1,\alpha}$, $R_{\alpha+2,\alpha+1}$, etc. To construct $R_{\beta\alpha}$ (when f is proper), note that for all sufficiently large positive integers n, there are unique closest points in $f^{-1}([-\infty, \alpha + (\beta - \alpha)i/n])$ to points in $f^{-1}([-\infty, \alpha + (\beta - \alpha)(i + 1)/n])$, for $i = 0, 1, \ldots, n - 1$. This follows from the existence of a unique closest point for points in a small neighborhood of a convex set, as noted earlier. This construct defines a retraction R of $f^{-1}([-\infty, \beta])$ onto $f^{-1}([-\infty, \alpha])$ by composition of the successive closest-point maps. Each of these closest-point maps is distance-nonincreasing modulo an error of order n^{-2}; this follows from a geometric argument that will be made explicit momentarily. A (sub)sequence of the R_n converges uniformly to a retraction that is distance-nonincreasing: the error terms are n in number and $O(n^{-2})$, so that their contribution vanishes in the limit.

Two further ingredients are needed to complete the geometric part of this argument, that is, to establish that projections onto the lower sublevels are distance-nonincreasing up to errors of order n^2. The first ingredient is to note an angle property of projection: Suppose that C is a compact (totally) convex set, and that x_1 and x_2 are points not in C but close enough to C to have unique closest points in C, say y_1 and y_2. Let γ_1 be minimal from x_1 to y_1, γ_2 from x_2 to y_2, and γ_3 from y_1 to y_2. Then the angle at y_1 between γ_1 and γ_3 (that is, $\angle x_1 y_1 y_2$) and the corresponding angle at y_2 are both $\geq \pi/2$. This is an obvious consequence of the first variation formula and the fact that $\gamma_3 \subset C$: If, say, $\angle x_1 y_1 y_2$ were acute, then moving along γ_3 from y_1 towards y_2 would give points (in C) closer to x_1 than is y_1, whereas y_1 is the closest point in C to x_1. In an obvious sense, this angle estimate shows that $\mathrm{dist}(x_1, x_2) \geq \mathrm{dist}(y_1, y_2)$ up to terms of second order in $\mathrm{dist}(x_1, C)$ and $\mathrm{dist}(y_1, C)$. (Here $\mathrm{dist}(x_2, y_2)$ is bounded uniformly if C is compact; otherwise, in case one wants to consider C only closed, convex, not necessarily compact, one still has a bound on $\mathrm{dist}(y_1, y_2)$ in terms of $\mathrm{dist}(x_1, x_2)$ and $\mathrm{dist}(x_1, C)$ and $\mathrm{dist}(x_2, C)$, which suffices.)

The second ingredient needed is the observation that, supposing the sublevel $f^{-1}((-\infty, b])$ to be compact and a to be greater than $\min_M f$, there is a constant C_0 independent of n such that

$$f(x) \in [a + i(b - a)/n, a + (b - a)(i + 1)/n] \quad \text{for } i = 0, 1, \ldots, n - 1 \implies$$
$$\mathrm{dist}(x, f^{-1}((-\infty, a + (b - a)i/n])) \leq C_0/n.$$

This estimate shows that "second-order" in terms of distance to the closest-point projection, as in the previous paragraph, implies second-order in the sense of order $1/n^2$. To establish this estimate, choose a number α such that $\alpha < a$ but $\alpha > \min_M f$. For x such that $f(x) \in [a + (b - a)i/n, a + (b - a)(i + 1)/n]$, choose $x_0 \in f^{-1}((-\infty, \alpha])$ such that

$$\mathrm{dist}(x, x_0) = \mathrm{dist}(x, f^{-1}((-\infty, \alpha])).$$

And choose a minimal, arclength parameter geodesic $\gamma : [0, D_x] \to M$ with $\gamma(0) = x_0$ and $\gamma(D_x) = x$.

Clearly $f(\gamma(t)) > \alpha$ for $t > 0$. Let t_0 be the smallest positive number such that $f(\gamma(t_0)) = a$. Now the function $t \mapsto f(\gamma(t))$ is convex. It follows that f is strictly increasing on $[0, D_x]$, and that if t_1 is the unique number in $[0, D_x]$ such that $f(\gamma(t_1)) = a + (b-a)i/n$, then

$$\frac{D_x - f(\gamma(t_1))}{D_x - t_1} \geq \frac{f(\gamma(t_0)) - f(\gamma(0))}{t_0 - 0}.$$

Thus

$$\mathrm{dist}(x, f^{-1}(-\infty, a + (b-a)i/n]) \leq D_x - t_1 \leq \frac{t_0}{a - \alpha} \left(f(x) - (a + (b-a)i/n) \right)$$

$$\leq \left(\frac{t_0}{a - \alpha} \right) \frac{1}{n} \leq \frac{d_\alpha}{a - \alpha} \frac{1}{n},$$

where $d_\alpha = \inf\{\mathrm{dist}(p, q) : f(p) = a, f(q) = \alpha\}$. Thus the required estimate holds, with $C_0 = d_\alpha(a - \alpha)$.

Retraction that is distance-nonincreasing onto $a = \min f$ follows by a limiting argument from the case of $a > \min f$. Also, if the sublevels are not compact, similar estimates hold on compact subsets and the desired constructions can be carried out globally by patching arguments, which are omitted in the interests of brevity.

A variant of this argument can be used to show that if $B > \alpha > \inf_M f$ then $f^{-1}([-\infty, \beta])$ is homeomorphic to $f^{-1}([-\infty, \alpha])$, and also that $f^{-1}((-\infty, \alpha))$ is homeomorphic to M. One does not map a point p in $f^{-1}([\alpha + (\beta - \alpha)i/n, \alpha + (\beta - \alpha)(i+1)/n])$ to its closest point in $f^{-1}([-\infty, \alpha + (\beta - \alpha)i/n])$. Rather, one takes the point p to the intersection with $\{x : f(p) = \alpha + (\beta - \alpha)i/n\}$ of the geodesic from p to the closest point to p in $f^{-1}(([-\infty, \alpha + (\beta - \alpha)(i-1)/n]))$— or, if $i = 0$, to $f^{-1}(([-\infty, \alpha - \varepsilon]))$, for some small $\varepsilon > 0$. This construction gives a sequence of "push-downs" that can be patched together to exhibit a product structure on $f^{-1}([\alpha, \beta])$, making it homeomorphic to $\{x \in M : f(x) = \alpha\} \times [\alpha, \beta]$. (See [Greene and Shiohama 1981a] for details, and [Greene and Shiohama 1981b] for a proof of the stronger result where "diffeomorphism" is substituted for "homeomorphism".) This process serves as a geometric substitute for the more analytic viewpoint of noncriticality of convex functions away from their minimum sets, discussed in Section 3.

The soul S of a complete manifold M of nonnegative sectional curvature is obtained as successive push-downs to the minimum set of convex functions, as explained in Section 4. Thus one obtains the following corollary (established in [Sharafutdinov 1977]) of the result on distance nonincreasing retractions:

RETRACTION ONTO THE SOUL. *If M is a complete Riemannian manifold with everywhere nonnegative sectional curvature, and if $S \subset M$ is a soul of M, then there is a distance-nonincreasing retraction $R : M \to S$.*

The retraction R is, as noted, obtained as a composition of retractions that are themselves limits of compositions of other retractions. This might seem to be a rather uncontrolled, even uncontrollable, construction, from the geometric point of view. Thus, the following result of Perelman [1994] is more than a little startling:

RIGIDITY OF RETRACTION ONTO THE SOUL. *If $R : M \to S$ is a distance-nonincreasing retraction of a complete manifold M of nonnegative curvature onto a soul S, then $R(\exp_x v) = x$ for each $x \in S$ and $v \in T_x M$ with $v \perp T_x S$.*

The exponential map on the normal bundle of S need not be a diffeomorphism onto M, even though M is indeed diffeomorphic to this normal bundle (see Section 3). For instance, if S is a point, so that M is diffeomorphic to $\mathbb{R}^n$, it may not be the case that M has any pole (recall that $x \in M$ is a *pole* if $\exp_x$ is a diffeomorphism). Even so, Perelman's rigidity result can well be thought of as saying that R is a one-sided inverse of exp of the normal bundle of S.

If M is a complete Riemannian manifold of everywhere positive sectional curvature, the supremum B of the Busemann functions is actually strongly convex in the support-function sense. That is, for each x in M and for σ a C^∞ function in a neighborhood of x with positive definite covariant differential at x, there is an $\varepsilon > 0$ such that $B - \varepsilon\sigma$ is convex in a neighborhood of x. The minimum set of a strongly convex function (in the support-function sense, which applies to not necessarily smooth functions) is either empty or consists of a single point. Thus the soul S of M is a point and M is diffeomorphic to $\mathbb{R}^n$, where $n = \dim M$. Homeomorphism to $\mathbb{R}^n$ was proved by Gromoll and Meyer [1969], prior to the general structure theorem for nonnegative curvature, and indeed this work introduced already the basic results about the suprema of the Busemann functions and so on. The explicit proof of diffeomorphism to $\mathbb{R}^n$, as opposed to just homeomorphism, came in [Poor 1974] (this is of course relevant only in dimensions 3 and 4). The same result was proved in [Greene and Wu 1976] by a particularly direct process: we first proved that the function B can be smoothed, that is, approximated by a strongly convex C^∞ function, which is also proper. That M is diffeomorphic to $\mathbb{R}^n$ then follows from standard smooth Morse theory techniques.

It was natural to ask whether the conclusions that S is a point and hence that M is diffeomorphic to $\mathbb{R}^n$ hold under the weaker hypotheses that M has nonnegative curvature everywhere and all sectional curvatures positive at a single point. This was known to be true in two particular cases: in dimension 2 [Cohn-Vossen 1935], and for manifolds M that arise as the boundaries of convex bodies in $\mathbb{R}^{n+1}$ (see the classification in [Busemann 1958]). Of course, for $n > 2$, it is atypical for a general Riemannian manifold to be isometric to a hypersurface, even locally, since local isometric embedding requires, generically, $\frac{1}{2}n(n + 1)$ dimensions at least (compare [Gromov and Rohlin 1970]). In any case, it was conjectured by Cheeger and Gromoll [1972] that sectional curvatures nonnegative everywhere and all positive at one point (on a complete, noncompact manifold) implies diffeomorphism to euclidean space.

Only very limited special cases of this conjecture were confirmed in the next twenty years: for example, when S has codimension 1 [Cheeger and Gromoll 1972] or 2 [Walschap 1988], and some cases involving detailed curvature assumptions [Elerath 1979]. Thus Perelman's proof of the conjecture, presented for the first time at the MSRI Workshop of which this volume is a record, caused excitement. In fact the solution follows from this general result [1994]:

FLAT RECTANGLES IN NONNEGATIVE CURVATURE. *Let M be a complete noncompact manifold of nonnegative curvature with soul S of positive dimension. Let γ be a geodesic in S, and let N be a parallel vector field along γ normal to S. Then the curves γ_t defined by*

$$\gamma_t(u) = \exp_{\gamma(u)} tN(u),$$

for $t \geq 0$, are geodesics in M and form a flat totally geodesic two-dimensional submanifold.

COROLLARY 5.1 [Perelman 1994]. *A complete, noncompact, n-dimensional manifold with everywhere nonnegative sectional curvature and with all sectional curvatures positive at some point is diffeomorphic to $\mathbb{R}^n$.*

The corollary follows from the theorem because every point p of M arises as $\exp_x tN$, where $x \in S$ and $N \perp T_xS$. This is easy to see, e.g., by taking for x a closest point in S to p and for N the tangent vector at x to a minimal-length geodesic from x to p. If S has positive dimension the theorem provides a zero sectional curvature at p. So if all sectional curvatures at some p are positive, S must be zero-dimensional, that is a point (S is convex, and therefore connected). Hence M is diffeomorphic in that case to $\mathbb{R}^n$.

The arguments used by Perelman to prove the rigidity theorem and the flat two-plane theorem are ingenious, but purely geometric and, in a sense, surprisingly elementary. The starting point is to define a function $f(r)$ as follows: if $R : M \to S$ is a distance-nonincreasing retraction, set

$$f(r) = \max \operatorname{dist}(x, R(\exp_x rv)),$$

where $x \in S$ and $v \in T_xM$ is orthogonal to S. Then, by a geometric argument, one shows that the upper (one-sided) derivative of f is nonnegative. Since f is clearly Lipschitz continuous, f is the integral of its derivative up to an additive constant. Since $f(0) = 0$ while $f \geq 0$ by definition, $f(r) \equiv 0$ for $r \geq 0$. This establishes the rigidity. The flat two-plane result then follows using comparison results applied to "rectangles" obtained by exponentiating the parallel vector field N along γ, applying R, and using the rigidity result.

It was also shown in [Perelman 1994] as part of the proofs of the results already quoted that Sharafutdinov's retraction R was a Riemannian submersion of class C^1. Recently, this submersion has been shown to be of class C^∞ by Guijarro [a].

6. Of How Little Curvature a Manifold of Nonnegative Curvature Can Have

In Section 1, the question was explored of how much positive curvature a noncompact manifold of nonnegative curvature could have. That there was some restriction was obvious, since if there were too much positive curvature, the manifold would close up, ceasing to be noncompact. But at first sight, there seems to be no restriction in the other direction, no lower bound on how much positivity of curvature can occur. After all, the most familiar examples of open manifolds of nonnegative curvature, euclidean spaces, have no positive curvatures at all; they are flat. It is thus rather surprising that, if one supposes in advance that the metric is not flat, there is a precise sense, in many cases, in which the manifold cannot be arbitrarily close to flat.

The first results of this type were discovered only relatively recently; they are not extensions of results from the classical period of Riemannian geometry, in the late nineteenth and early twentieth centuries. The absence of any such results classically is probably just a consequence of the fact that for surfaces there are no results of the type. It is easy, for instance, to construct complete metrics (from surfaces of revolution) on $\mathbb{R}^2$ that have zero curvature outside some compact set, nonnegative curvature everywhere, and positive curvature somewhere. The corresponding situation for metrics on $\mathbb{R}^n$, for $n \geq 3$, cannot occur, as we shall see momentarily. And this nonoccurrence is the basic instance of the restrictions on near-flatness without total flatness with which we shall be concerned.

The first results of this general type were actually obtained for complete, simply connected Kähler manifolds of nonpositive curvature and "faster-than-quadratic curvature decay", in the sense that, for some $C > 0$ and $\varepsilon > 0$,

$$\sup |K(\sigma)| \leq Cr^{-(2+\varepsilon)},$$

where the supremum of absolute value of sectional curvature is taken over all two-planes σ at distance r from a fixed point. Such manifolds had been proposed as objects of study in [Greene and Wu 1977], where, in particular, it was conjectured that they should be necessarily biholomorphic to $\mathbb{C}^n$. This conjecture was proved in [Siu and Yau 1977]. But except in complex dimension 1, no examples were known except the standard metric on $\mathbb{C}^n$. And in [Mok et al. 1981], it was shown that indeed such manifolds had to be isometric to $\mathbb{C}^n$ if their complex dimension n were ≥ 2. (The biholomorphism theorem was extended in the same paper to allow small amounts of positive curvature, so that then nonflat examples occur). This work on Kähler geometry motivated H. Wu and myself to consider the corresponding, more general Riemannian situation to be discussed now.

The most basic and prototypical Riemannian-geometric result has to do with manifolds that are flat outside some compact set:

THEOREM 6.1 [Greene and Wu 1982]. *Let M be a complete noncompact manifold such that* (a) *there is a compact set K such that all sectional curvatures are 0* on $M - K$, and (b) M is connected at infinity and simply connected at infinity. Then there are compact sets K_1 in M and K_2 in $\mathbb{R}^n$ with $M - K_1$ isometric to $M - K_2$.

COROLLARY 6.2 [Greene and Wu 1982; Greene and Wu 1993]. *If a manifold satisfying the hypotheses of the Theorem also has the property of everywhere nonnegative Ricci curvature, it is isometric to $\mathbb{R}^n$.*

COROLLARY 6.3 [Greene and Wu 1982]. *If a complete, simply connected Riemannian manifold of dimension $n \geq 3$ has everywhere nonpositive sectional curvature and has zero curvature outside some compact set, it is isometric to $\mathbb{R}^n$.*

The first corollary follows from the theorem by considering a (large) region in M containing K_1, the boundary of which is isometric to the boundary of a cube in $\mathbb{R}^n$. Identifying opposite faces gives a manifold of dimension n with nonnegative Ricci curvature and with first homology of rank n. Such a manifold must be flat, by the Bochner technique, so M itself is flat.

The second corollary follows from the Theorem by noting that M has exactly euclidean volume growth, that is,

$$\lim_{r \to +\infty} \operatorname{vol} B(p,r)/\operatorname{vol}_e B(r) = 1$$

for each $p \in M$, where $\operatorname{vol}_e B(r)$ is the volume of the euclidean ball of radius T. Then M must be flat by the Bishop volume comparison result [Bishop and Crittenden 1964]: if any sectional curvature of M at a point p, say, were negative, the volume growth limit would be greater than 1 from p.

It is an important fact that no analogue of the first corollary holds for nonpositive Ricci curvature: There are compactly supported perturbations of the standard euclidean metric on $\mathbb{R}^n$, for $n \geq 3$, which have Ricci curvature nonpositive everywhere and negative somewhere [Lohkamp 1992].

The proof of the Theorem on manifolds flat outside a compact set is obtained by considering the locally isometric "developing map" of M outside a compact set to $\mathbb{R}^n$. Using the convexity results of [Greene and Wu 1973; 1974], one can take the compact set to have C^∞ boundary and to be convex in M. Simple connectivity gives a well-defined developing map which maps the boundary of the compact convex set in M to an immersed, locally convex hypersurface in $\mathbb{R}^n$. Since $n \geq 3$, this immersion is in fact an embedding [Sacksteder 1960; van Heijenoort 1952], and the consideration of exterior normal maps gives an isometric map from the exterior of this compact convex hypersurface in $\mathbb{R}^n$ to the exterior of the corresponding set in M. This technique can be extended to classify the possible structures of manifolds flat outside a compact set even when the manifolds are not assumed simply connected at infinity ([Greene and Wu 1993]; this classification was in fact obtained earlier by a different method in [Schroeder and Ziller 1989]).

It is natural to try to replace the condition of flatness outside a compact set by some condition of rapid decay of curvature to zero with increasing distance. The Kähler-geometric situation suggests that faster than quadratic decay is the appropriate choice. So does the relationship between faster-than-quadratic curvature decay and quasi-isometry of the exponential map established in [Greene and Wu 1979]. In [Greene and Wu 1982], the faster-than-quadratic-decay condition was analyzed as a substitute for flat-outside-a-compact-set in the case of manifolds admitting a pole.

In summary form, it was proved in [Greene and Wu 1982] that a manifold with a pole, with faster than quadratic curvature decay, and with sectional curvature of one sign (either everywhere ≤ 0 or everywhere ≥ 0) was necessarily flat, if the dimension of the manifold was ≥ 3, except when the dimension is 4 or 8 and the curvature is nonnegative.

These results were called "gap theorems" because in effect they showed the existence of a gap between flat $\mathbb{R}^n$ and other metrics of signed curvature on $\mathbb{R}^n$: such metrics cannot be too close to the flat metric.

The basic proof technique for these results of [Greene and Wu 1982] is to estimate the $(n-1)$-dimensional volume of distance spheres with a view to showing that M has euclidean volume growth. Then the flatness follows from the Bishop volume comparison [Bishop and Crittenden 1964], as already mentioned. For curvature ≤ 0, one wishes to estimate volume from above. One can estimate the sectional curvature of distance spheres from below with a positive lower bound depending on the radius, using comparison methods such as the Hessian comparison of [Greene and Wu 1979]. The volume estimate above follows. (This argument, simpler than the ones used in [Greene and Wu 1982], was communicated to the author by M. Gromov.) But for curvature ≥ 0, one seeks estimates of $(n-1)$-dimensional volume of the distance spheres from below, while the information available directly from comparison theory is an upper (and lower) curvature bound on the spheres' intrinsic metrics. In general, such curvature information fails to yield a lower bound on volume, as the phenomenon of "collapse with bounded curvature" shows (e.g., Berger spheres).

In odd dimensions one does get a lower volume estimate on the (even-dimensional) distance spheres by means of the generalized Gauss–Bonnet formula of [Allendoerfer and Weil 1943]. A bound on the absolute value of sectional curvature provides a bound on the generalized Gauss–Bonnet integrand and, since the integral must be the Euler characteristic of the even-dimensional sphere, namely 2, a lower volume bound follows. This method was introduced in [Greene and Wu 1982] for the odd-dimensional pole case. The same approach was used successfully in the case of manifolds not necessarily diffeomorphic to $\mathbb{R}^n$ (soul not a point) in [Eschenburg et al. 1989] to generate "gap theorems" in that situation again in odd dimensions, and to deal with the odd-dimensional case of diffeomorphism to $\mathbb{R}^n$, but with no pole.

When the manifold itself is even-dimensional, the estimation of distance sphere volume from below must use a different method. The idea used in [Greene and Wu 1982] is to replace the generalized Gauss–Bonnet integrand by a Gauss–Kronecker curvature computed relative to an almost parallel frame. (The almost parallel frame can be constructed in a neighborhood of a given (large) distance sphere because of the nearly-zero curvature of the manifold at large distances.) The Gauss–Kronecker integrand can be estimated above by comparison methods as before. Because the Gauss–Kronecker integrand is the determinant of the Gauss map relative to the almost parallel frame, its integral is, up to a normalization constant, equal to the degree of the Gauss map. Since the Gauss map is a classifying map for the tangent bundle of the sphere, the integral is bounded away from 0 (the degree being nonzero and in fact 1, *provided that the tangent bundle of S^{n-1} is nontrivial*, that is, provided that $n \neq 2$, 4, or 8). This method introduced in [Greene and Wu 1982] for the pole case was extended in [Drees 1994] to apply to curvature ≥ 0, no pole assumed. (The cases $n = 4$, 8 with a pole were treated in [Kasue and Sugahara 1987], but the method there does not generalize to the nonpole case.)

That $n = 4$ really is a special case was made a matter not just of restricted proof technique but also of concrete example in [Unnebrink a]. There a metric is constructed on $\mathbb{R}^4$ with cubic curvature decay and volume growth of cubic order, $\mathrm{vol}\, B(p,r) = O(r^3)$—an order of magnitude less than one might hope. Somewhat mysteriously, as yet no example of a similar nature on $\mathbb{R}^8$ has been found.

Meanwhile, recently, the author, P. Petersen, and S. Zhu have shown that no such example can exist (on any $\mathbb{R}^n$, for $n \geq 3$) with curvature decay faster than cubic; that is, if, for some C and $\varepsilon > 0$, one has

$$\sup |K(\sigma)| \leq Cr^{-(3+\varepsilon)},$$

where the sup is over sectional curvatures at points of distance r from a fixed $p \in M$, then the volume growth is euclidean:

$$\lim_{r \to +\infty} \mathrm{vol}\, B(p,r)/\mathrm{vol}_{eu}\, B(r) = 1.$$

(No hypothesis is made about signed curvature. The corresponding result with faster-than-quartic decay was given in [Greene et al. 1994]; the faster-than-cubic result here is a recent improvement by the same authors.) A corresponding result with faster-than-cubic decay also holds without topological hypotheses (of simple connectivity at infinity), the conclusion being in that case that the manifold admits a new metric, equivalent to the original one of fast curvature decay, with the new metric's sectional curvature being zero outside some compact set (the possibilities for metrics of that type having been already determined in [Schroeder and Ziller 1989] and [Greene and Wu 1993]).

7. Of Convergence and Behavior at Infinity

The rather specific proof techniques of the previous section can be put advan-
tageously into a general picture via the concept of Gromov–Hausdorff convergence (hereafter, GH convergence). Suppose (M, g) is a complete, noncompact
manifold with the property that

$$\liminf_{r \to +\infty} (r^2 \inf K(\sigma)) > -\infty,$$

where as usual the inf $K(\sigma)$ means the infimum of sectional curvature $K(\sigma)$
over two-planes σ at distance r from a fixed base point p_0 in M. A manifold
of faster-than-quadratic curvature decay has this property, for example. Then
the family of Riemannian manifolds $\{(M, \lambda^{-1}g) : \lambda \geq 1, \lambda \in \mathbb{R}\}$ is precompact
in the sense of GH convergence. (This precompactness property really requires
only a corresponding estimate on Ricci curvature as a function of distance, but
this greater generality would not be relevant here.) That is, every sequence
$\{(M, \lambda_i^{-1}g) : \lim \lambda_i = +\infty\}$ has a subsequence that converges in the GH sense
for pointed metric spaces, all the $(M, \lambda^{-1}g)$ being taken to have p_0 as base point.
(See [Gromov 1981b] for the relevant convergence definitions.) The role here of
the curvature hypothesis is to guarantee that the family $\{(M, \lambda^{-1}g) : \lambda \geq 1\}$
has curvature bounded below λ, away from the base point. (In this set-up,
the base point p_0 in general becomes a singular point in a limit as $\lambda_i \to +\infty$,
corresponding to the curvature at p_0 "blowing up" in the rescaling.) This lower
bound on curvature arises from the fact that distance is rescaling by $1/\lambda$ and the
curvature by λ^2. So a lower bound on $\liminf r^2 K(\sigma)$ as $r \to +\infty$ gives a lower
curvature bound at points fixed distance from p_0 that is uniform over variation
of λ, for $r > 0$ fixed.

According to [Grove and Petersen 1991], a space arising as a GH limit of
Riemannian manifolds (of fixed dimension) with curvature bounded uniformly
below has constant dimension: The GH limit of a sequence of such manifolds
of dimension n is either again of dimension n, with some metric singularities
perhaps, or alternatively "collapsing" occurs everywhere, and the limit space
has everywhere a fixed dimension $k < n$. In either case the limit space is an
Alexandrov space. This result applies, in particular, to the GH limit of a GH-
convergent sequence $\{(M, \lambda_i^{-1}g)\}$, where $\lim \lambda_i = +\infty$. The collapsing case
(into a nonmanifold) can indeed occur, as one sees for instance from a capped-
off two-dimensional half-infinite cylinder, with curvature ≥ 0 everywhere and
curvature $\neq 0$ outside a compact set: the limit space in this case is isometric to
$\{t \in \mathbb{R} : t \geq 0\}$.

In the case of GH limits of sequences $\{(M_i, g_i)\}$ of manifolds with curvature
bounded both above and below, stronger conclusions hold than in the more
general situation of [Grove and Petersen 1991], where the curvature is bounded
below only. Collapsing can still occur, as the example of Berger spheres shows.

But with bounds both above and below and with *no* collapsing, the limit space is (identifiable with) an n-dimensional Riemannian manifold with a $C^{1,\alpha}$ metric.

In this no collapsing situation, [Grove and Petersen 1991] guarantees that injectivity radius is bounded away from zero uniformly (or uniformly on compact sets, in case the (M_i, g_i) are noncompact, pointed). Thus the $C^{1,\alpha}$ statement is just a restatement of the standard $C^{1,\alpha}$ Convergence Theorem proposed by Gromov [1981b] and established in detail in [Greene and Wu 1988], and independently in [Peters 1987], using results from [Jost and Karcher 1982] (compare [Nikolaev 1980]).

This line of reasoning applies in particular to a family of the form $\{(M, \lambda^{-1}g) : \lambda \geq 1\}$ if M is a manifold of faster than quadratic curvature decay. The family has (sectional) curvature bounded both below and above (away from the point p_0, that is, on the complement in each $(M, \lambda^{-1}g)$ of a ball in that $(M, \lambda^{-1}g)$ around p_0 of any fixed positive radius $\varepsilon > 0$, the curvature bounds depending on ε). This follows by the reasoning of the first paragraph of this section.

Suppose that $\{(M, \lambda_i^{-1}g)\}$, with $\lim \lambda_i = +\infty$, is a GH convergent sequence in this setting and that the convergence occurs without collapsing. (The occurrence of the single singular point p_0 at which curvature is possibly unbounded causes no difficulty in the applications of [Grove and Petersen 1991] and the $C^{1,\alpha}$ convergence theorem that is used here.) Then the limit space (with p_0 ignored) is a Riemannian manifold with $C^{1,\alpha}$ Riemannian metric, and this metric is flat, in the sense for instance that it is the limit of C^∞ metrics with curvature going to zero, so that it is also flat in the triangle comparison sense. Thus the limit must be in fact a C^∞ Riemannian manifold of zero curvature. If M is connected at infinity and simply connected at infinity, this limit flat manifold with the isolated singularity p_0 removed is isometric to $\mathbb{R}^n$ with a single point removed (compare Theorem 6.1). Note that the metric space structure limit includes p_0 so that p_0 is metrically isolated too, that is the limit space minus p_0 is isometric to $\mathbb{R}^n$ minus a point, not to $\mathbb{R}^n$ minus a set with more than one point). Since $n \geq 3$, the limit space is in fact simply isometric to $\mathbb{R}^n$.

In this situation, one concludes that M is "euclidean at infinity", not only in the sense that M minus some compact set is topologically $\mathbb{R}^n$ minus some compact set, but also that the metric structure of M converges (in $C^{1,\alpha}$) to that of $\mathbb{R}^n$ in the following sense: There are coordinates on M that amount to a map of $\mathbb{R}^n$ minus a (large) ball into M in which coordinates the metric g converges to the Euclidean $g_{ij} = \delta_{ij}$ metric with increasing distance from p_0. This program is carried out in detail in [Bando et al. 1989].

Continuing in this faster-than-quadratic decay setting, one notes that the required "no collapsing" can be guaranteed by assuming Euclidean volume growth in the sense already discussed, that is, that $r^{-n} \operatorname{vol} B(p_0, r)$ bounded away from 0 as $r \to +\infty$. In this set-up, the assumption of simple connectivity made earlier for convenience of exposition is not in fact needed: The limit flat Riemannian manifold with an isolated singular point p_0 must have finite fundamental group (otherwise it could not have euclidean volume growth, as one sees from the metric classification of manifolds flat outside a compact set in [Schroeder and Ziller 1989] or [Greene and Wu 1993]). So by passing to a finite cover, one can reason as in the simply connected case to conclude that M "at infinity" converges to $(S^{n-1}/\Gamma^1) \times \mathbb{R}^+$, where Γ is a finite group of fixed-point-free isometries of S^{n-1}: see [Bando et al. 1989] for details.

This perspective explains the repeated occurrence of volume estimation as the vital point in the proofs of the "gap theorems" of the previous section: euclidean volume growth is the natural guarantee of "good", that is, essentially euclidean, behavior at infinity.

The convergence viewpoints so far discussed in relation to the gap theorems and asymptotically locally euclideanness are also related to the concept of ideal boundaries for open manifolds defined in terms of geodesic geometry. This concept was first introduced in substantial form by Eberlein and O'Neill [1973] for manifolds of nonpositive curvature. Two rays emanating from a base point p_0 were to be regarded as equivalent if they remained a bounded distance apart; and the ideal boundary, what is now known as the (boundary of) the Eberlein–O'Neill compactification, was obtained by introducing an appropriate metric on the equivalence classes of rays.

For manifolds of nonnegative curvature, a corresponding idea was suggested in [Ballmann et al. 1985] with some "exercises" pointing the way to detailed development. Further development in detail indeed was carried out by Kasue [1988], Shioya [1988], and Shiohama [a]. In summary form, the ideal boundary or "space at infinity" consists of equivalence classes of rays emanating from a base point p_0. Two rays γ_1 and γ_2 are taken to be equivalent if $\lim t^{-1} \operatorname{dist}(\gamma_1(t), \gamma_2(t)) = 0$ as $t \to +\infty$.

Then the distance between two equivalence classes, represented by, say, $\alpha(t)$ and $\beta(t)$, is defined to be $\lim t^{-1} \operatorname{dist}_t(A_t, B_t)$, where A_t and B_t are the intersections of the rays α and β with the radius-t sphere around p_0, and dist_t is the distance obtained by using the intrinsic metric of the radius-t sphere around p_0. (Of course, this intrinsic metric distance differs considerably, in general, from the distance in M between the two points.)

For manifolds of nonnegative curvature, the space at infinity or ideal boundary in this geodesic sense is related to the GH limiting ideas as follows [Shiohama a]: If (M, g) is a complete noncompact manifold of nonnegative sectional curvature, with fixed base point p_0, the GH limit of the family $(M, \lambda^{-1}g)$ as $\lambda \to +\infty$ exists and is a cone over a compact (metric) space $M(\infty)$. The space $M(\infty)$ is isometric to the space at infinity in the geodesic sense.

At first sight, the space $M(\infty)$ might seem to be quite arbitrary. It has the property that it is an Alexandrov space of curvature ≥ 1 (see [Guijarro and Kapovitch 1995] for history), but no other restrictions spring immediately to mind. Thus the following result of Luis Guijarro and Vitali Kapovitch [1995] is rather surprising:

THEOREM 7.1. *If (M, g) is a complete, noncompact, n-dimensional manifold of nonnegative curvature and if $M(\infty)$ is connected, there is a locally trivial fibration $f : S^k \to M(\infty)$ for some $k \leq n - 1$.*

From this line of thought, one can deduce that there are compact manifolds of curvature ≥ 1 that do not occur as the ideal boundary $M(\infty)$ of any complete noncompact M of nonnegative curvature.

References

[Abresch 1985]   U. Abresch, "Lower curvature bounds, Toponogov's theorem, and bounded topology I", *Ann. scient. Éc. Norm. Sup.* (4) **18** (1985), 651–670.

[Abresch and Gromoll 1990]   U. Abresch and D. Gromoll, "On complete manifolds with nonnegative Ricci curvature", *J. Amer. Math. Soc.* **3** (1990), 355–374.

[Allendoerfer and Weil 1943]   C. Allendoerfer and A. Weil, "The Gauss–Bonnet theorem for Riemannian polyhedra", *Trans. Amer. Math. Soc.* **53** (1943), 101–129.

[Anderson and Cheeger 1992]   M. Anderson and J. Cheeger, "C^α compactness for manifolds with Ricci curvature and injectivity radius bounded from below", *J. Diff. Geom.* **35** (1992), 265–281.

[Bando et al. 1989]   S. Bando, A. Kasue, and H. Nakajima, "A construction of coordinates at infinity on manifolds with fast curvature decay and maximal volume growth", *Invent. Math.* **97** (1989), 313–349.

[Ballmann et al. 1985]   W. Ballmann, M. Gromov, and V. Schroeder, *Manifolds of Nonpositive Curvature*, Birkhäuser, Boston, 1985.

[Becker et al. 1980]   J. Becker, C. W. Henson, and L. Rubel, "First-order conformal invariants", *Ann. of Math.* **112** (1980), 123–178.

[Bishop and Crittenden 1964]   R. Bishop and R. Crittenden, *Geometry of Manifolds*, Academic Press, New York, 1964.

[Burago et al. 1992]   Y. Burago, M. Gromov, and G. Perelman, "A. D. Aleksandrov spaces with curvatures bounded below", *Uspehi Mat. Nauk* **47**:2 (1992), 3–51, 222; translation in *Russian Math. Surveys* **47**:2 (1992), 1–58.

[Busemann 1958]   H. Busemann, *Convex Surfaces*, Tracts in Pure and Appl. Math. **6**, Interscience, New York, 1958.

[Calabi 1957]   E. Calabi, "An extension of E. Hopf's maximum principal with an application to geometry", *Duke Math. J.* **25** (1957), 45–56.

[Cai 1991]   M. Cai, "Ends of Riemannian manifolds with nonnegative Ricci curvature outside a compact set", *Bull. Amer. Math. Soc.* **24** (1991), 371–377.

[Cai et al. 1994]   M. Cai, G. Galloway and Z. Liu, "Local splitting theorems for Riemannian manifolds", *Proc. Amer. Math. Soc.* **120** (1994), 1231–1239.

[Cheeger and Gromoll 1971]   J. Cheeger and D. Gromoll, "The splitting theorem for manifolds of nonnegative Ricci curvature", *J. Differential Geom.* **6** (1971), 119–129.

[Cheeger and Gromoll 1972]   J. Cheeger and D. Gromoll, "On the structure of complete manifolds with nonnegative curvature", *Ann. of Math.* **96** (1972), 413–443.

[Cheeger and Tian 1994] J. Cheeger and G. Tian, "On the cone structure at infinity of Ricci flat manifolds with Euclidean volume growth and quadratic curvature decay", *Invent. Math.* **118** (1994), 493–571.

[Cohn-Vossen 1935]   S. Cohn-Vossen, "Kurzeste Wege und Totalkrümmung auf Flächen", *Compositio Math.* **2** (1935), 69–133.

[Drees 1994]   G. Drees, "Asymptotically flat manifolds of nonnegative curvature", *Differential Geom. Appl.* **4** (1994), 77–90.

[Eberlein and O'Neill 1973] P. Eberlein and B. O'Neill, "Visibility manifolds", *Pacific J. Math.* **46** (1973), 45–109.

[Elencwajg 1975] G. Elencwajg, "Pseudoconvexité locale dans les variétés kahlériennes, *Ann. Inst. Fourier* **25** (1975), 295–314.

[Elerath 1979]   D. Elerath, "Open nonnegatively curved 3-manifolds with a point of positive curvature", *Proc. Amer. Math. Soc.* **75** (1979), 92–94.

[Eschenburg et al. 1989] J. Eschenburg, V. Schroeder, and M. Strake, "Curvature at infinity of open nonnegatively curved manifolds", *J. Differential Geom.* **30** (1989), 155–166.

[Gao 1990] L. Gao, "$L^{n/2}$ curvature pinching", *J. Diff. Geo.* **32** (1990), 713–774.

[Greene 1989] R. E. Greene, "Some recent developments in Riemannian geometry", pp. 1–30 in *Recent Developments in Riemannian Geometry* (Los Angeles, 1987), edited by S.-Y. Cheng et al., Contemp. Math. **101**, Amer. Math. Soc., Providence, 1989.

[Greene and Shiohama 1981a] R. E. Greene and K. Shiohama, "Convex functions on complete noncompact manifolds: topological structure", *Invent. Math.* **63** (1981), 129–157.

[Greene and Shiohama 1981b] R. E. Greene and K. Shiohama, "Convex functions on complete noncompact manifolds: differentiable structure", *Ann. scient. Éc. Norm. Sup.* (4) **14** (1981), 357–367.

[Greene and Wu 1973]   R. E. Greene and H. Wu, "On the subharmonicity and plurisubharmonicity of geodesically convex functions", *Indiana Univ. Math. J.* **22** (1973), 641–653.

[Greene and Wu 1974] R. E. Greene and H. Wu, "Integrals of subharmonic functions on manifolds of nonnegative curvature", *Invent. Math.* **27** (1974), 265–298.

[Greene and Wu 1976] R. E. Greene and H. Wu, "C^∞ Convex functions and manifolds of positive curvature", *Acta Math.* **137** (1976), 209–245.

[Greene and Wu 1977] R. E. Greene and H. Wu, "Analysis on noncompact Kähler manifolds", pp. 69–100 in *Several complex variables* (Williams College, 1975), edited by R. O. Wells, Jr., Proc. Sym. Pure Math. **30**, Amer. Math. Soc., Providence, 1977.

[Greene and Wu 1978] R. E. Greene and H. Wu, "On Kähler manifolds of positive bisectional curvature and a theorem of Hartogs", *Abh. Math. Sem. Univ. Hamburg* **47** (1978), 171–185.

[Greene and Wu 1979] R. E. Greene and H. Wu, *Function theory on manifolds which possess a pole*, Lecture Notes in Math. **699**, Springer, Berlin, 1979.

[Greene and Wu 1982] R. E. Greene and H. Wu, "Gap theorems for noncompact Riemannian manifolds", *Duke Math. J.* **49** (1982), 731–756.

[Greene and Wu 1988] R. E. Greene and H. Wu, "Lipschitz convergence of Riemannian manifolds", *Pacific J. Math.* **131** (1988), 119–141.

[Greene and Wu 1993] R. E. Greene and H. Wu, "Nonnegatively curved manifolds which are flat outside a compact set", pp. 327–335 in *Differential geometry: Riemannian geometry* (Los Angeles, 1990), edited by R. Greene and S. T. Yau, Proc. Symp. Pure Math. **54** (3), Amer. Math. Soc., Providence, 1993.

[Greene et al. 1994] R. E. Greene, P. Petersen, and S. H. Zhu, "Riemannian manifolds of faster-than-quadratic curvature decay", *Internat. Math. Res. Notices* **1994**, no. 9, 363–377.

[Gromoll and Meyer 1969] D. Gromoll and W. Meyer, "On complete manifolds of positive curvature", *Ann. of Math.* **90** (1969), 75–90.

[Gromov 1981a] M. Gromov, "Curvature, diameter and Betti numbers", *Comment. Math. Helv.* **56** (1981), 179–195.

[Gromov 1981b] M. Gromov, *Structures métriques pour les variétés riemanniennes*, edited by J. Lafontaine and P. Pansu, CEDIC, Paris, 1981.

[Gromov 1986] M. Gromov, *Partial Differential Relations*, Springer, Berlin, 1986.

[Gromov and Rohlin 1970] M. Gromov and V. Rohlin, "Imbeddings and immersions in Riemannian geometry", *Uspehi Mat. Nauk* **25**:5 (1970), 3–62; translation in *Russian Math. Surveys* **25**:5 (1970), 1–58.

[Grove and Petersen 1991] K. Grove and P. Petersen, "Manifolds near the boundary of existence", *J. Differential Geom.* **33** (1991), 379–394.

[Grove and Shiohama 1977] K. Grove and K. Shiohama, "A generalized sphere theorem", *Ann. of Math.* **106** (1977), 201–211.

[Guijarro a] L. Guijarro, in preparation.

[Guijarro and Kapovitch 1995] L. Guijarro and V. Kapovitch, "Restrictions on the geometry at infinity of nonnegatively curved manifolds", *Duke Math. J.* **78** (1995), 257–276.

[Guijarro and Petersen a] L. Guijarro and P. Petersen, "Rigidity in nonnegative curvature", preprint.

[van Heijenoort 1952] J. van Heijenoort, "On locally convex manifolds", *Comm. Pure Appl. Math.* **5** (1952), 223–242.

[Hartmann and Nirenberg 1951] Hartmann and L. Nirenberg, "On the spherical image maps whose Jacobians do not change signs", *Amer. J. Math.* **73** (1951).

[Jost and Karcher 1982] J. Jost and H. Karcher, "Geometrische methoder zur Gewinnung von a-priori-Schranken für harmonische Abbildungen", *Manuscripta Math.* **40** (1982), 27–77.

[Karcher 1989] H. Karcher, "Riemannian comparison constructions", pp. 170–222 in *Global differential geometry*, edited by S.-S. Chern, MAA Stud. Math. **27**, Math. Assoc. America, Washington, DC, 1989.

[Kasue 1988] A. Kasue, "A compactification of a manifold with asymptotically nonnegative curvature", *Ann. scient. Éc. Norm. Sup.* (4) **21** (1988), 593–622.

[Kasue and Sugahara 1987] A. Kasue and K. Sugahara, "Gap theorems for certain submanifolds of euclidean spaces and hyperbolic space forms", *Osaka J. Math.* **24** (1987), 679–704.

[Li and Tam 1992] P. Li and L.-F. Tam, "Harmonic functions and the structure of complete manifolds", *J. Diff. Geom.* **35** (1992), 359–383.

[Lohkamp 1992] J. Lohkamp, "Negatively Ricci curved manifolds", *Bull. Amer. Math. Soc.* (2) **27** (1992), 288–291.

[Marenich 1981] V. B. Marenich, "Metric structure of open manifolds of nonnegative curvature" (Russian), *Dokl. Akad. Nauk SSSR* **261** (1981), 801–804.

[Mok et al. 1981] N. Mok, Y. T. Siu, and S. T. Yau, "The Poincaré–Lelong equation on complete Kähler manifolds", *Compositio Math.* **44** (1981), 183–218.

[Nikolaev 1980] I. G. Nikolaev, "Parallel translation and smoothness of the metric of spaces of bounded curvature", *Dokl. Akad. Nauk SSSR* **250** (1980), 1056–1058.

[Özaydin and Walschap 1994] M. Özaydin and G. Walschap, "Vector bundles with no soul", *Proc. Amer. Math. Soc.* **120** (1994), 565–567.

[Perelman 1994] G. Perelman, "Proof of the soul conjecture of Cheeger and Gromoll", *J. Differential Geom.* **40** (1994), 209–212.

[Peters 1987] S. Peters, "Convergence of Riemannian manifolds", *Compositio Math.* **62** (1987), 3–16.

[Poor 1974] W. A. Poor, "Some results on nonnegatively curved manifolds", *J. Differential Geom.* **9** (1974), 583–600.

[Rushing 1973] T. Rushing, *Topological embeddings*, Academic Press, New York, 1973.

[Sacksteder 1960] R. Sacksteder, "On hypersurfaces with no negative curvature", *Amer. J. Math.* **82** (1960), 609–630.

[Schroeder and Ziller 1989] V. Schroeder and W. Ziller, "Rigidity of convex domains in manifolds with nonnegative Ricci and sectional curvature", *Comment. Math. Helv.* **64** (1989), 173–186.

[Sharafutdinov 1977] V. Sharafutdinov, "Pogorelov–Klingenberg theorem for manifolds homeomorphic to $\mathbb{R}^n$", *Sibirsk. Math. Zh.* **18** (1977), 915–925.

[Shiohama a] K. Shiohama, "An introduction to the geometry of Alexandrov spaces", lecture notes, Seoul National University.

[Shioya 1988] T. Shioya, "Splitting theorems for nonnegatively curved manifolds", *Manuscripta Math.* **61** (1988), 315–325.

[Siu and Yau 1977] Y. T. Siu and S. T. Yau, "Complete Kähler manifolds of faster than quadratic curvature decay", *Ann. Math.* **105** (1977), 225–264.

[Strake 1988] M. Strake, "A splitting theorem for open nonnegatively curved manifolds", *Manuscripta Math.* **61** (1988), 315–325.

[Toponogov 1959] V. Toponogov, "Riemann spaces with curvature bounded below" (Russian), *Uspehi Mat. Nauk* **14** (1959), 85, 87–130.

[Toponogov 1964] V. Toponogov, "Spaces with straight lines", *Amer. Math. Soc. Transl.* **37** (1964), 287–290.

[Unnebrink a] S. Unnebrink, "Asymptotically flat 4-manifolds", preprint.

[Walschap 1988] G. Walschap, "Nonnegatively curved manifolds with souls of codimension 2", *J. Differential Geom.* **27** (1988), 525–537.

[Wu 1970] H. Wu, "An elementary method in the study of nonnegative curvature", *Acta Math.* **142** (1970), 57–78.

[Wu 1987] H. Wu, "On manifolds of partially positive curvature", *Indiana Univ. Math. J.* **36** (1987), 525–548.

[Yang 1992] D. Yang, "Convergence of Riemannian manifolds with integral bounds on curvature I–II", *Ann. Sci. École Norm. Sup.* **25** (1992), 77–105, 179–199.

[Yim 1988] J. W. Yim, "Distance nonincreasing retraction on a complete open manifold of nonnegative sectional curvature", *Ann. Global Anal. Geom.* **6** (1988), 191–206.

ROBERT E. GREENE
DEPARTMENT OF MATHEMATICS
UNIVERSITY OF CALIFORNIA
LOS ANGELES, CA 90024
greene@math.ucla.edu

Support, Comparison and Splitting

Second variation support functions (Calabi et al.)

Toponogov triangle comparison [1959]
(Second variation $\Rightarrow$ Toponogov: H. Karcher [1989])

|

Toponogov splitting [1964]
(second curvature ≥ 0 and existence of a line $\Rightarrow$ isometric to $N \times \mathbb{R}$)

|

Cheeger–Gromoll splitting [1971]
(Ricci ≥ 0 and existence of a line $\Rightarrow$ isometric to $N \times \mathbb{R}$)

|

Greene and Wu [1974], Greene and Shiohama [1981a]
(sectional curvature ≥ 0 outside a compact $\Rightarrow$ finite number of product ends)

|

P. Li and L. Tam [1992]
(Ricci curvature ≥ 0 outside a compact $\Rightarrow$ finite number of ends)
(Geometric proof by M. Cai [1991])

|

Cai, Galloway, Liu [1994]　(localized version)

Related support function applications

Elenczwag [1975], Greene and Wu [1978] (Kähler manifolds of nonnegative
　　holomorphic bisectional curvature, plurisubharmonic functions, etc.)

H. Wu [1987]　(mixed curvature conditions, q-convexity)

Convex Function Theory

Greene and Wu [1973; 1976]
(approximation of convex functions by almost convex C^∞ functions; existence
　　of C^∞ strictly convex exhaustion function under positive curvature)

|　　　　　　　　　|

　　　　　　Sharafutdinov [1977] (distance nonincreasing retraction onto soul)

|　　　　　　　　　|

Greene and Shiohama [1981a; 1981b]　(structure of convex functions,
　　distance-nonincreasing retraction in general convex function case)

|　　　　　　　　　|

　　　　　　　　　　Yim [1988]　(space of souls)

　　　　　G. Perelman [1994] (rigidity of retraction onto soul)

Open Manifolds of Nonnegative Curvature

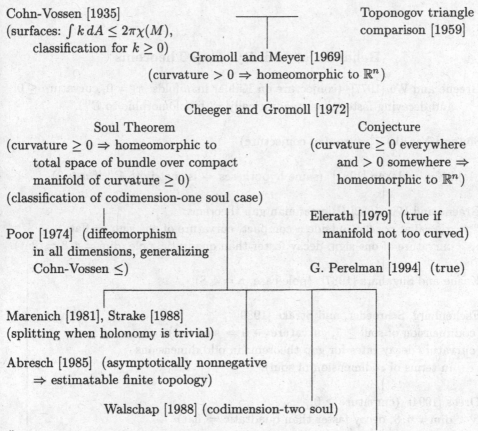

Cohn-Vossen [1935]
(surfaces: $\int k\,dA \le 2\pi\chi(M)$,
   classification for $k \ge 0$)

Toponogov triangle
comparison [1959]

Gromoll and Meyer [1969]
(curvature $> 0 \Rightarrow$ homeomorphic to $\mathbb{R}^n$)

Cheeger and Gromoll [1972]

Soul Theorem
(curvature $\ge 0 \Rightarrow$ homeomorphic to
   total space of bundle over compact
   manifold of curvature ≥ 0)
(classification of codimension-one soul case)

Conjecture
(curvature ≥ 0 everywhere
   and > 0 somewhere $\Rightarrow$
   homeomorphic to $\mathbb{R}^n$)

Elerath [1979] (true if
   manifold not too curved)

Poor [1974] (diffeomorphism
   in all dimensions, generalizing
   Cohn-Vossen $\le$)

G. Perelman [1994] (true)

Marenich [1981], Strake [1988]
(splitting when holonomy is trivial)

Abresch [1985] (asymptotically nonnegative
   $\Rightarrow$ estimatable finite topology)

Walschap [1988] (codimension-two soul)

Özaydin and Walschap [1994] (not all bundles over compact manifolds of
   curvature ≥ 0 occur)

Rescaling and Convergence

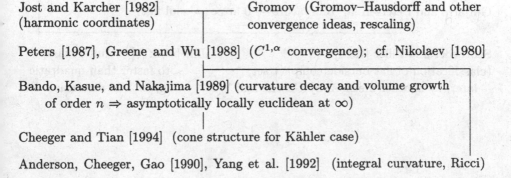

Jost and Karcher [1982]
(harmonic coordinates)

Gromov (Gromov–Hausdorff and other
   convergence ideas, rescaling)

Peters [1987], Greene and Wu [1988] ($C^{1,\alpha}$ convergence); cf. Nikolaev [1980]

Bando, Kasue, and Nakajima [1989] (curvature decay and volume growth
   of order $n \Rightarrow$ asymptotically locally euclidean at ∞)

Cheeger and Tian [1994] (cone structure for Kähler case)

Anderson, Cheeger, Gao [1990], Yang et al. [1992] (integral curvature, Ricci)

Behavior at Infinity: Gap Theorems

Greene and Wu [1977] (conjecture on Kähler manifolds: $\pi_1 = 0$, curvature ≤ 0
   and decaying faster than quadratically $\Rightarrow$ biholomorphic to $\mathbb{C}^n$)

Siu and Yau [1977] (proof of conjecture)

Mok, Siu, and Yau [1981] (same hypotheses $\Rightarrow$ isometric to $\mathbb{C}^n$ if $n \geq 2$)

Greene and Wu [1982] (Riemannian gap theorems:
   $\pi_1 = 0$ at ∞, flat outside a compact, curvature of one sign $\Rightarrow$ flat;
   curvature of one sign, decay faster than quadratic, pole, dim $\neq 4, 8 \Rightarrow$ flat)

Kasue and Sugahara [1987] (pole case, $n = 4, 8$)

Eschenburg, Schroeder, and Strake [1989]
(codimension of soul ≥ 3, curvature $\to 0 \Rightarrow$ soul flat)
(curvature decay rates for gap theorems in odd dimensions
   in terms of codimension of soul)

Drees [1994] (curvature ≥ 0,
   dim $\neq 4, 8$, decay faster than quadratic $\Rightarrow$ flat)

Greene, Petersen, and Zhu [1994]
(π_1 finite at ∞, decay faster than cubic (if dim $\neq 4, 8$) or quartic (else)
   $\Rightarrow$ quotient of asymptotically euclidean manifold)
(π_1 infinite, faster than cubic $\Rightarrow$ equivalent metric flat outside a compact)

Guijarro and Petersen [a] (curvature $\to 0$ at $\infty \Rightarrow$ soul flat)

Schroeder and Ziller [1989]   Unnebrink [a] (counterexample
(classification of flat outside compact set;   to faster than quadratic
   later, by another method, by Greene and Wu)   growth in dimension 4)

Comparison Geometry
MSRI Publications
Volume 30, 1997

Differential Geometric Aspects
of Alexandrov Spaces

YUKIO OTSU

ABSTRACT. We summarize the results on the differential geometric structure of Alexandrov spaces developed in [Otsu and Shioya 1994; Otsu 1995;
Otsu and Tanoue a]. We discuss Riemannian and second differentiable
structure and Jacobi fields on Alexandrov spaces of curvature bounded below or above.

1. Introduction

Let $f : \mathbb{R}^n \to \mathbb{R}$ be a convex function, $B := \{(u, t) \in \mathbb{R}^n \times \mathbb{R} : t > f(u)\}$,
and $\Gamma = \Gamma_f := \{(u, f(u)) \in \mathbb{R}^n \times \mathbb{R}\} = \partial B$. Then B is an open convex set. A
hyperplane L in $\mathbb{R}^{n+1}$ is a *support* at $x \in \Gamma$ if $x \in L$ and $L \cap B = \varnothing$. We say
that $x \in \Gamma$ is a *singular point* if supports at x are not unique. Let S_Γ be the set
of singular points in Γ, and $S_f := \{u \in \mathbb{R}^n : (u, f(u)) \in S_\Gamma\}$. If $u \notin S_f$, then f
is differentiable and the differential is continuous at u.

THEOREM 1.1 [Reidemeister 1921]. *Let $f : \mathbb{R}^n \to \mathbb{R}$ be a convex function.
Then f is a.e. differentiable: more precisely, f is C^1 on $\mathbb{R}^n \setminus S_f \subset \mathbb{R}^n$, and the
Hausdorff dimension* $\dim_H S_f = \dim_H S_\Gamma$ *is at most $n - 1$.*

If $n = 1$, the map $\mathbb{R} \to \mathbb{R}$ given by $u \mapsto df_u$ is monotone by convexity
and, therefore, df_u is a.e. differentiable, that is, f is a.e. twice differentiable. In
general, we have [Busemann and Feller 1935; Alexandrov 1939]:

THEOREM 1.2 (ALEXANDROV'S THEOREM). *Let $f : \mathbb{R}^n \to \mathbb{R}$ be a convex function. Then f is a.e. twice differentiable in the sense of Stolz, that is, for a.a.
$x \in \mathbb{R}^n$ there exist $A \in \mathbb{R}^n$ and an $n \times n$ symmetric matrix $B \in \mathrm{Sym}(n)$ such
that*

$$f(u + h) = f(u) + Ah + \tfrac{1}{2}{}^t hBh + o(|h|^2)$$

1991 *Mathematics Subject Classification.* 53C20, 53C23, 53C45, 26B25.

Key words and phrases. Alexandrov space, differentiable structure, comparison geometry.

for $h \in \mathbb{R}^n$.

Consider the inner geometry of Γ. The length of a path $c : [a, b] \to \Gamma$ is

$$|c| := \sup_{a=a_0 < \ldots < a_l = b} \sum_{i=1}^{l} |c(a_{i-1})c(a_i)|,$$

where $|\ |$ on the right denotes the Euclidean metric. The intrinsic distance d on Γ is defined by

$$d(p, q) = |pq| = \inf_c |c|,$$

where c is a path from p to q. Then Γ is a geodesic space, that is, for any p, $q \in \Gamma$ there is a path from p to q whose length is equal to $|pq|$; this is called a minimal segment or minimal geodesic (from p to q), and denoted by pq. Any convex function can be written as a limit of a sequence of C^∞ convex functions $\{f_i\}_{i=1}^\infty$; equivalently,

$$d_{pH}((\Gamma, p), (\Gamma_i, p_i)) \to 0 \quad \text{as } i \to \infty, \tag{1.1}$$

where d_{pH} denotes the pointed Hausdorff distance, $\Gamma_i = \Gamma_{f_i}$, $p = (0, f(0))$, and $p_i = (0, f_i(0))$. Note that Γ_i is a Riemannian manifold with sectional curvature ≥ 0.

Let X be a geodesic space. For p, q, $r \in X$ the *triangle* $\triangle pqr$ is a triad of segments pq, qr, rp. A *comparison triangle* for $\triangle pqr$ is a triangle $\tilde{\triangle}pqr = \triangle \tilde{p}\tilde{q}\tilde{r}$ in $\mathbb{R}^2$ with $|\tilde{p}\tilde{q}| = |pq|$, $|\tilde{r}\tilde{p}| = |rp|$, $|\tilde{r}\tilde{q}| = |qr|$. If X is a Riemannian manifold with sectional curvature ≥ 0, it has the following property:

PROPERTY 1.3 (ALEXANDROV CONVEXITY). *Given a triangle $\triangle pqr$ in X, there is a comparison triangle $\tilde{\triangle}pqr$ such that, for $s \in qr$, we have*

$$|ps| \geq |\tilde{p}\tilde{s}|,$$

where $\tilde{s} \in \tilde{q}\tilde{r}$ with $|\tilde{s}\tilde{q}| = |sq|$.

This property is equivalent to Toponogov's comparison theorem. Since (1.1) implies that any segment on Γ is approximated by segments on Γ_i, Alexandrov convexity is also valid on Γ.

For $k \leq 0$ a metric space X is an *Alexandrov space of curvature $\geq k$* if X is a locally compact, complete length space of dimension $\dim_H X < \infty$, satisfying the Alexandrov convexity property with $\mathbb{R}^2$ replaced by $H^2(k)$, the simply connected space form of curvature k. Then the dimension of X is an integer, say n, and the n-dimensional Hausdorff measure V_H^n satisfies

$$0 < V_H^n(B(p, r)) < \infty$$

for any $B(p, r) := \{x \in X : |px| < r\}$.

EXAMPLES. (1) A complete Riemannian manifold of sectional curvature $\geq k$ is an Alexandrov space of curvature $\geq k$.

(2) It follows from the preceding discussion that the graph Γ of a convex function is an Alexandrov space of curvature ≥ 0 and has a natural a.e. twice differentiable structure.

(3) As the above argument illustrates, the Hausdorff limit of a sequence of Riemannian manifolds $\{M_i\}$ of curvature $\geq k$ is an Alexandrov space of curvature $\geq k$. Gromov's convergence theorem states that if, for each M_i, the absolute value of the sectional curvature is bounded above by a constant and the injectivity radius is bounded below by a positive constant, then X is a $C^{1,\alpha}$ Riemannian manifold for $0 < \alpha < 1$, that is, X has a $C^{2,\alpha}$ differentiable structure and a $C^{1,\alpha}$ Riemannian metric.

(4) Let X be an n-dimensional Alexandrov space and let $p \in X$. Then there exists a pointed Hausdorff limit of (iX, p), as $i \to \infty$; it is called a *tangent cone at p*, and is denoted by K_p. (In Section 2 we will give another definition of a tangent cone.) The point p is called *singular* if K_p is not isometric to $\mathbb{R}^n$. The tangent cone K_p is an Alexandrov space of curvature ≥ 0.

In some of these examples we find that there exists a second differentiable structure. In general Reidemeister's and Alexandrov's theorems are generalized as follows:

THEOREM 1.4 [Otsu and Shioya 1994; Otsu 1995]. *Let X be an n-dimensional Alexandrov space of curvature bounded below, and let S_X be the set of singular points on X.*

(i) *The complement $X \setminus S_X \subset X$ has a C^1 differentiable and Riemannian structure, and $\dim_H S_X \leq n - 1$. The induced metric from the Riemannian structure coincides with the original metric on X.*

(ii) *There is a set $X_0 \subset X \setminus S_X$ of full measure with respect to V_H^n such that $X_0 \subset X \setminus S_X$ has an approximately second differentiable structure in the sense of Stolz.*

In this paper we do not give precise definitions for the various structures mentioned in this theorem. Instead, we give in Section 2 a rough sketch of the proof of the theorem, and show how to develop differential geometry on Alexandrov spaces of curvature bounded below. In Section 3 we discuss differential geometry on Alexandrov space of curvature bounded above.

2. Elements of Differential Geometry on Alexandrov Spaces of Curvature Bounded Below

The exponential map. For simplicity we assume that X is an n-dimensional Alexandrov space of curvature ≥ 0. Note that no segment on X branches, by

Alexandrov convexity. For $p, q, r \in X$ and segments $pq = \gamma : [0, a] \to X$ and $pr = \sigma : [0, b] \to X$, put

$$\omega(t, s) = \angle \tilde{\gamma}(t) \tilde{p} \tilde{\sigma}(s)$$

for $0 < t \leq a$ and $0 < s \leq b$. Then, by Alexandrov convexity, $(t, s) \mapsto \omega(t, s)$ is a monotone nonincreasing function; thus the angle between γ and σ,

$$\angle qpr := \lim_{(t,s) \searrow (0,0)} \omega(t, s),$$

is well-defined. It follows easily that X has the following property:

PROPERTY 2.1 (TOPONOGOV CONVEXITY). *For any triangle $\triangle pqr$ there is a comparison triangle $\triangle \tilde{p}\tilde{q}\tilde{r}$ such that*

$$\angle rpq \geq \angle \tilde{r}\tilde{p}\tilde{q}, \quad \angle qrp \geq \angle \tilde{q}\tilde{r}\tilde{p}, \quad \angle pqr \geq \angle \tilde{p}\tilde{q}\tilde{r}.$$

Set $\tilde{W}_p = \{pq : q \in X\}$ and $\tilde{\Sigma}_p = (\tilde{W}_p \setminus o_p)/\sim$, where o_p is the trivial segment pp and $pq \sim pr$ implies $pq \subset pr$ or $pr \subset pq$. Denote by v_{pq} the equivalence class of pq. The *space of directions* Σ_p is the completion of $(\tilde{\Sigma}_p, \angle)$, and is a compact Alexandrov space of curvature ≥ 1 [Burago et al. 1992]. The *tangent cone* K_p is obtained from $[0, \infty) \times \Sigma_p$ by identifying together all elements of the form $(0, u_0)$, and its elements are denoted by tu_0 for $t \geq 0$ and $u_0 \in \Sigma_p$, or simply by u. We introduce a distance on K_p by setting

$$|tu_0 \, sv_0| := \sqrt{t^2 + s^2 - 2st \cos \angle u_0 v_0}$$

for $tu_0, sv_0 \in K_p$. Then $(K_p, |\ |)$ is an Alexandrov space of curvature ≥ 0. By considering $\tilde{W}_p \subset K_p$, we define the exponential map $\mathrm{Exp}_p : \tilde{W}_p \to X$ by

$$\mathrm{Exp}_p|pq|v_{pq} = q.$$

Toponogov convexity implies

$$|\mathrm{Exp}_p u \, \mathrm{Exp}_p v| \leq |uv|. \qquad (2.1)$$

For $\delta > 0$, let W_p^δ be the set of $x \in X$ for which there exists $y \in X$ such that $x \in py$ and $|py| = (1 + \delta)|px|$. In this case y is unique, and we define a map $E_p^\delta : W_p^\delta \to X$ by setting

$$E_p^\delta(x) = y.$$

By Alexandrov convexity we have

$$|E_p^\delta(x) E_p^\delta(y)| \leq (1 + \delta)|xy|, \qquad (2.2)$$

and clearly

$$E_p^\delta \circ \mathrm{Exp}_p u = \mathrm{Exp}_p(1 + \delta)u. \qquad (2.3)$$

Define the *cut locus* of p by $\mathrm{Cut}_p = X \setminus \bigcup_{\delta > 0} W_p^\delta$. If $x \in \mathrm{Cut}_p$, then x is not an interior point of any segment from p, so this definition coincides with that of Riemannian manifold.

PROPOSITION 2.2. *We have $V_H^n(\mathrm{Cut}_p) = 0$.*

PROOF. Because the map $E_p^\delta : W_p^\delta \cap B(p, R) \to B(p, (1+\delta)R)$ is surjective for $R > 0$, we have

$$V_H^n(B(p, (1+\delta)R)) \le (1+\delta)^n V_H^n(W_p^\delta \cap B(p, R))$$

by (2.2). As $\delta \searrow 0$ we have

$$V_H^n(B(p, R)) \le V_H^n((X \setminus \mathrm{Cut}_p) \cap B(p, R)). \qquad \square$$

Then, since X is separable, $V_H^n(S_X) = 0$ by Toponogov's splitting theorem. A more careful argument will give us $\dim_H(S_X) \le n - 1$.

The first variation formula. Suppose given a Riemannian manifold M, a point $p \in M$, and a segment $\sigma : [0, a] \to M$. Applying the first variation formula for $s \mapsto |p\sigma(s)|$, we have

$$\frac{d}{ds}\bigg|_{s=0} |p\sigma(s)| = - \cos \min_{p\sigma(0)} \angle p\sigma(0)\sigma(a).$$

In the case of a point p in an Alexandrov space X of curvature ≥ 0, we also have

$$d_p(y) = d_p(x) - |xy| \cos \min_{px} \angle pxy + o(|xy|) \qquad (2.4)$$

by the Lipschitz continuity of $x \mapsto d_p(x) = |px|$ and the compactness of Σ_x.

For $p_1, \ldots, p_n \in X$ we define a map $\psi : X \to \mathbb{R}^n$ by

$$\psi(x) = (|p_1 x|, \ldots, |p_n x|)$$

and $g_\psi : X \setminus \bigcup_{i=1}^n \mathrm{Cut}_{p_i} \to \mathrm{Sym}(n)$ by

$$g_\psi(x) = (\cos \angle p_i x p_j).$$

If $x_0 \in X \setminus S_X$, we can choose points $p_1, \ldots, p_n \in X$ so that $g_\psi(x_0) > 0$. Because the angle is continuous at a nonsingular point, as the differential is continuous at regular points of Γ, there is a neighborhood U_ψ of x_0 such that $g_\psi(x) > 0$ on U_ψ and $\psi : U_\psi \to \mathbb{R}^n$ is a homeomorphism onto an open set in $\mathbb{R}^n$. Since

$$\psi(y) = \psi(x) + |xy|(- \cos \angle p_i xy) + o(|xy|)$$

by (2.4), the set $\{(\psi, U_\psi, g_\psi)\}$ gives us an a.e. C^1 structure and Riemannian structure on $X \setminus S_X \subset X$.

Jacobi fields. For simplicity we choose $p \in X \setminus S_X$, that is, $K_p = \mathbb{R}^n$. Since $\mathrm{Exp}_p : \tilde{W}_p \to X$ is a Lipschitz map by (2.1), using the differentiable and Riemannian structure on X we conclude that Exp_p is differentiable a.e. by extending Rademacher's theorem, which states that a Lipschitz map between Euclidean spaces is differentiable a.e. Let $x \in X \setminus S_X$ be such that Exp_p is a.e. differentiable on $[0, |px|]v_{px}$. For a segment $\sigma : [0, a] \to X$ starting at x we want to construct a Jacobi field $J(t)$ from $\alpha(s, t) = \gamma_s(t)$, where γ_s is a segment from p to $\sigma(s)$ whose parameter is scaled for $[0, |px|]$. Let $\tilde{\alpha}$ be a lift of α by Exp_p

and let $w \in K_{|px|v_{px}} K_p = \mathbb{R}^n$ be such that $d\mathrm{Exp}_p|_{|px|v_{px}} w = \dot{\sigma}(0)$. Then $\tilde{\alpha}$ is differentiable at $s = 0$ and

$$\frac{d}{ds}\bigg|_{s=0} \tilde{\alpha}(s,t) = tw,$$

which we denote by $\tilde{J}_w(t)$. For a.a. $t \in (0, |px|]$ there exists

$$\frac{d}{ds}\bigg|_{s=0} \alpha(s,t) = d\mathrm{Exp}_p|_{tv_{px}} \tilde{J}_w(t).$$

This is the desired Jacobi field $J_w(t)$, that is,

$$J_w(t) = d\mathrm{Exp}_p|_{tv_{px}} \tilde{J}_w(t). \tag{2.5}$$

Then $w \mapsto J_w(t)$ is a well-defined linear map on $K_{|px|v_{px}} K_p = \mathbb{R}^n$.

Second differential of the distance function. As mentioned on the preceding page, the first differentiable structure is deduced from the first variation formula. Similarly, the second differentiable structure is determined by the second variation of the distance function. First we examine the second variation of the distance function on a Riemannian manifold M. For $p \in M$ and a segment $\sigma : [0, a] \to M$ with $\sigma(0) \notin \mathrm{Cut}_p$, we have

$$\frac{d^2}{ds^2}\bigg|_{s=0} |p\sigma(s)| = \left\langle J, \frac{d}{ds}\nabla d_p \right\rangle = \left\langle J, \nabla_{t=|p\sigma(0)|} J_w^{\perp}(t) \right\rangle, \tag{2.6}$$

where ∇d_p denotes the gradient vector of d_p and $J^{\perp}$ the orthogonal component of J with respect to ∇d_p. Therefore

$$|py| = |px| + |xy|\langle v_{xy}, \nabla d_p \rangle + \tfrac{1}{2}|xy|^2 \langle J_{v_{xy}}^{\perp}(t), \nabla J_{v_{xy}}^{\perp}(t) \rangle + o(|xy|^2). \tag{2.7}$$

Here we present facts that assure us that (2.7) is valid. Let $p \in X$ and let $\sigma : [0, a] \to X$ be a segment.

PROPOSITION 2.3. *The function $s \mapsto \angle p\sigma(s)\sigma(a)$ is of bounded variation; in particular, it is a.e. differentiable. For any s_0 where it is differentiable, we have*

$$|p\sigma(s)| = |p\sigma(s_0)| - (s - s_0) \cos \angle p\sigma(s_0)\sigma(a)$$
$$+ \frac{1}{2}(s - s_0)^2 \sin \angle p\sigma(s)\sigma(a) \frac{d}{ds}\bigg|_{s=s_0} \angle p\sigma(s)\sigma(a) + o(|s - s_0|^2)$$

and

$$\frac{d}{ds}\bigg|_{s=s_0} \angle p\sigma(s)\sigma(a) \leq \frac{1}{|p\sigma(s_0)|} \sin \angle p\sigma(s)\sigma(a).$$

PROOF. By Toponogov convexity,

$$\angle p\sigma(s')\sigma(a) = \pi - \angle p\sigma(s')\sigma(s) \leq \pi - \angle \widetilde{p\sigma(s')}\widetilde{\sigma(s)} = \angle \widetilde{p\sigma(s)}\widetilde{\sigma(a)} + \angle \widetilde{\sigma(s)}\widetilde{p\sigma(a)}$$
$$< \angle p\sigma(s)\sigma(a) + (s' - s)\frac{1}{|p\sigma(s)|} \sin \angle p\sigma(s)\sigma(a) + O(|s - s_0|^2)$$

for $s < s'$. (See Figure 1.) □

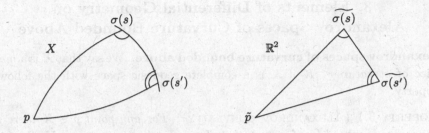

Figure 1. Comparison triangles in the proof of Proposition 2.3.

It is difficult to deduce an expansion like (2.7) from Proposition 2.3, since the above argument is restricted to segments; the estimation is not uniform on the direction; it is not clear that the second differential is a quadratic form, etc. We show here that the last term of (2.6) exists a.e. and that it is a quadratic form.

PROPOSITION 2.4. *The function $t \mapsto |J(t)|$ has bounded variation. In particular, for a.a. t there is a first differential for $|J(t)|$, satisfying*

$$\frac{d}{dt}|J(t)| \leq \frac{1}{t}|J(t)|,$$

and the map

$$w \mapsto \frac{d}{dt}|J_w(t)|^2$$

on $K_{|px|v_{px}}K_p$ is a quadratic form.

PROOF. As in Section 2.3, we consider a family of segments $\gamma_s(t)$ as a variation. By (2.2), we have

$$|\gamma_0((1+\delta)t)\gamma_s((1+\delta)t)| \leq (1+\delta)|\gamma_0(t)\gamma_s(t)|.$$

By taking $s \searrow 0$ we have $|J((1+\delta)t)| \leq (1+\delta)|J(t)|$ for a.a. t and δ, that is, $t \mapsto |J(t)|$ has bounded variation. Then

$$\frac{1}{\delta t}(|J((1+\delta)t)| - |J(t)|) \leq \frac{|J(t)|}{t}. \qquad \square$$

Although it is quite interesting to show the second differential of the distance function coincides with $d|J(t)^{\perp}|/dt$, we omit this proof because it is complicated. Then we have twice differentiability of d_p a.e., and a second differentiable structure on X a.e.

3. Elements of Differential Geometry on Alexandrov Spaces of Curvature Bounded Above

Alexandrov spaces of curvature bounded above. We say that X is a *metric space of curvature $\leq K$* if X is a complete geodesic space with the following property:

PROPERTY 3.1 (ALEXANDROV CONVEXITY). *For any point $p \in X$ there is a convex neighborhood U of p such that, for any triangle $\triangle xyz$ in U, there is a comparison triangle $\tilde{\triangle}xyz$ such that if $s \in yz$, then*

$$|xs| \leq |\tilde{x}\tilde{s}|$$

for $\tilde{s} \in \tilde{y}\tilde{z}$ with $|\tilde{s}\tilde{y}| = |sy|$. Here the comparison triangle $\triangle \tilde{x}\tilde{y}\tilde{z}$ is taken in the space form $H^2(K)$; if $K > 0$, we assume that $|xy| + |yz| + |zx| \leq 2\pi/\sqrt{K}$.

Notice that the comparison inequality is valid only on U; the situation is completely different from the case of curvature bounded below. If $x \in U$, then x is not a cut point of $p \in U$.

EXAMPLES. (1) A complete Riemannian manifold of sectional curvature $\leq K$.

(2) A finite graph.

(3) Let X_i be a binary tree where each edge has length 2^{-i}. Then X_i can be isometrically embedded in X_{i+1}; the figure on the right shows in thick lines a finite approximation to X_i, embedded in a finite approximation to X_{i+1}. From the sequence $X_1 \subset X_2 \subset \ldots$ we can construct an inductive limit X_∞, which is also a metric space of curvature ≤ 0. This is not locally compact.

(4) Next we consider another binary tree X such that the length of each edge at height i is 2^{-i}, as in the figure below. Then X is locally compact but not geodesically complete. Notice that the Hausdorff dimension of its boundary is 1.

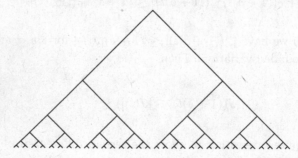

(5) A two-dimensional cone at whose vertex the total angle is greater than 2π.

(6) A simplicial complex which branches, as in the figure on the right.

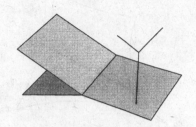

(7) For (s,t) and $(s',t') \in \mathbb{R}^2$, define the distance between them as $|s-s'|$ if $t = t'$, $s+s'+|t-t'|$ if $s, s' \geq 0$ and $t \neq t'$, $s + \sqrt{s'^2 + |t - t'|^2}$ if $s \geq 0$ and $s' \leq 0$, and as the Euclidean metric if $s, s' \leq 0$. This makes $\mathbb{R}^2$ into a metric space of curvature ≤ 0 that is not locally compact.

Angles and related concepts are defined in a similar way as for Alexandrov spaces of curvature bounded below. In the absence of additional restrictions, it is difficult to deduce for metric spaces of curvature bounded above a topological structure and a result similar to Theorem 1.4, as the examples illustrate. We therefore make the following definition: A geodesic space X is an *Alexandrov space of curvature bounded above* if X is a locally compact, geodesically complete metric space of curvature bounded above.

THEOREM 3.2 [Otsu and Tanoue a]. *Let X be an Alexandrov space of curvature bounded above. Then, for any $p \in X$ and $r > 0$, there is an integer n such that $0 < V_H^n(B(p,r)) < \infty$. Let S_X^n be the set of $x \in B(p,r)$ such that K_x is not isometric to $\mathbb{R}^n$.*

(i) *There exists a C^1 differentiable and a C^0 Riemannian structure for $B(p,r) \setminus S_X^n \subset B(p,r)$, and S_X^n is a set of V_H^n null measure.*

(ii) *There exists an a.e. second differentiable structure in the sense of Stolz on $B(p,r)$.*

We now give a description of differential geometric properties in the absence of the above restriction.

The Jacobi norm. As we know from the examples given earlier, we cannot define an exponential map. Thus we define the Jacobi norm from variation. For simplicity we assume that X is a metric space of curvature ≤ 0 and $U = X$. Let $\sigma : [0,a] \to X$ be a segment and let $\alpha(s,t) = \gamma_s(t)$ be a family of normal segments from p to $\sigma(s)$. It follows from Alexandrov convexity that

$$(1 + \delta) |\alpha(0,t)\, \alpha(s,t)| \leq |\alpha(0, (1 + \delta)t)\, \alpha(s, (1 + \delta)t)|.$$

Hence for a.a. t there is

$$|J|(t) = |J|_\alpha(t) := \limsup_{h \to 0} \frac{1}{h} |\alpha(0,t)\alpha(h,t)|.$$

As in Proposition 2.4, for a.a. t there is a first differential of $|J|$ at t, and

$$\frac{d}{dt}|J|(t) \geq |J|(t) \times \frac{1}{t}.$$

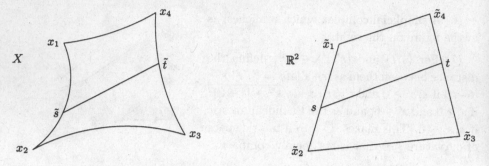

Figure 2. For a space with the Wald convexity property, $|st| \le |\tilde{s}\tilde{t}|$.

Here we examine again the Riemannian case: Let $J(t)$ be a Jacobi field on M along a segment γ. Then

$$\frac{d^2}{dt^2}|J(t)| = \frac{d}{dt}\frac{1}{|J|}\langle J(t), \nabla_t J(t)\rangle$$

$$= \frac{1}{|J|}\left(|\nabla_t J(t)|^2 - \frac{1}{|J|^2}\langle J(t), \nabla_t J(t)\rangle^2\right) + \left\langle\frac{1}{|J|}J(t), R(\dot{\gamma}(t), J(t))\dot{\gamma}(t)\right\rangle$$

$$= \frac{1}{|J|}\left(|\nabla_t J(t)|^2 - \left(\frac{d}{dt}|J(t)|\right)^2\right) - K_{\langle\dot{\gamma}(t),J(t)\rangle}|\dot{\gamma}(t) \wedge J(t)|, \qquad (3.1)$$

where $K_{\langle\dot{\gamma}(t),J(t)\rangle}$ denotes the sectional curvature of the section spanned by $\dot{\gamma}(t)$ and $J(t)$, and $\wedge$ denotes the exterior product of vectors. If the sectional curvature of M is nonpositive, we have

$$\frac{d^2}{dt^2}|J(t)| \ge \frac{1}{|J|}\left(|\nabla_t J(t)|^2 - \left\langle\frac{1}{|J|}J(t), \nabla_t J(t)\right\rangle^2\right) \ge 0. \qquad (3.2)$$

This argument does not hold for the case of curvature bounded below, and this is one reason why the treatment of Alexandrov spaces of curvature bounded below is difficult.

The Jacobi equation. The geometric expression of (3.2) is the following (see Figure 2):

THEOREM 3.3 (WALD CONVEXITY [Reshetnyak 1968]). *Let X be a metric space of curvature ≤ 0. For any $x_1, \dots, x_4 \in X$ there exist $\tilde{x}_1, \dots, \tilde{x}_4 \in \mathbb{R}^2$ such that*

$$|x_1 x_2| = |\tilde{x}_1 \tilde{x}_2|, \quad |x_2 x_3| = |\tilde{x}_2 \tilde{x}_3|, \quad |x_3 x_4| = |\tilde{x}_3 \tilde{x}_4|, \quad |x_4 x_1| = |\tilde{x}_4 \tilde{x}_1|,$$

and that for $s \in x_i x_j$ and $t \in x_{i'} x_{j'}$ we have

$$|st| \le |\tilde{s}\tilde{t}|,$$

where $\tilde{s} \in \tilde{x}_i \tilde{x}_j$ and $\tilde{t} \in \tilde{x}_{i'} \tilde{x}_{j'}$ satisfy $|sx_i| = |\tilde{s}\tilde{x}_i|$ and $|tx_{i'}| = |\tilde{t}\tilde{x}_{i'}|$.

Applying this to $\alpha(0,t)$, $\alpha(0,t')$, $\alpha(h,t')$, $\alpha(h,t)$, we conclude that $t \mapsto |J|(t)$ is a convex function; in particular, it is continuous on $[0, |px|)$ and for a.a. t there is a second differential of $|J|$ at t, satisfying

$$\frac{d^2}{dt^2}|J|(t) \geq 0.$$

PROPOSITION 3.4. *Set* $\varphi = \angle\alpha(h,t)\alpha(0,t)\sigma(0)$, $\psi = \angle\alpha(0,t)\alpha(h,t)\sigma(h)$, $\varphi' = \angle\alpha(0,t+k)\alpha(h,t+k)p$, *and* $\psi' = \angle\alpha(h,t+k)\alpha(0,t+k)p$ *(see Figure 3). Then* $|J|$ *has a first and second differentials a.e., and*

$$\frac{d}{dt}|J|(t) = \lim_{h\to 0}\frac{-1}{h}(\cos\varphi + \cos\psi),$$

$$\frac{d^2}{dt^2}|J|(t) = \lim_{k\to 0}\lim_{h\to 0}\frac{1}{hk}(\cos\varphi + \cos\psi + \cos\varphi' + \cos\psi').$$

In the case of a Riemannian manifold, the covariant derivative of J is written as

$$\nabla J(t) = \lim_{h,k\to 0}\frac{1}{hk}(c - b - a),$$

where

$$a = |\alpha(h,t)\alpha(0,t)|v_{\alpha(0,t)\alpha(h,t)},$$

$$b = |\alpha(h,t+k)\alpha(0,t)|v_{\alpha(0,t)\alpha(h,t+k)},$$

$$c = |\alpha(0,t+k)\alpha(0,t)|v_{\alpha(0,t)\alpha(h,t+k)};$$

its existence is not clear in our case. If we set

$$|\nabla J|(t) = \lim_{k\to 0}\lim_{h\to 0}\frac{1}{hk}|c - b - a|,$$

then we have

$$\lim_{k\to 0}\lim_{h\to 0}\frac{1}{hk}(\theta + \eta - \varphi) = \frac{1}{2|J|}|\nabla J|^2,$$

where $\theta = \angle\alpha(h,t+k)\alpha(0,t)\alpha(0,t+k)$ and $\eta = \angle\alpha(h,t+k)\alpha(0,t)\alpha(h,t)$.

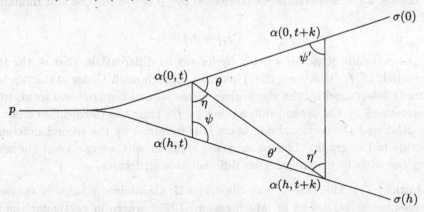

Figure 3. Angles in Proposition 3.4. In addition, $\varphi = \angle\alpha(h,t)\alpha(0,t)\sigma(0)$ and $\varphi' = \angle\alpha(0,t+k)\alpha(h,t+k)p$.

Next we consider the sectional curvature of M. If $\dim M = 2$, Gauss–Bonnet's theorem implies

$$\angle pqr + \angle qrp + \angle qrp - \pi = \int_{\triangle pqr} G\,dA,$$

where G denotes Gaussian curvature and dA the area element. Thus

$$G_x = \lim_{\triangle pqr \to x} \frac{1}{\text{area of } \triangle \tilde{p}\tilde{q}\tilde{r}}(\angle pqr + \angle qrp + \angle rpq - \pi). \tag{3.3}$$

In higher dimensions we have a similar result.

In the case of metric spaces of curvature bounded above we define the *connection norm* of $|J|$ as

$$|\nabla_t J|(t) = \left(|J| \limsup_{k \to 0} \limsup_{h \to 0} \frac{1}{hk}(\theta + \eta - \varphi + \theta' + \eta' - \varphi')\right)^{\frac{1}{2}},$$

and the *sectional curvature* by

$$K_\alpha(t) = \limsup_{k \to 0} \limsup_{h \to 0} \frac{1}{\tilde{h}k}(\psi + \eta + \theta' - \pi + \psi' + \eta' + \theta - \pi),$$

where $\theta' = \angle\alpha(0,t)\alpha(h,t+k)\alpha(h,t)$, $\eta' = \angle\alpha(0,t)\alpha(h,t+k)\alpha(0,t+k)$, and $\tilde{h} = |\alpha(0,t)\alpha(h,t)|$.

Then:

THEOREM 3.5 [Otsu and Tanoue a]. *The second differential of the norm of Jacobi field, the sectional curvature, and $|\nabla_t J|(t)$ are well defined on α a.e. If they are well defined at $(0,t)$, then*

$$\frac{d^2}{dt^2}|J|(t) = \frac{1}{|J|}|\nabla_t J|(t)^2 - K_\alpha(t) \times |J|(t).$$

Consider a C^∞ hypersurface determined by $f : \mathbb{R}^n \to \mathbb{R}$. The first fundamental form is

$$g_{ij} = \delta_{ij} + \partial_i f \partial_j f,$$

and the curvature tensor is written by its second differentials, that is, the third differentials of f. However, the Theorema Egregium of Gauss states the curvature is determined by the eigenvalues of the second fundamental form, which is determined by the second differentials of f. Thus the third differentials kill each other and the sectional curvature is determined by the second differential structure of the graph. This observation explains why we can treat the above quantities without invoking higher differentiable structures.

REMARKS. (1) The case of two-dimensional Alexandrov spaces of curvature bounded below is studied in [Machigashira 1995], where in particular the sectional curvature and a generalization of Gauss–Bonnet theorem are given. Note that here parallel translation along a segment is trivially determined.

We also mention Petrunin's work on parallel translation in Alexandrov spaces of curvature bounded below [Petrunin a; Berestvskii and Nikolaev 1993].

(2) Recently there have been several studies on harmonic maps on simply connected Alexandrov spaces of curvature ≤ 0 [Gromov and Schoen 1992; Jost a; Korevaar and Schoen 1993]. These results would seem to have close connections with ours.

References

[Alexandrov 1939]   A. D. Alexandrov, "Almost everywhere existence of second differentials of convex functions and convex surfaces connected with it", *Leningrad State Univ. Ann. (Uchenye Zapiski) Math. Ser.* **6** (1939), 3–35.

[Alexandrov 1951] A. D. Alexandrov, "A theorem on triangles in a metric space and some of its applications", *Trudy Mat. Inst. Steklov* **38** (1951), 4–23.

[Alexander et al. 1993]   S. Alexander, I. D. Berg, and R. L. Bishop, "Geometric curvature bounds in Riemannian manifolds with boundary", *Trans. Amer. Math. Soc.* **339** (1993), 703–716.

[Berestvskii and Nikolaev 1993]   V. N. Berestvskii and I. G. Nikolaev, "Multidimensional generalized Riemannian spaces", pp. 165–243 in *Geometry IV: nonregular Riemannian geometry*, edited by Y. G. Reshetnyak, Encycl. Math. Sciences **70**, Springer, Berlin, 1993.

[Burago et al. 1992]   Y. Burago, M. Gromov, and G. Perelman, "A. D. Aleksandrov spaces with curvatures bounded below", *Uspehi Mat. Nauk* **47**:2 (1992), 3–51, 222; translation in *Russian Math. Surveys* **47**:2 (1992), 1–58.

[Busemann 1958] H. Busemann, *Convex Surfaces*, Tracts in Pure and Appl. Math. **6**, Interscience, New York, 1958.

[Busemann and Feller 1935]   H. Busemann and W. Feller, "Krümmungseigenschaften Konvexer Flächen", *Acta Math.* **66** (1935), 1–47.

[Evans and Gariepy 1992] L. Evans and R. Gariepy, *Measure theory and fine properties of functions*, CRC Press, Boca Raton (FL), 1992.

[Gromov 1981]   M. Gromov, *Structures métriques pour les variétés riemanniennes*, edited by J. Lafontaine and P. Pansu, CEDIC, Paris, 1981.

[Gromov and Schoen 1992] M. Gromov and R. Schoen, "Harmonic maps into singular spaces and p-adic superrigidity for lattices in groups of rank one", *Pub. Math. IHES* **76** (1992), 165–246.

[Jost a] J. Jost, "Equilibrium Maps between metric spaces", preprint.

[Korevaar and Schoen 1993] N. Korevaar and R. Schoen, "Sobolev spaces and harmonic maps for metric space targets", *Comm. Anal. Geom.* **1** (1993), 561–659.

[Machigashira 1995]   Y. Machigashira, "The Gaussian curvature of Alexandrov surfaces", preprint, Saga University, 1995.

[Otsu 1995] Y. Otsu, "Almost everywhere existence of second differentiable structure of Alexandrov spaces", preprint, University of Tokyo, 1995.

[Otsu and Shioya 1994] Y. Otsu and T. Shioya, "The Riemannian structure of Alexandrov spaces", *J. Diff. Geom.* **39** (1994), 629–658.

[Otsu and Tanoue a] Y. Otsu and H. Tanoue, "The Riemannian structure of Alexandrov spaces with curvature bounded above", in preparation.

[Petrunin a] A. Petrunin, "Parallel transportation and second variation", preprint.

[Rademacher 1919] H. Rademacher, "Über partielle und totale Differenzierbarkeit von Funktionen mehrerer Variabeln und über die Transformation der Doppelintegrale", *Math. Ann.* **79** (1919), 340–359.

[Reidemeister 1921] K. Reidemeister, "Über die singulären Randpunkte eines konvexen Körpers", *Math. Ann.* **83** (1921), 116–118.

[Reshetnyak 1968] Y. G. Reshetnyak, "Inextensible mapping in a space of curvature no greater than K", *Sib. Math. Zhur.* **9** (1968), 918–927.

[Shiohama 1993] K. Shiohama, *An introduction to the geometry of Alexandrov spaces*, Lecture Notes Series **8**, Seoul National University, 1993.

YUKIO OTSU
UNIVERSITY OF TOKYO
DEPARTMENT OF MATHEMATICAL SCIENCES
HONGO, TOKYO 113
JAPAN

Comparison Geometry
MSRI Publications
Volume 30, 1997

Collapsing with No Proper Extremal Subsets

G. PERELMAN

ABSTRACT. This is a technical paper devoted to the investigation of collapsing of Alexandrov spaces with lower curvature bound. In a previous paper, the author defined a canonical stratification of an Alexandrov space by the so-called extremal subsets. It is likely that if the limit of a collapsing sequence has no proper extremal subsets, then the collapsing spaces are fiber bundles over the limit space. In this paper a weaker statement is proved, namely, that the homotopy groups of those spaces are related by the Serre exact sequence. A restriction on the ideal boundary of open Riemannian manifolds of nonnegative sectional curvature is obtained as a corollary.

We assume familiarity with the basic notions and results about Alexandrov spaces with curvature bounded below [Burago et al. 1992; Perelman 1994; Perelman and Petrunin 1994], and with earlier results on collapsing with lower curvature bound [Yamaguchi 1991]. For motivation for the collapsing problem, see [Cheeger et al. 1992] and references therein.

1. Background

Notation. Throughout the paper we denote by M a fixed m-dimensional compact Alexandrov space with curvature $\geq k$; by N a variable n-dimensional compact Alexandrov space with curvature $\geq k$, where $n > m$; by $\Phi : M \to N$ a ν-approximation. For $p \in M$ we denote by $\bar{p}$ the image $\Phi(p)$. We set

$$I_a^l(v) = \{x \in \mathbb{R}^l : |x_i - v_i| < a \text{ for all } i\}.$$

For a map $f : M \to \mathbb{R}^{l_1}$ and $1 \leq l \leq l_1$ we denote by $f_{[l]}$ a map from M to $\mathbb{R}^l$ whose coordinate functions are the first l coordinate functions of f.

Admissible functions, maps, and regular points. A function $f : M \to \mathbb{R}$ is called *admissible* if $f(x) = \sum_\alpha \phi_\alpha(\text{dist}_{q_\alpha}(x))$, where $q_\alpha \in M$, the ϕ_α are smooth, increasing, concave functions, and the set of indices α is finite.

This paper was prepared while the author was a Miller fellow at the University of California at Berkeley.

A map $\hat{f} : M \to \mathbb{R}^l$ is said to be *admissible* in a domain $U \subset M$ if it can be represented there as $\hat{f} = H \circ f$, where H is a homeomorphism of $\mathbb{R}^l$ and the coordinate functions f_i of f are admissible.

A map $\hat{f} = H \circ f$ admissible in U is *regular* at $p \in U$ if the coordinate functions $f_i = \sum_{\alpha_i} \phi_{\alpha_i} \circ \mathrm{dist}_{q_{\alpha_i}}$ satisfy

(1) $\sum_{\alpha_i, \alpha_j} \phi'_{\alpha_i}(\mathrm{dist}_{q_{\alpha_i}}(p)) \, \phi'_{\alpha_j}(\mathrm{dist}_{q_{\alpha_j}}(p)) \cos \tilde{\angle} q_{\alpha_i} p q_{\alpha_j} < 0$ for any $i \neq j$, and
(2) there exists $\xi \in \Sigma_p$ such that $f'_i(\xi) > 0$ for all i.

(Note that $\hat{f}$ has no regular points if $l > m$.)

An admissible map $\hat{f} = H \circ f$ is said to be regular in a domain $V \subset U$ if it is regular at all points $p \in V$.

Conditions (1) and (2) are slightly more restrictive than those given in the definitions in [Perelman 1994]; however, it is easy to check that the arguments of [Perelman 1994] go through with this modified definition. The reason for this modification is that our new definition is stable: if f_i satisfies (1) and (2) then $\bar{f}_i = \sum_{\alpha_i} \phi_{\alpha_i}(\mathrm{dist}_{\bar{q}_{\alpha_i}}(x))$ satisfies the same conditions at $\bar{p} \in N$, provided that $\nu > 0$ is small enough. Therefore if $\hat{f} = H \circ f$ is regular at p then $\bar{\hat{f}} = H \circ \bar{f}$ is regular at $\bar{p}$. Now if $\hat{f} = H \circ f$ is regular in U and, for some subset $\tilde{K} \subset \mathbb{R}^l$, the set $K = \hat{f}^{-1}(\tilde{K}) \cap U$ has compact closure in U, then, assuming $\nu > 0$ to be small enough, we can unambiguously define $\bar{K} \subset N$ as a union of those components of $\bar{\hat{f}}^{-1}(\tilde{K})$ that are Hausdorff-close to $\Phi(K)$.

Canonical neighborhoods. Let $\hat{f} = H \circ f : M \to \mathbb{R}^{l+1}$ be admissible in U; let $p \in U$, and $a > 0$. Suppose that

(i) $\hat{f}_{l+1}(U) \subset (-\infty, 0]$ and $\hat{f}_{l+1}(p) = 0$;
(ii) the coordinate functions of $f_{[l]}$ satisfy conditions (1) and (2) in U, and the coordinate functions of f satisfy the same conditions in $U \backslash \hat{f}_{l+1}^{-1}(0)$;
(iii) the set $K_p(a)$ of points $x \in U$ such that $|\hat{f}_i(x) - \hat{f}_i(p)| < a$ for $i = 1, \ldots l+1$ has compact closure in U; and
(iv) $\hat{f}$ is one-to-one on $K_p(a) \cap \hat{f}_{l+1}^{-1}(0)$.

Then $K_p(a)$ is called a canonical neighborhood of p with respect to $\hat{f}$. (In particular we allow $l = m$ and $\hat{f}_{l+1} \equiv 0$.)

It has been proved in [Perelman 1994] that if $f : M \to \mathbb{R}^l$ is regular at p, then f is an open map near p and there exists a canonical neighborhood of p with respect to some map $g : M \to \mathbb{R}^{l_1}$ with $l_1 > l$, such that $g_{[l]} = f$.

Regular fibers and fiber data. Let $f : M \to \mathbb{R}^m$ be regular at p. Then f is a local homeomorphism near p. If $\nu > 0$ is small enough, $\bar{f}$ is regular in a neighborhood of $\bar{p}$ of size independent of N. Therefore, according to [Perelman 1994, Main Theorem (B)], we have a trivial bundle $\bar{f} : \bar{f}^{-1}(I_a^m(f(p))) \to I_a^m(f(p))$ for small $a \gg \nu$. In this case the fiber $\bar{f}^{-1}(f(p))$ is called regular and the pair (f, p) is called *fiber data*. It is easy to see that a regular fiber is connected.

It is easy to see that if (f_1, p) and (f_2, p) are fiber data then the corresponding regular fibers F_1, F_2 are homotopy equivalent, provided that $\nu > 0$ is small enough. Since a point $p \in M$ is a part of fiber data if and only if Σ_p contains $m + 1$ directions making obtuse angles with each other, and since the set of such points is open, dense and convex (convexity follows from Petrunin's result on parallel translation), we can conclude that the statement about homotopy equivalence is valid for arbitrary fiber data (f_1, p_1) and (f_2, p_2).

If N is a smooth manifold, a regular fiber is a topological manifold. In general one can check that a regular fiber has $\mathbb{Z}/2\mathbb{Z}$-fundamental class in $(n - m)$-dimensional singular homology (relative to the boundary if $\partial N \neq \varnothing$).

Homotopy groups. Fix an integer $l \geq 0$. Let (f, p) be fiber data and let $F = \bar{f}^{-1}(f(p))$ be the corresponding regular fiber. We can assume that $\bar{p} \in F$. We can try to define a homomorphism $\pi_l(N, F, \bar{p}) \to \pi_l(M, p)$ in the following way. Given a spheroid in $\pi_l(N, F, \bar{p})$, consider its fine triangulation and for each vertex x_α find a point $y_\alpha \in M$ so that $|x_\alpha \Phi(y_\alpha)| \leq \nu$; now span the corresponding spheroid in $\pi_l(M, p)$ using vertices y_α. The local geometric contractibility of M implies that this procedure gives a correctly defined homomorphism provided that $\nu > 0$ is small enough.

The main result of this paper is that the constructed homomorphism has an inverse, namely the lifting map constructed after Proposition 2.4 below, provided that M has no proper extremal subsets.

2. The Lifting Map

A point $p \in M$ is called *good* if it satisfies the following condition.

CONDITION PN. For any $R > 0$ there exists a number $\rho = \rho(p, R) > 0$ such that, for any fiber data (f, q) with $q \in B_p(\rho)$, one can find $\bar{\nu} = \bar{\nu}(p, R, f, q) > 0$ so that, if $\nu \leq \bar{\nu}$, then N contains a *product neighborhood*, that is, a domain U with $B_{\bar{p}}(\rho) \subset U \subset B_{\bar{p}}(R)$ such that the inclusion $\bar{f}^{-1}(f(q)) \hookrightarrow U$ induces isomorphisms of homotopy and homology groups.

Let NPN denote the set of all bad points of M.

PROPOSITION 2.1. *The closure of NPN, if nonempty, is a proper extremal subset of M.*

Observe that the closure of NPN is not all of M, since p is definitely good if Σ_p contains $m+1$ directions making obtuse angles with each other. The extremality of $\mathrm{clos}(NPN)$ follows from two lemmas.

LEMMA 2.2. *Let K be a canonical neighborhood with respect to $f : M \to \mathbb{R}^{l+1}$. Assume that $p, q \in K \cap f_{l+1}^{-1}(0)$, and that p is good. Then q is also good.*

PROOF. Fix $R > 0$. Find $a > 0$ so that $K_p(3a) \subset K$ and $K_q(3a) \subset B_q(R) \cap K$. Take $r > 0$ such that $B_p(r) \subset K_p(a)$. Choose $d > 0$ so that $K_p(3d) \subset B_p(\rho(p,r))$. Finally, define $\rho(q, R) > 0$ so that $B_q(\rho(q, R)) \subset K_q(d)$.

To check Condition PN, let $L = K \cap f_{l+1}^{-1}(-d, 0]$ and observe that, according to [Perelman 1994, Main Theorem (A)], the map $\bar{f}_{[l]}$ is a (topological) submersion in $\bar{K}$, whereas $\bar{f}$ is a submersion in $\bar{K} \setminus \bar{L}$, if $\nu > 0$ is small enough. Therefore, applying the results of [Siebenmann 1972], we can construct a homeomorphism $\psi : F \times I^l \to \bar{K}$ that respects $\bar{f}_{[l]}$ and, in addition, respects $\bar{f}_{l+1}$ over $\bar{K} \setminus \bar{L}$. Thus we can use the homeomorphism $\theta = \psi \circ \text{transl}(\overline{f_{[l]}(p) f_{[l]}(q)}) \circ \psi^{-1}$ to transfer $\bar{K}_p(\rho)$ to $\bar{K}_q(\rho)$ for $d \leq \rho \leq 3a$, and it is easy to see that if U is a product neighborhood, $\bar{K}_p(2d) \subset U \subset \bar{K}_p(2a)$, then $\theta(U)$ is also a product neighborhood that satisfies Condition PN with respect to given q, R and chosen $\rho(q, R)$. □

LEMMA 2.3. *Let* $f : M \to \mathbb{R}^l$ *be regular in a neighborhood* U *of* p, *and let* $L = \bigcap_{i>1} f_i^{-1}(f_i(p))$. *Then* f_1 *restricted to* $\text{clos}(NPN \cap L) \cap U$ *cannot attain its minimum.*

PROOF. We use reverse induction on l. If $l = m$, our assertion is clear since all points in U are good in this case. Assume that $l < m$. If the minimum is attained, we can assume that it is attained at p. Consider a canonical neighborhood $K_p(a) \subset V \subset U$ with respect to a map $g : M \to \mathbb{R}^{l_1}$, where $l_1 > l$, such that $g_{[l]} = f$ in V. Since p is the point of minimum, we have $K_p(a) \cap L \cap g_{l_1}^{-1}(0) \not\subset NPN$. Therefore Lemma 2.2 implies that $K_p(a) \cap NPN \cap g_{l_1}^{-1}(0) = \varnothing$. On the other hand, since $p \in \text{clos}(L \cap NPN)$, we can find $v \in \mathbb{R}^{l_1}$ such that $v_i = f_i(p)$ for $1 < i \leq l$, $|v_i - f_i(p)| < a$ for $l + 1 \leq i \leq l_1$, $v_{l_1} \neq 0$, and $K_p(a) \cap NPN \cap L_1 \neq \varnothing$, where $L_1 = \bigcap_{1 < i \leq l_1} g_i^{-1}(v_i) \subset L$. Let $U_1 = K_p(a) \setminus g_{l_1}^{-1}(0)$, and $p_1 \in U_1 \cap L_1$. Then g is regular in $U_1 \ni p_1$, and $g_1(= f_1)$ restricted to $\text{clos}(NPN \cap L_1) \cap U_1$ attains its minimum value because $f_1(p) \leq f_1(x) < f_1(p) + a$ for all $x \in \text{clos}(NPN \cap L_1) \cap U_1$, whereas $f_1(L_1 \cap \partial U_1) \subset \{f_1(p) + a, f_1(p) - a\}$. This proves the induction step. □

The extremality of $\text{clos}(NPN)$ follows immediately from the case $l = 1$ of the lemma above and the definition.

Now assume that M has no proper extremal subsets. According to Proposition 2.1, all points of M are good. Moreover, the compactness of M implies that we can define a function $\rho(R)$ satisfying Condition PN and independent of p. (Choose a finite covering of M by balls $B(p_\alpha, \rho(p_\alpha, R_\alpha/10)/10)$ and let $\rho(R) = \min_\alpha \rho(p_\alpha, R/10)/10$).

Fix a positive integer l and fiber data (f_0, p_0). Choose $R_{2l+3} > 0$ so small that f_0 is regular in a R_{2l+3}-neighborhood of p_0 and for any finite simplicial complex K of dimension at most l and its subcomplex L, any two maps of (K, L) into (M, p_0) that are uniformly R_{2l+3}-close are homotopic relative to L. Define $R_i > 0$ for $0 \leq i \leq 2l + 2$ inductively, in such a way that $\rho(R_{i+1}/10) \geq 10R_i$ for all i. Take a finite family of fiber data (f_α, q_α) such that q_α form an R_0-net in M. Repeating the compactness argument, we can choose a universal $\bar{\nu}$, with

$0 < \bar{\nu} < R_0$, so that Condition PN is satisfied for all $p \in M$, $R = R_i$, $f = f_\alpha$, $q = q_\alpha$. From now on we assume $\nu \leq \bar{\nu}$.

PROPOSITION 2.4. *Let K be a finite simplicial complex of dimension $\leq l$. Suppose $\phi : K \times I \to M$ and $\bar{\phi} : K \to N$ satisfy $|\bar{\phi}(x)\Phi \circ \phi(x)| < R_{i+1}$ for $x \in$ skel$_i K$, for $0 \leq i \leq l$, and that $\mathrm{diam}\phi(\Delta) < R_0$ for all simplices $\Delta \subset K$. Then $\bar{\phi}$ can be extended to a map from $K \times I$ to N such that $|\bar{\phi}(x)\Phi \circ \phi(x)| < R_{i+1}$ for $x \in$ skel$_i(K \times I)$, for $0 \leq i \leq l+1$.*

PROOF. A standard argument reduces our extension problem to the case when $K = \Delta^i$, $0 \leq i \leq l$, $\mathrm{diam}(\phi(\Delta^i \times I)) < R_0$, and $\bar{\phi}$ has already been defined on $\Delta^i \times \{0\} \cup \partial\Delta^i \times I$. Take any $p \in \phi(\Delta^i \times I)$ and let U, V be the product neighborhoods such that $B_{\bar{p}}(10R_{i+1}) \subset U \subset B_{\bar{p}}(R_{i+2}/10)$ and $B_{\bar{p}}(10R_i) \subset V \subset B_{\bar{p}}(R_{i+1}/10)$. Clearly $\bar{\phi}(\Delta^i \times \{0\} \cup \partial\Delta^i \times I) \subset U$ and $\bar{\phi}(\partial\Delta^i \times \{1\}) \subset V$. Since there exists a regular fiber F such that $F \hookrightarrow V$ and $F \hookrightarrow U$ induce isomorphisms of homotopy groups, we conclude that $V \hookrightarrow U$ has the same property; in particular, $\pi_{i+1}(U, V) = 0$. Therefore we can easily extend $\bar{\phi}$ so that $\bar{\phi}(\Delta^i \times I) \subset U$ and $\bar{\phi}(\Delta^i \times \{1\}) \subset V$. $\square$

Now we can define a lifting map $\pi_l(M, p_0) \to \pi_l(N, F_0, \bar{p}_0)$: Given a spheroid ϕ in $\pi_l(M, p_0)$, let $\bar{\phi}$ be its image if $|\bar{\phi}(x)\Phi \circ \phi(x)| < R_{l+1}$ for all x. Existence of such $\bar{\phi}$ follows from Proposition 2.4 and the fact that the inclusion of F_0 into its appropriate neighborhood of size R_{2l+3} is a homotopy equivalence; correctness follows from Proposition 2.4 with R_{l+i+1} substituted for R_{i+1} in the assumption and the conclusion. It is clear that this lifting homomorphism is an inverse of the one described at the end of Section 1.

3. Corollaries

COROLLARY 3.1. *Let N be a complete noncompact Riemannian manifold of nonnegative sectional curvature that does not admit isometric splitting and is not diffeomorphic to $\mathbb{R}^n$. Then its asymptotic cone M has proper extremal subsets. In particular, the radius of its ideal boundary is at most $\pi/2$.*

REMARK. The last assertion was conjectured by Shioya [1993]. Recently Sérgio Mendonça independently obtained a direct proof of the same result.

PROOF OF COROLLARY 3.1. The manifold N with rescaled metrics collapses to M. If M has no proper extremal subsets, then, in particular, its apex is a good point. Therefore we can easily construct neighborhoods $U_1 \supset V \supset U_2 \supset S$ of the soul S, such that U_1, U_2 are product neighborhoods, whereas V is a convex neighborhood. Thus S must have the homology groups of the regular fiber, whence $\dim S = n - m$. Now an easy packing argument shows that the normal bundle of S has finite holonomy, which is therefore a quotient of the fundamental group of S and N. Thus a finite cover $\tilde{N}$ of N has normal bundle with trivial holonomy, and, according to [Strake 1988], $\tilde{N}$ splits isometrically. It follows that

M is the quotient of the asymptotic cone of $\tilde{N}$ by an isometric action of a finite group, which fixes the apex. Thus, according to [Perelman and Petrunin 1994, § 4.2], M has proper extremal subsets, unless the group action is trivial and $\tilde{N} = N$.

The second statement of the corollary follows from [Perelman and Petrunin 1994, Proposition 1.4.1]. □

COROLLARY 3.2. *If Σ^{m-1} is a limit of a collapsing sequence of compact $(n-1)$-dimensional Alexandrov spaces with curvature ≥ 1, then either the diameter of Σ^{m-1} is at most $\pi/2$ or Σ^{m-1} has proper extremal subsets.*

In particular, according to [Perelman and Petrunin 1994, Proposition 1.4.1], the radius of Σ^{m-1} cannot exceed $\pi/2$. This conclusion was obtained earlier by Petrunin (in his unpublished Master's Thesis), and it also follows immediately from [Grove and Petersen 1993, Theorem 3(3)].

PROOF. Consider the collapsing of the corresponding cones. If the conclusion is false, then $M = \text{cone}(\Sigma^{m-1})$ has no proper extremal subsets; in particular, its apex is a good point. Thus we can construct neighborhoods $U_1 \supset V \supset U_2$ in the collapsing cone, such that U_1, U_2 are product neighborhoods, whereas V is a spherical neighborhood of the apex. Therefore the inclusion $U_2 \hookrightarrow U_1$ factors through a contractible space—a contradiction. □

References

[Burago et al. 1992] Y. Burago, M. Gromov, and G. Perelman, "A. D. Aleksandrov spaces with curvatures bounded below", *Uspehi Mat. Nauk* **47**:2 (1992), 3–51, 222; translation in *Russian Math. Surveys* **47**:2 (1992), 1–58.

[Cheeger et al. 1992] J. Cheeger, K. Fukaya, and M. Gromov, "Nilpotent structures and invariant metrics on collapsed manifolds", *J. Amer. Math. Soc.* **5** (1992), 327–372.

[Grove and Petersen 1993] K. Grove and P. Petersen, "A radius sphere theorem", *Invent. Math.* **112** (1993), 577–583.

[Perelman 1994] G. Perelman, "Elements of Morse theory on Alexandrov spaces" (Russian), *Algebra i analiz* **5**:1 (1993), 232–241; translation in *St. Petersburg Math. J.* **5**:1 (1994), 207–214.

[Perelman and Petrunin 1994] G. Perelman and A. Petrunin, "Extremal subsets in Alexandrov spaces and the generalized Liberman theorem", *St. Petersburg Math. J.* **5**/1 (1994), 215–227.

[Siebenmann 1972] L. C. Siebenmann, "Deformation of homeomorphisms on stratified sets", *Comment. Math. Helv.* **47** (1972), 123–163.

[Shioya 1993] T. Shioya, "Splitting theorems for nonnegatively curved open manifolds with large ideal boundary", *Math. Zeit.* **212** (1993), 223–238.

[Strake 1988] M. Strake, "A splitting theorem for open nonnegatively curved manifolds", *Manuscripta Math.* **61** (1988), 315–325.

[Yamaguchi 1991] T. Yamaguchi, "Collapsing and pinching under a lower curvature bound", *Ann. of Math.* (2) **113** (1991), 317–357.

G. PERELMAN
ST. PETERSBURG BRANCH
V. A. STEKLOV INSTITUTE OF MATHEMATICS (POMI)
RUSSIAN ACADEMY OF SCIENCES
FONTANKA 27
ST. PETERSBURG, 191011
RUSSIA
  perelman@pdmi.ras.ru

Comparison Geometry
MSRI Publications
Volume 30, 1997

Construction of Manifolds of
Positive Ricci Curvature with
Big Volume and Large Betti Numbers

G. PERELMAN

ABSTRACT. It is shown that a connected sum of an arbitrary number of
complex projective planes carries a metric of positive Ricci curvature with
diameter one and, in contrast with the earlier examples of Sha–Yang and
Anderson, with volume bounded away from zero. The key step is to construct complete metrics of positive Ricci curvature on the punctured complex projective plane, which have uniform euclidean volume growth and
almost contain a line, thus showing topological instability of the splitting
theorem of Cheeger–Gromoll, even in the presence of the lower volume
bound. In the absence of such a bound, the topological instability was
earlier shown by Anderson; metric stability holds, even without the volume
bound, by the recent work of Colding–Cheeger.

1. Outline

We start from a singular space of positive curvature, namely the double spherical suspension of a small round two-sphere. The size of that sphere can be
estimated explicitly and is fixed in our construction. The singular points of our
space fill a circle of length 2π. We smooth our space near the singular circle
in a symmetric way. Then we remove a collection of disjoint small metric balls
centered at the former singular points, and glue in our "building blocks" instead.
A building block is a metric on $\mathbb{C}P^2 \setminus$ ball, having positive Ricci curvature and
strictly convex boundary. We arrange that the boundary of the building block
is isometric to the boundary of the removed ball and is "more convex". This
allows us to smooth the resulting space to get a manifold of positive Ricci curvature. Its volume is close to the volume of the double suspension we started
from, whereas its second Betti number equals the number of building blocks it

This paper was prepared while the author was a Miller fellow at the University of California
at Berkeley.

contains. This number can be made arbitrarily large by localizing the initial smoothing of the double suspension to smaller neighborhoods of the singular circle, and by removing a larger number of smaller balls.

The main difficulty is to construct building blocks. This is carried out in Section 2. In Section 3 we construct the ambient space, i.e., we make precise the relation between the number and the size of balls to be removed and the smoothing of our double suspension. In Section 4 we explain how to smooth the result of gluing two manifolds with boundaries, so as to retain positive Ricci curvature.

Notation. $K(X \wedge Y)$ denotes the sectional curvature in the plane spanned by X, Y. We abbreviate $\operatorname{Ric}(X,Y)/|X||Y|$ as $\operatorname{Ri}(X,Y)$.

2. Construction of the Building Block

Our building block is glued from two pieces: the core and the neck. The core is a metric on $\mathbb{C}P^2 \setminus$ ball with positive Ricci curvature and strictly convex boundary; moreover, the boundary is isometric to a round sphere. The neck is a metric on $S^3 \times [0,1]$ with positive Ricci curvature, such that the boundary component $S^3 \times \{0\}$ is concave and is isometric to a round sphere, whereas the boundary component $S^3 \times \{1\}$ is convex and looks like a lemon, that is, like a smoothed spherical suspension of a small round two-sphere. There are some additional conditions on the normal curvatures of the boundary of the neck; they will be made clear below.

Construction of the core. It is well known that the canonical metric of $\mathbb{C}P^2$ in a neighborhood of $\mathbb{C}P^1$ can be expressed as

$$ds^2 = dt^2 + A^2(t)\,dx^2 + B^2(t)\,dy^2 + C^2(t)\,dz^2,$$

where t is the distance from $\mathbb{C}P^1$ and X, Y, Z is the standard invariant framing of S^3, satisfying $[X,Y] = 2Z$, $[Y,Z] = 2X$, $[Z,X] = 2Y$. The canonical metric has $A = \sin t \cos t$ and $B = C = \cos t$, but we are going to consider general A, B, C.

The curvature tensor in this presentation is listed on page 165 of this volume. Assuming now that $B \equiv C$, we deduce from those formulas that

$$\operatorname{Ri}(X,X) = (-A''/A - 2A'B'/AB + 2A^2/B^4),$$
$$\operatorname{Ri}(Y,Y) = \operatorname{Ric}(Z,Z) = (-B''/B - A'B'/AB - B'^2/B^2 + (4B^2 - 2A^2)/B^4),$$
$$\operatorname{Ri}(T,T) = (-A''/A - 2B''/B),$$

and all off-diagonal entries vanish.

Now let $A = \sin t \cos t$, $B = \frac{1}{100}\cosh(t/100)$. Clearly this choice gives a smooth metric. We have, for $0 < t < \frac{1}{10}$,

$$-A''/A = 4, \quad -B''/B = -10^{-4}, \quad 0 < A'B'/AB < \tfrac{1}{10}, \quad 0 < B'/B < \tfrac{1}{100},$$

and $(4B^2 - 2A^2)/B^4 > 100$ provided that $A \le B$.

Thus if $0 < t_0 < \frac{1}{10}$ is such that $A(t_0) = B(t_0)$, $A(t) < B(t)$ for $t < t_0$, then Ricci curvatures are positive for $0 \le t \le t_0$. Such t_0 exists because $A(0) = 0 < B(0)$, whereas $A(\frac{1}{10}) > \frac{1}{20} > B(\frac{1}{10})$. Therefore the core is constructed.

Construction of the neck. We first prove the following result.

ASSERTION. *Let (S^n, g) be a rotationally symmetric metric of sectional curvature > 1, distance between the poles πR and waist $2\pi r$; that is, g can be expressed as $ds^2 = dt^2 + B^2(t)\, d\sigma^2$, where $d\sigma^2$ is the standard metric of S^{n-1}, $t \in [0, \pi R]$, and $\max_t B(t) = r$. Let $\rho > 0$ be such that $\rho < R$ and $r^{n-1} < \rho^n$. Then there exists a metric of positive Ricci curvature on $S^n \times [0,1]$ such that (a) the boundary component $S^n \times \{1\}$ has intrinsic metric g and is strictly convex; moreover, all its normal curvatures are > 1; (b) the boundary component $S^n \times \{0\}$ is concave, with normal curvatures equal to $-\lambda$, and is isometric to a round sphere of radius $\rho\lambda^{-1}$, for some $\lambda > 0$.*

(Note that, if ρ is small enough, the core constructed earlier in this section can be glued, after rescaling, to the neck along $S^n \times \{0\}$, so that the resulting space can be smoothed with positive Ricci curvature; see Section 4.)

To prove the assertion, we start by rewriting our metric g in the form

$$ds^2 = r^2 \cos^2 x \, d\sigma^2 + A^2(x)\, dx^2,$$

where $-\pi/2 \le x \le \pi/2$ and A is a smooth positive function satisfying $A(\pm\pi/2) = r$, $A'(\pm\pi/2) = 0$. Clearly $A(x) \ge R > r$ for some x, so we can write

$$A(x) = r(1 - \eta(x) + \eta(x) \cdot a_\infty),$$

where $\max_x \eta(x) = 1$ and $a_\infty \ge R/r$, $\eta(\pm\pi/2) = 0$, and $\eta'_x(\pm\pi/2) = 0$.

Consider a metric on $S^n \times [t_0, t_\infty]$ of the form

$$ds^2 = dt^2 + A^2(t,x)\, dx^2 + B^2(t,x)\, d\sigma^2,$$

where $B = tb(t)\cos x$, $A = tb(t)\bigl(1 - \eta(x) + \eta(x)a(t)\bigr)$, $a(t_0) = 1$, $a'(t_0) = 0$, $b(t_0) = \rho$, $b'(t_0) = 0$, $a(t_\infty) = a_\infty > 1$, $b(t_\infty) > r$. That metric will satisfy the conditions of our assertion after rescaling by a multiple $r/(t_\infty \cdot b(t_\infty))$, provided that the functions $a(t)$ and $b(t)$ are chosen appropriately.

The curvatures of our metric can be computed as follows:

$$K(T \wedge X) = -A_{tt}/A$$
$$K(T \wedge \Sigma) = -B_{tt}/B$$
$$K(X \wedge \Sigma) = -A_t B_t/AB + A_x B_x/A^3 B - B_{xx}/BA^2$$
$$K(\Sigma_1 \wedge \Sigma_2) = 1/B^2 - B_x^2/A^2 B^2 - B_t^2/B^2,$$

and the Ricci tensor has only one nonzero off-diagonal term, namely

$$\text{Ri}(T, X) = (n-1)(A_t B_x/A^2 B - B_{xt}/AB).$$

Further computations give

$$\text{Ri}(T,T) = -n\left(\frac{b''}{b} + \frac{2b'}{tb}\right) - \left(\frac{\eta a''}{1-\eta+\eta a} + \frac{2\eta a'}{t(1-\eta+\eta a)} + 2\frac{b'}{b}\frac{\eta a'}{1-\eta+\eta a}\right),$$

$$\text{Ri}(T,X) = -(n-1)\,\text{tg}\,x\,\frac{\eta a'}{tb(1-\eta+\eta a)^2}$$

for the Ricci tensor, and

$$K(X \wedge \Sigma) = \frac{1}{t^2 b^2}\left(\frac{1}{(1-\eta+\eta a)^2} - \frac{\eta_x\,\text{tg}\,x\,(a-1)}{(1-\eta+\eta a)^3}\right)$$

$$- \left(\frac{1}{t} + \frac{b'}{b}\right)\left(\frac{1}{t} + \frac{b'}{b} + \frac{\eta a'}{1-\eta+\eta a}\right),$$

$$K(\Sigma_1 \wedge \Sigma_2) = \frac{1}{t^2 b^2}\left(\frac{1}{\cos^2 x} - \frac{\text{tg}^2 x}{(1-\eta+\eta a)^2}\right) - \left(\frac{1}{t} + \frac{b'}{b}\right)^2$$

for the sectional curvatures. Note that the first terms in the last two formulas are the intrinsic curvatures of the hypersurfaces $t = \text{const}$, $K_i(X \wedge \Sigma)$, $K_i(\Sigma_1 \wedge \Sigma_2)$.

We construct the functions $b(t)$ and $a(t)$ as follows:

$$b'/b = -\beta(t-t_0)/2t_0^2 \log 2t_0 \quad \text{for } t_0 \leq t \leq 2t_0,$$

$$b'/b = -\beta \log 2t_0/t \log^2 t \quad \text{for } t \geq 2t_0 \quad (b \text{ must be smoothed near } t = 2t_0),$$

$$a'/a = -\alpha b'/b \quad \text{for } t \geq t_0.$$

Here $\beta = (1-\varepsilon)(\log \rho - \log r)/(1 + 1/(4 \log 2t_0))$ and $\alpha = (1+\delta)\beta^{-1} \log a_\infty/(1 + 1/(4 \log 2t_0))$, so that $\int_{t_0}^\infty b'/b = (1-\varepsilon)(\log r - \log \rho)$ and $\int_{t_0}^\infty a'/a = (1+\delta)\log a_\infty$; the small positive numbers ε, δ are to be determined later.

At this point we still have freedom in the choice of t_0, ε, δ, and t_∞. Note, however, that t_∞ is determined by δ from the relation $a(t_\infty) = a_\infty$. We choose $\varepsilon > 0$ first in such a way that the metric $(\rho/r)^\varepsilon \cdot g$ still has sectional curvatures > 1. In this situation we prove below that the curvatures $K_i(X \wedge \Sigma)$ and $K_i(\Sigma_1 \wedge \Sigma_2)$ can be estimated from below by c/t^2 for some $c > 1$ independent of t_0, t_∞, and δ. Then we show that the Ricci curvatures of our metric are strictly positive if t_0 is chosen large enough, independently of δ, if δ is small. Finally, we choose δ in such a way that the normal curvatures of $S^n \times \{t_\infty\}$ are sufficiently close to t_∞^{-1}.

First of all we need to check that $1 < \alpha < n$. Indeed, at the maximum point of $\eta(x)$ we have $\text{tg}\,x\,\eta_x = 0$, so $K_g(X \wedge \Sigma) = 1/r^2 a_\infty^2$. Thus $a_\infty < 1/r$ and $\log a_\infty < n(\log \rho - \log r)$ since $r^{n-1} < \rho^n$. On the other hand $a_\infty \geq R/r > \rho/r$, so $\log a_\infty > \log \rho - \log r$.

Now we are in a position to check that $K_i(X \wedge \Sigma) \geq c/t^2$ and $K_i(\Sigma_1 \wedge \Sigma_2) \geq c/t^2$ for some $c > 1$. Let $\psi = \log(t^2 K_i(\Sigma_1 \wedge \Sigma_2))$. We know that $\psi|_{t=t_0} > 0$ and

$\psi|_{t=t_\infty} > 0$. A computation gives

$$\psi_t = -\frac{2b'}{b}\left(1 + \frac{\alpha\eta a \sin^2 x}{(1 - \eta + \eta a)((1 - \eta + \eta a)^2 - \sin^2 x)}\right).$$

This expression is positive if $\eta \geq 0$, and it is decreasing in a if $\eta < 0$. Therefore ψ cannot have a local minimum on (t_0, t_∞) for any fixed x, so $\psi > 0$ as required.

Now let $\phi = \log(t^2 K_i(X \wedge \Sigma))$. Again consider the behavior of this function for $t \in [t_0, t_\infty]$ and fixed x. Suppose that $\operatorname{tg} x \, \eta_x \geq 0$. A computation gives

$$\phi_t = -\frac{b'}{b}\frac{1}{1 - \eta + \eta a}\left[2(1 - \eta) + 2\eta a(1 - \alpha) - \frac{\alpha a \operatorname{tg} x \, \eta_x}{1 - \eta + \operatorname{tg} x \, \eta_x + a(\eta - \operatorname{tg} x \, \eta_x)}\right]$$

$$= -\frac{b'}{b}\left(2 - \frac{\alpha a}{1 - \eta + \eta a}\left[2\eta + \frac{\operatorname{tg} x \, \eta_x}{1 - \eta + \operatorname{tg} x \, \eta_x + a(\eta - \operatorname{tg} x \, \eta_x)}\right]\right).$$

If $\eta \geq 0$ the first expression in large brackets is clearly decreasing in a, whereas if $\eta < 0$ the second expression in brackets, and also $\alpha a/(1 - \eta + \eta a)$, are clearly increasing in a, so ϕ cannot have a local minimum in (t_0, t_∞), and therefore $\phi > 0$ since $\phi|_{t=t_0} > 0$ and $\phi|_{t=t_\infty} > 0$. Now suppose that $\operatorname{tg} x \, \eta_x < 0$. Then $\phi > -2 \log ab > 0$, because $a(t_\infty)b(t_\infty) < 1$ and $\log(ab)$ is increasing in t (note that $\log(ab)' = -(b'/b)(\alpha - 1) > 0$). The estimate for $K_i(X \wedge \Sigma)$ and $K_i(\Sigma_1 \wedge \Sigma_2)$ is proved.

Now we can estimate the Ricci curvatures of our metric. Note that the normal curvatures $1/t + b'/b$ and $1/t + b'/b + \eta a'/(1 - \eta + \eta a)$ can be estimated by $1/t + O(1/t \log t)$. It follows that $K(X \wedge \Sigma) \geq c/t^2$ and $K(\Sigma_1 \wedge \Sigma_2) \geq c/t^2$, for $c > 0$. It is also clear that $|\operatorname{Ri}(T, X)|$, $|K(X \wedge T)|$, and $|K(\Sigma \wedge T)|$ are all bounded above by $c \log t_0/(t^2 \log^2 t)$; hence, in particular, $\operatorname{Ri}(X, X) \geq c/t^2$ and $\operatorname{Ri}(\Sigma, \Sigma) \geq c/t^2$, with $c > 0$.

To estimate $\operatorname{Ri}(T, T)$, write it as

$$\operatorname{Ri}(T, T) = \left(\frac{\alpha\eta a}{1 - \eta + \eta a} - n\right)\left((b'/b)' + \frac{2}{t}\frac{b'}{b}\right)$$

$$- n(b'/b)^2 - \frac{\eta a}{1 - \eta + \eta a}\left(2\frac{b'}{b}\frac{a'}{a} + \left(\frac{a'}{a}\right)^2\right).$$

Note that, since $\eta \leq 1$ and $\alpha < n$, the first multiple is negative and bounded away from 0. Therefore, the first term is $\geq c \log t_0/(t^2 \log^2 t)$, for $c > 0$. The remaining terms are of order $c \log^2 t_0/(t^2 \log^4 t)$, so we get

$$\operatorname{Ri}(T, T) \geq \frac{c \log t_0}{t^2 \log^2 t}.$$

Since $|\operatorname{Ri}(T, X)| \leq c\operatorname{Ri}(T, T) \ll \operatorname{Ri}(X, X)$, we conclude that Ricci curvature is positive, if t_0 was chosen large enough.

To complete the construction it remains to choose $\delta > 0$ so small, and correspondingly t_∞ so large, that normal curvatures of $S^n \times \{t_\infty\}$ are $> (r/\rho)^\varepsilon t_\infty^{-1}$; this is possible since they are estimated by $1/t + O(1/t \log t)$.

3. Construction of the Ambient Space

The metric of the double suspension of a sphere of radius R_0 near singular circle can be written as

$$ds^2 = dt^2 + \cos^2 t\, dx^2 + R^2(t)\, d\sigma^2,$$

where $t \geq 0$ is the distance from the singular circle, $R(t) = R_0 \sin t$. We smooth this metric by modifying $R(t)$ in such a way that $R(0) = 0$, $R'(0) = 1$, and $R''(0) = 0$. The curvatures of this metric are easily computed to be $K(T \wedge X) = 1$, $K(T \wedge \Sigma) = -R''/R$, $K(X \wedge \Sigma) = (R'/R)\,\mathrm{tg}\,t$, $K(\Sigma_1 \wedge \Sigma_2) = (1 - R'^2)/R^2$, and all mixed curvatures in this basis vanish. We will choose $R(t)$ in such a way that $1 > R' > 0$ and $R'' < 0$ when $t > 0$, thus making all sectional curvatures positive.

Consider a metric ball of small radius r_0 centered on the axis $t = 0$. We assume that $-R'' \geq R$ for $0 \leq t \leq r_0$, so that $K(T \wedge \Sigma) \geq 1$. In this case the normal curvatures of the boundary of this ball do not exceed $\mathrm{ctg}\,r_0$. We'll show that the function $R(t)$ can be chosen so that the intrinsic curvatures of the boundary are strictly bigger than $\mathrm{ctg}^2 r_0$. It would follow that for sufficiently small $R_0 > 0$, we can construct a building block that, after rescaling by $\mathrm{tg}\,r_0$, can be glued (using the results of Section 4) in our ambient space instead of the metric ball.

At a point of the boundary of our metric ball, which is at a distance t from the axis, we can compute the intrinsic curvatures as follows:

$$K_i(\Sigma_1 \wedge \Sigma_2) = (1/R^2)(1 - R'^2 \sin^2 \phi),$$

$$K_i(Y \wedge \Sigma) = -\frac{R''}{R}\sin^2 \phi + \frac{R'}{R}\,\mathrm{tg}\,t\cos^2 \phi + \mathrm{ctg}\,r_0\frac{R'}{R}\cos \phi,$$

where $Y \in X \wedge T$ is a tangent vector to the boundary, ϕ is the angle between T and the normal vector, $\cos \phi = \mathrm{ctg}\,r_0\,\mathrm{tg}\,t$.

Since $K(T \wedge \Sigma) \geq 1$, it follows from the comparison theorem that

$$K_i(\Sigma_1 \wedge \Sigma_2) \geq 1/\sin^2 t - \mathrm{ctg}^2 t(1 - \cos^2 \phi) = 1 + \mathrm{ctg}^2 r_0 > \mathrm{ctg}^2 r_0.$$

To make sure that $K_i(Y \wedge \Sigma) > \mathrm{ctg}^2 r_0$, we have to choose $R(t)$ more carefully. Namely, let $R(t) = R_0 \sin\big(t + \delta\gamma(t/r_0 - 1)\big)$ for $t \geq \varepsilon$, where γ is a standard smooth function interpolating between $\gamma(x) = 1$ for $x \leq 0$ and $\gamma(x) = 0$ for $x \geq 1$. Extend $R(t)$ to the segment $[0, \varepsilon]$ in such a way that $R(0) = 0$, $R'(0) = 1$, and $-R''/R \geq 2r_0^{-2}$ on $[0, \varepsilon]$. (This is possible with an appropriate choice of ε and δ—for example, if $\varepsilon = \frac{1}{2}r_0^2$ and $\delta = \frac{1}{4}r_0^4$.) Now it is clear that already the first term in the expression for $K_i(Y \wedge \Sigma)$ is bigger than $\mathrm{ctg}^2 r_0$ if $0 \leq t \leq \varepsilon$. For $\varepsilon \leq t \leq r_0$ we have

$$K_i(Y \wedge \Sigma) = \sin^2 \phi + \mathrm{ctg}(t + \delta)\,\mathrm{tg}\,t\cos^2 \phi + \mathrm{ctg}^2 r_0\,\mathrm{ctg}(t + \delta)\,\mathrm{tg}\,t$$

$$\geq \mathrm{ctg}(t + \delta)\,\mathrm{tg}\,t\,(\mathrm{ctg}^2 r_0 + 1).$$

So we need to check that $\operatorname{tg}(t+\delta)/\operatorname{tg} t < 1 + \operatorname{tg}^2 r_0$. This is true for $t \geq \varepsilon = \frac{1}{2} r_0^2$ and $\delta = \frac{1}{4} r_0^4$ and for small r_0. Thus, both $K_i(\Sigma_1 \wedge \Sigma_2)$ and $K_i(X \wedge \Sigma)$ are bigger than $\operatorname{ctg}^2 r_0$, and the construction of the ambient space is complete.

4. Gluing and Smoothing

To justify gluing and smoothing in our construction we need the following fact.

Let M_1, M_2 be compact smooth manifolds of positive Ricci curvature, with isometric boundaries $\partial M_1 \simeq \partial M_2 = X$. Suppose that the normal curvatures of ∂M_1 are bigger than the negatives of the corresponding normal curvatures of ∂M_2. Then the result $M_1 \cup_X M_2$ of gluing M_1 and M_2 can be smoothed near X to produce a manifold of positive Ricci curvature.

To prove this, express the metric of $M_1 \cup_X M_2$ in normal coordinates with respect to X; let t be the normal coordinate. Introduce a small parameter $\tau > 0$, and for arbitrary coordinate vectors X_1, X_2 tangent to X replace our given function $\langle X_1, X_2 \rangle(x, t)$ by its interpolation on the segment $[-\tau, \tau]$. At first we can take a C^1 interpolation given by a cubic polynomial in t, whose coefficients are linear functions of $\langle X_1, X_2 \rangle(x, \pm\tau)$ and $\langle X_1, X_2 \rangle'_t(x, \pm\tau)$. Clearly this procedure is independent of the choice of coordinates in X, and it gives a C^1 metric, which is C^2 outside two hypersurfaces $X_\tau, X_{-\tau}$, corresponding to $t = \pm\tau$. It is easy to see that in the segment $[-\tau, \tau]$ we get all $K(T \wedge X)$ positive of order $c\tau^{-1}$ and all $K(X_1 \wedge X_2)$ and $\operatorname{Ri}(T, X)$ bounded, so the Ricci curvature is positive. Now we can use a similar procedure to smooth our manifold near hypersurfaces $t = \pm\tau$. This time we choose another $\tau' \ll \tau$ and construct a C^2-interpolation. It is clear that only the components $R(T, \cdot, \cdot, T)$ of the curvature tensor were discontinuous on, say, X_τ, and, up to an error of order τ', these components now interpolate linearly between their original values on the different sides of X_τ. Since positivity of Ricci curvature is open and convex condition, the smoothed manifold has positive Ricci curvatures.

G. Perelman
St. Petersburg Branch
V. A. Steklov Institute of Mathematics (POMI)
Russian Academy of Sciences
Fontanka 27
St. Petersburg, 191011
Russia
  perelman@pdmi.ras.ru

Comparison Geometry
MSRI Publications
Volume 30, 1997

A Complete Riemannian Manifold of Positive Ricci Curvature with Euclidean Volume Growth and Nonunique Asymptotic Cone

G. PERELMAN

Consider the metric $ds^2 = dt^2 + A^2(t)\,dx^2 + B^2(t)\,dy^2 + C^2(t)\,dz^2$, where t is the radial coordinate and x, y, z are "spherical coordinates" with $[X, T] = [Y, T] = [Z, T] = 0$, $[X, Y] = 2Z$, $[Y, Z] = 2X$, and $[Z, X] = 2Y$. (Taking $A(t) = B(t) = C(t) = t$ we get the standard Euclidean metric.) A straightforward computation gives

$$\langle R(X, T)T, X \rangle = -\frac{A''}{A}\|X\|^2\|T\|^2,$$

$$\langle R(X, Y)Y, X \rangle = \|X\|^2\|Y\|^2$$
$$\times \left(-\frac{A'B'}{AB} + \frac{1}{A^2B^2C^2}(A^4 + B^4 - 3C^4 + 2A^2C^2 + 2B^2C^2 - 2A^2B^2) \right)$$

and similar equalities obtained by permutation of the pairs (X, A), (Y, B), (Z, C); similarly

$$\langle R(X, Y)Z, T \rangle = \|X\|\,\|Y\|\,\|Z\|\,\|T\|$$
$$\times \frac{1}{ABC} \left(-\frac{A'}{A}(C^2 + A^2 - B^2) + \frac{B'}{B}(A^2 - B^2 - C^2) + 2C'C \right),$$

while $\langle R(X, T)T, Y \rangle = \langle R(T, Y)T, Z \rangle = \langle R(Z, T)T, X \rangle = \langle R(X, Y)Y, Z \rangle = \langle R(Y, Z)Z, X \rangle = \langle R(Z, X)X, Y \rangle = \langle R(T, X)X, Y \rangle = \langle R(T, X)X, Z \rangle = \langle R(T, Y)Y, X \rangle = \langle R(T, Y)Y, Z \rangle = \langle R(T, Z)Z, X \rangle = \langle R(T, Z)Z, Y \rangle = 0$. In particular, the matrix of Ricci curvature in these coordinates is diagonal. Now take

$$A(t) = \tfrac{1}{10}\,t\big(1 + \phi(t)\sin(\ln\ln t)\big),$$
$$B(t) = \tfrac{1}{10}\,t\big(1 + \phi(t)\sin(\ln\ln t)\big)^{-1},$$
$$C(t) = \tfrac{1}{10}\,t\big(1 - \gamma(t)\big),$$

This note was prepared when the author was visiting the Courant Institute of Mathematical Sciences.

where $\phi(t)$ is a smooth function such that $\phi(t) = 0$ for $t \in [0, T]$ (where $T > 0$ is a sufficiently big number), $\phi(t) > 0$ for $t > T$, $0 \le \phi'(t) \le t^{-2}$, and $|\phi''(t)| \le t^{-3}$; and $\gamma(t)$ is a smooth function such that $\gamma(t) = 0$ for $t \in [0, T/2]$, $\gamma'(t), \gamma''(t) > 0$ for $t \in (T/2, T)$, and $\gamma'(t) = (t \ln^{3/2} t)^{-1}$ for $t > T$.

Computation shows that $\|\mathrm{Rm}\| = O(t^{-2})$ and $\mathrm{Ric}(T, T) \ge C/(t^2 \ln^{3/2} t)$, while $\mathrm{Ric}(X, X)$, $\mathrm{Ric}(Y, Y)$, $\mathrm{Ric}(Z, Z)$ are all $\ge C/t^2$. It is also clear that the asymptotic cone is not unique. It remains only to smooth off the vertex ($t = 0$), where our space is isometric to a cone over a sphere of constant curvature 100.

REMARK. Mike Anderson has pointed out to me that a similar construction was used earlier by Brian White, in the context of surfaces in euclidean spaces.

G. PERELMAN
ST. PETERSBURG BRANCH
V. A. STEKLOV INSTITUTE OF MATHEMATICS (POMI)
RUSSIAN ACADEMY OF SCIENCES
FONTANKA 27
ST. PETERSBURG, 191011
RUSSIA
   perelman@pdmi.ras.ru

Comparison Geometry
MSRI Publications
Volume 30, 1997

Convergence Theorems in Riemannian Geometry

PETER PETERSEN

ABSTRACT. This is a survey on the convergence theory developed first by
Cheeger and Gromov. In their theory one is concerned with the compactness of the class of riemannian manifolds with bounded curvature and lower
bound on the injectivity radius. We explain and give proofs of almost all
the major results, including Anderson's generalizations to the case where
all one has is bounded Ricci curvature. The exposition is streamlined by the
introduction of a norm for riemannian manifolds, which makes the theory
more like that of Hölder and Sobolev spaces.

1. Introduction

This paper is an outgrowth of a talk given in October 1993 at MSRI and
a graduate course offered in the Spring of 1994 at UCLA. The purpose is to
introduce readers to the convergence theory of riemannian manifolds not so much
through a traditional survey article, but by rigorously proving most of the key
theorems in the subject. For a broader survey of this subject, and how it can be
applied to various problems, we refer the reader to [Anderson 1993].

The prerequisites for this paper are some basic knowledge of riemannian geometry, Gromov–Hausdorff convergence and elliptic regularity theory. In particular,
the reader should be familiar with the comparison geometry found in [Karcher
1989], for example. For Gromov–Hausdorff convergence, it suffices to read Section 6 in [Gromov 1981a] or Section 1 in [Petersen 1993]. In regard to elliptic
theory, we have an appendix that contains all the results we need, together with
proofs of those theorems that are not explicitly stated in [Gilbarg and Trudinger
1983].

In Section 2 we introduce the concept of (pointed) $C^{k+\alpha}$ convergence of riemannian manifolds. This introduces a natural topology on (pointed) riemannian
manifolds and immediately raises the question of which "subsets" are precompact. To answer this, we use, for the first time in the literature, the idea that

Partially supported by NSF and NYI grants.

a riemannian manifold in a natural way has a $C^{k+\alpha}$ norm for each fixed scale $r > 0$. This norm is a quantitative version of the definition of a manifold. It is basically computed by finding atlases of charts from r-balls in $\mathbb{R}^n$ into the manifold and then, for each of these atlases, by computing the largest $C^{k+\alpha}$ norm of the metric coefficients in these charts, and for compatibility reasons, also the largest $C^{k+1+\alpha}$ norm of the transition functions. The scale r is an integral part of the definition, so that $\mathbb{R}^n$ becomes the only space that has zero norm on all scales, while flat manifolds have zero norm on small scales, and nonflat manifolds have the property that the norm goes to zero as the scale goes to zero. The norm concept is dual to the usual radii concepts in geometry, but has nicer properties.

We have found this norm concept quite natural to work with: It gives, for instance, a very elegant formulation of what we call The Fundamental Theorem of Convergence Theory. This theorem is completely analogous to the classical Arzela–Ascoli theorem, and says that, for each fixed scale r, the class of riemannian n-manifolds of $C^{k+\alpha}$ norm $\leq Q$ is compact in the $C^{k+\beta}$ topology for $\beta < \alpha$. The only place where we use Gromov–Hausdorff convergence is in the proof of this theorem. Aside from the new concept of norm and the use of Gromov–Hausdorff convergence, the proof of this fundamental theorem is essentially contained in [Cheeger 1970].

In Section 3 we use the Fundamental Theorem to prove the Compactness Theorem of Cheeger and Gromov for manifolds of bounded curvature, as it is stated in [Gromov 1981b]. In addition we give a new proof by contradiction of Cheeger's lemma on the injectivity radius, using convergence techniques. Finally, we give S.-H. Zhu's proof of the compactness of the class of n-manifolds with lower sectional curvature bounds and lower injectivity radius bounds.

The theory really picks up speed in Section 4, where we introduce $L^{p,k}$ norms on the scale of r, using harmonic coordinates. It is at this point that we need to use elliptic regularity theory. The idea of using harmonic coordinates, rather than just general coordinates, makes a big difference in the theory, since it makes the norm a locally realizable number. This is basically a consequence of the fact that we don't have to worry about the norms of transition functions. Our "harmonic" norm is, of course, dual to Anderson's harmonic radius, but we again find that the norm idea has some important technical advantages over the use of harmonic radius.

Most of Anderson's convergence results are proved in Section 5. They are all generalizations of the Cheeger–Gromov Compactness Theorem, but the proofs are, in contrast, all by contradiction.

The final Section 6 has some applications of convergence theory to pinching problems in riemannian geometry. Some of the results in this section are extensions of work in [Gao 1990].

2. The Fundamentals

For a function $f : \Omega \to \mathbb{R}^h$, where $\Omega \subset \mathbb{R}^n$ the Hölder C^α-constant $0 < \alpha \leq 1$ is defined as:

$$\|f\|_\alpha = \sup_{x,y \in \Omega} \frac{|f(x) - f(y)|}{|x - y|^\alpha}.$$

In other words, $\|f\|_\alpha$ is the smallest constant C such that $|f(x) - f(y)| \leq C |x - y|^\alpha$. Notice that $\|f\|_1$ is the best Lipschitz constant for f. If $\Omega \subset \mathbb{R}^n$ is open, $k \geq 0$ is an integer, and $0 < \alpha \leq 1$, we define the $C^{k+\alpha}$ norm of f as

$$\|f\|_{k+\alpha} = \|f\|_{C^k} + \max_{|j|=k} \|\partial^j f\|_\alpha.$$

Here $\|f\|_{C^k}$ is the usual C^k-norm and $\partial^j = \partial_1^{j_1} \cdots \partial_n^{j_n}$, where $\partial_i = \partial/\partial x^i$ and $j = (j_1, \ldots, j_n)$ is a multi-index. Note that the norm $\|f\|_{k+1}$ is not the same as $\|f\|_{C^{k+1}}$.

We denote by $C^{k+\alpha}(\Omega)$ the space of functions with finite $(k + \alpha)$-norm. This norm makes $C^{k+\alpha}(\Omega)$ into a Banach space. The classical Arzela–Ascoli Theorem says that for $k + \alpha > 0$, $0 < \alpha \leq 1$, and $l + \beta < k + \alpha$, any sequence satisfying $\|f_i\|_{k+\alpha} \leq \kappa$ has a subsequence that converges in the $l + \beta$-topology to a function f, with $\|f\|_{k+\alpha} \leq \kappa$. Note that if $\alpha = 0$ it is not necessarily true that f is C^k, but it will be C^{k-1+1}. We shall therefore always look at $(k + \alpha)$-topologies with $0 < \alpha \leq 1$.

We can now define $C^{k+\alpha}$ convergence of tensors on a given manifold M. Namely, a sequence T_i of tensors on M is said to converge to T in the $C^{k+\alpha}$-topology if we can find a covering $\varphi_s : U_s \to \mathbb{R}^n$ of coordinate charts so that the overlap maps are at least $C^{k+1+\alpha}$ and all the components of the tensors T_i converge in the $C^{k+\alpha}$ topology to the components of T in these coordinate charts, considered as functions on $\varphi_\alpha(U_\alpha) \subset \mathbb{R}^n$. This convergence concept is clearly independent of our choice of coordinates. Note that it is necessary for the overlaps to be $C^{k+1+\alpha}$, since the components of tensors are computed by evaluating these tensors on the $C^{k+\alpha}$ fields $\partial/\partial x^i$ and dx^i.

In the sequel we shall restrict our attention to complete or closed riemannian manifolds. Some of the theory can, with modifications, be generalized to incomplete manifolds and manifolds with boundary. These generalizations, while useful, are not very deep and can be found in the literature.

A pointed sequence (M_i, g_i, p_i) of riemannian n-manifolds with metrics g_i, and $p_i \in M_i$, is said to converge to a riemannian manifold (M, g, p) in the pointed $C^{k+\alpha}$ topology if, for each $R > 0$, there is a domain Ω in M containing the open ball $B(p, R)$ and embeddings $f_i : \Omega \to M_i$, for large i, such that $f_i(\Omega) \supset B(p_i, R)$ and the $f_i^* g_i$ converge to g in the $C^{k+\alpha}$ topology on Ω. If we can choose $\Omega = M$ and $f_i(\Omega) = M_i$, we say that (M_i, g_i) converges to (M, g) in the $C^{k+\alpha}$-topology. There is obviously no significant difference between pointed and unpointed topologies when all manifolds in use are closed.

It is very important to realize that this concept of $C^{k+\alpha}$ convergence is not the same as the one we just defined for tensors on a given manifold M, even when we are considering a sequence of riemannian metrics g_i on the same space M. This is because one can have a sequence of metrics g_i and diffeomorphisms $f_i : M \to M$ such that $\{f_i^* g_i\}$ converges while $\{g_i\}$ does not converge.

We say that a collection of riemannian n-manifolds is precompact in the (pointed) $C^{k+\alpha}$ topology if any sequence in this collection has a subsequence that is convergent in the (pointed) $C^{k+\alpha}$-topology—in other words, if the (pointed) $C^{k+\alpha}$-closure is compact.

The rest of the paper is basically concerned with finding reasonable (geometric) conditions that imply precompactness in some of these topologies. In order to facilitate this task and ease the exposition a little, we need to introduce some auxiliary concepts.

For a riemannian n-manifold we define the $C^{k+\alpha}$-norm on the scale of $r > 0$, denoted $\|(M,g)\|_{k+\alpha,r}$, as the infimum over all numbers $Q \geq 0$ such that we can find coordinates charts $\varphi_s : B(0,r) \subset \mathbb{R}^n \to U_s \subset M$ with these properties:

(n1) Every ball of radius $\delta = \frac{1}{10} e^{-Q} r$ is contained in some U_s.
(n2) $|d\varphi_s^{-1}| \leq e^Q$ and $|d\varphi_s| \leq e^Q$ on $B(0,r)$.
(n3) $r^{|j|+\alpha}\|\partial^j g_{s..}\|_\alpha \leq Q$ for all multi-indices j with $0 \leq |j| \leq k$.
(n4) $\|\varphi_s^{-1} \circ \varphi_t\|_{k+1+\alpha} \leq (10+r)\, e^Q$ on the domain of definition.

Here $g_{s..}$ represents the metric components in the coordinates φ_s, considered as functions on $B(0,r) \subset \mathbb{R}^n$. The first condition is equivalent to saying that δ is a Lebesgue number for the covering $\{U_s\}$. The second condition can be rephrased as saying that the eigenvalues of $g_{s..}$ with respect to the standard euclidean metric lie between e^{-Q} and e^Q. This gives both a C^0 bound for $g_{s..}$ and a uniform positive definiteness for $g_{s...}$.

If A is a subset of M, we can define $\|(A,g)\|_{k+\alpha,r}$ in a similar way, only changing (n1) to say that all δ-balls centered on A are contained in some U_s. In (n1) it would perhaps have been more natural to merely assume that the sets U_s cover M (or A). That, however, is not a desirable state of affairs, because this would put us in a situation where the norm $\|\ \ \|_{k+\alpha,r}$ wouldn't necessarily be realized by some Q. A fake but illustrative example is the sphere, covered by two balls of radius $r > \pi/2$ centered at antipodal points. As $r \to \pi/2$ the sets will approach a situation where they no longer cover the space.

From the definition it is clear that $\|(M,g)\|_{k+\alpha,r}$ must be finite for all r, when M is closed. For open manifolds it would be more natural to have some weight function $f : M \to \mathbb{R}$, which allows for Q to get bigger as we go farther and farther out on the manifold. We'll say a few more words about this later in this section.

EXAMPLE. If $M = \mathbb{R}^n$ with the canonical euclidean metric, then $\|M\|_{k+\alpha,r} = 0$ for all r. More generally, we can easily prove that $\|(M,g)\|_{k+\alpha,r} = 0$ for $r \leq$

$\operatorname{inj\,rad}(M, g)$ if (M, g) is a flat manifold. We shall see later in this section that these properties characterize $\mathbb{R}^n$ and flat manifolds in general.

With this example in mind, it is pretty clear that conditions (n1)–(n3) say that on a scale of r the metric on (M, g) is Q-close to the euclidean metric in the $C^{k+\alpha}$ topology. Condition (n4) is a compatibility condition that ensures that $C^{k+\alpha}$-closeness to euclidean space means the same in all coordinates.

For a given Q and r we can in the usual fashion consider maximal atlases of all charts satisfying (n1)–(n4), but we won't use this much. Let us now turn to some of the properties of this norm.

PROPOSITION 2.1. *Let (M, g) be a C^∞ riemannian n-manifold.*

(i) $\|(A, \lambda^2 g)\|_{k+\alpha, \lambda r} = \|(A, g)\|_{k+\alpha, r}$; *in other words the norm is scale invariant, for all $A \subset M$.*

(ii) *If M is compact, then $\|(M, g)\|_{k+\alpha, r}$ is finite for all r; moreover, this number depends continuously on r, and it tends to 0 as $r \to 0$.*

(iii) *If $(M_i, g_i, p_i) \to (M, g, p)$ in the pointed $C^{k+\alpha}$-topology, then for any bounded domain $B \subset M$ we can find domains $B_i \subset M_i$ such that*

$$\|(B_i, g_i)\|_{k+\alpha, r} \to \|(B, g)\|_{k+\alpha, r} \quad \text{for every } r > 0.$$

When all spaces involved are closed manifolds we can set $B_i = M_i$ and $B = M$.

PROOF. (i) If we change g to $\lambda^2 g$ we can change charts $\varphi^s : B(0, r) \to M$ to $\varphi^s_{\lambda^{-1}}(x) = \varphi(\lambda^{-1} x) : B(0, \lambda r) \to M$. Since we are also scaling the metric, this means that conditions (n1)–(n4) still hold with the same Q.

(ii) If, as in (i), we only change g to $\lambda^2 g$, but do not scale in euclidean space, we get

$$\text{for } \lambda < 1, \quad \|(M, \lambda^2 g)\|_{k+\alpha, r} \le Q - \log \lambda$$
$$\text{for } \lambda > 1, \quad \|(M, \lambda^2 g)\|_{k+\alpha, r} \le \max\{\lambda Q, Q + \log \lambda\},$$

where $\|(M, g)\|_{k+\alpha, r} < Q$. Thus

$$\|(M, g)\|_{k+\alpha, \lambda^{-1} r} \le \max\{\lambda Q, Q - \log \lambda, Q + \log \lambda\}.$$

If, therefore, the norm is finite for some r, it will be finite for all r. Furthermore, $f(r) = \|(M, g)\|_{k+\alpha, r}$ is a function with the property that, for each r,

$$f(\lambda^{-1} r) \le h(\lambda, f(r)) = \max\{\lambda f(r), f(r) - \log \lambda, f(r) + \log \lambda\},$$

and $f(r) = h(1, f(r))$. If, therefore, $r_i \to r$, we clearly have $\limsup f(r_i) \le f(r)$, since $f(r_i) = f(r(r_i/r)) \le g(r/r_i, f(r))$. Conversely, $f(r) = f(r_i r/r_i) \le h(r_i/r, f(r_i)) = \max\{f(r_i) r_i/r, f(r_i) - \log(r_i/r), f(r_i) + \log(r_i/r)\} \le f(r_i) + \varepsilon$, for any $\varepsilon > 0$ as $r_i \to r$. Thus $f(r) \le \liminf f(r_i)$, and we have established continuity of $f(r)$.

To see that $\|(M, g)\|_{k+\alpha, r} \to 0$ as $r \to 0$ just use exponential coordinates $\exp_{p_i} : B(0, r) \to B(p_i, r)$.

(iii) Fix $r > 0$ and $B \subset M$. Then choose embeddings $f_i : \Omega \to M_i$ such that $f_i^* g_i$ converge to g and $B \subset \Omega$. Define $B_i = f_i(B)$.

For $Q > \|(B, g)\|_{k+\alpha, r}$, choose appropriate charts $\varphi^s : B(0, r) \to M$ satisfying (n1)–(n4), with $B = A$. Then $\varphi_i^s = f_i \circ \varphi^s : B(0, r) \to M_i$ will satisfy (n1) and (n4). Now, since $f_i^* g_i \to g$ in the $C^{k+\alpha}$-topology, (n2) and (n3) will hold with a Q_i satisfying $Q_i \to Q$ as $i \to \infty$. Thus $\limsup \|(B_i, g)\|_{k+\alpha, r} \le \|(B, g)\|_{k+\alpha, r}$.

Conversely, for large i and $Q > \liminf \|(B_i, g_i)\|_{k+\alpha, r}$. We can choose charts $\varphi_i^s : B(0, r) \to M_i$ satisfying (n1)–(n4) with $B_i = A$. Now consider $\varphi^s = f_i^{-1} \circ \varphi_i^s : B(0, r) \to M$. These charts will satisfy n1 and n4, and again n2 and n3 for some $Q_i \ge Q$, which can be chosen to converge to Q as $i \to \infty$. Thus $\liminf \|(B_i, g_i)\|_{k+\alpha, r} \ge \|(B, g)\|_{k+\alpha, r}$. $\square$

A few comments about our definition of $\|(M, g)\|_{k+\alpha, r}$ are in order at this point. From a geometric point of view, it would probably have been nicer to assume that our charts $\varphi^s : B(p, r) \subset M \to \Omega \subset \mathbb{R}^n$, were defined on metric balls in M rather than on $\mathbb{R}^n$. The reason for not doing so is technical, but still worth pondering. With revised charts, $\|(M, g)\|_{k+\alpha, r}$ would definitely be ∞ for large r, unless (M, g) were euclidean space; even worse, our proof of (iii) would break down, as the charts $\varphi_i \circ f_i$ and $\varphi \circ f_i^{-1}$ would not be defined on metric r-balls but only on slightly smaller balls. We could partially avert these problems by maximizing r for fixed Q, rather than minimizing Q for fixed r, and thereby define a $(Q, k + \alpha)$ radius $r_{k+\alpha, Q}(M)$. This is exactly what was done in Anderson's work on convergence, and it follows more closely the standard terminology in riemannian geometry. But even with this definition, we still cannot conclude that this radius concept varies continuously in the $C^{k+\alpha}$ topology for fixed Q. All we get is that $r_{k+\alpha, Q_i}(M_i, g_i) \to r_{k+\alpha, Q}(M, g)$ for some sequence $Q_i \downarrow Q$, where Q is chosen in advance. Our concept of norm, therefore, seems to behave in a nicer manner from an analytic point of view. However, when it really matters, the two viewpoints are more or less equivalent. We also think that the norm concept is more natural, since it is a quantitative version of the qualitative coordinate definition of a manifold with a riemannian structure.

Another issue is exactly how to define the norm for noncompact manifolds. The most natural thing to do would be to consider pointed spaces (M, g, p), choose a nondecreasing weight function $f(R)$, and assume $\|B(p, R))\|_{k+\alpha, r} \le f(R)$. Our whole theory can easily be worked out in this context, but for the sake of brevity, we won't do this.

For noncompact spaces one can also use the norm concept to define what it should mean for a space to be asymptotically locally euclidean of order $k + \alpha$— namely, that it should satisfy $\|(M - B(p, R), g)\|_{k+\alpha, r} \to 0$ as $R \to \infty$. Here one can also refine this and assume that r should, in some way, depend on R.

Our next theorem is, in a way, as basic as our proposition, and in fact it uses none of our basic properties. The proof, however, is a bit of a mouthful. Nevertheless, the reader is urged to go through it carefully, because many of the subsequent corollaries are only true corollaries in the context of some of the constructions in the proof.

THEOREM 2.2 (FUNDAMENTAL THEOREM OF CONVERGENCE THEORY). *For given $Q > 0$, $n \geq 2$ integer, $k+\alpha > 0$, and $r > 0$, consider the class $\mathcal{H}^{k+\alpha}(n, Q, r)$ of complete, pointed riemannian n-manifolds (M, g, p) with $\|(M, g)\|_{k+\alpha, r} \leq Q$. Then $\mathcal{H}^{k+\alpha}(n, Q, r)$ is compact in the pointed $C^{k+\beta}$-topology for all $k+\beta < k+\alpha$.*

PROOF. We proceed in stages. First we make some general comments about the charts we use. We then show that $\mathcal{H} = \mathcal{H}^{k+\alpha}(n, Q, r)$ is precompact in the pointed Gromov–Hausdorff topology. Next we prove that $\mathcal{H}$ is compact in the Gromov–Hausdorff topology. Finally we consider the statement of the theorem.

Setup: First fix $K > Q$. Whenever we select an $M \in \mathcal{H}$ we shall assume that it comes equipped with an atlas of charts satisfying (n1)–(n4) with K in place of Q. Thus we implicitly assume that all charts under consideration belong to these atlases. We will, in consequence, only prove that limit spaces (M, g, p) satisfy $\|(M, g)\|_{k+\alpha, r} \leq K$, but since K was arbitrary we still get $(M, g, p) \in \mathcal{H}$.

We proceed by establishing several simple facts.

FACT 1. *Every chart $\varphi : B(0, r) \to U \subset M \in \mathcal{H}$ satisfies*

(a) $d(\varphi(x_1), \varphi(x_2)) \leq e^K |x_1 - x_2|$ *and*
(b) $d(\varphi(x_1), \varphi(x_2)) \geq \min\{e^{-K} |x_1 - x_2|, e^{-K}(2r - |x_1| - |x_2|)\}$,

where d is distance measured in M, and $|\cdot|$ is the usual euclidean norm. The condition $|d\varphi| \leq e^K$ together with convexity of $B(0, r)$ immediately implies (a). For (b), first observe that if any segment from x_1 to x_2 lies in U, then $|d\varphi| \geq e^{-K}$ implies that $d(\varphi(x_1), \varphi(x_2)) \geq e^{-K}|x_1 - x_2|$. So we may assume that $\varphi(x_1)$ and $\varphi(x_2)$ are joined by a segment $\sigma : [0, 1] \to M$ that leaves U. Split σ into $\sigma : [0, t_1) \to U$, and $\sigma : (t_2, 1) \to U$ such that $\sigma(t_i) \notin U$. Then we clearly have

$$d(\varphi(x_1), \varphi(x_2)) = L(\sigma) \geq L(\sigma|_{[0,t_1)}) + L(\sigma|_{(t_2,1]})$$
$$\geq e^{-K}\left(L(\varphi^{-1} \circ \sigma|_{[0,t_1)}) + L(\varphi^{-1} \circ \sigma|_{(t_2,1]})\right)$$
$$\geq e^{-K}(2r - |x_1| - |x_2|).$$

The last inequality follows from the fact that $\varphi^{-1}\sigma(0) = x_1, \varphi^{-1} \circ \sigma(1) = x_2$ and that $\varphi^{-1} \circ \sigma(t)$ approaches the boundary of $B(0, r)$ as $t \nearrow t_1$, or $t \searrow t_2$.

FACT 2. *Every chart $\varphi : B(0, r) \to U \subset M \in \mathcal{H}$, and hence any δ-ball in M, where $\delta = \frac{1}{10}e^{-K}r$, can be covered by at most $N(\delta/4)$-balls, where N depends only on n, K, r.* Clearly there exists an $N(n, K, r)$ such that $B(0, r)$ can be covered by at most Ne^{-K} $(\delta/4)$-balls. Since $\varphi : B(0, r) \to U$ is a Lipschitz map with Lipschitz constant $\leq e^K$, we get the desired covering property.

FACT 3. *Every ball $B(x, l\delta/2) \subset M$ can be covered by $\leq N^l$ $\delta/4$-balls.* For $l = 1$ we just proved this. Suppose we know that $B(x, l\delta/2)$ is covered by $B(x_1, \delta/4)$, ..., $B(x_{N^l}, \delta/4)$. Then $B(x, l\delta/2 + \delta/2) \subset \bigcup_{i=1}^{N} B(x_i, \delta)$. Now each $B(x_i, \delta)$

can be covered by at most $N\delta/4$-balls; hence $B(x, (l+1)\delta/2)$ can be covered by $\leq N N^l = N^{l+1} \, \delta/4$-balls.

FACT 4. $\mathcal{H}$ *is precompact in the pointed Gromov–Hausdorff topology.* This is equivalent to asserting that, for each $R > 0$, the family of metric balls $B(p, R) \subset (M, g, p) \in \mathcal{H}$ is precompact in the Gromov–Hausdorff topology. This claim is equivalent to showing that we can find a function $N(\varepsilon) = N(\varepsilon, R, K, r, n)$ such that each $B(p, R)$ can contain at most $N(\varepsilon)$ disjoint ε-balls. To check this, let $B(x_1, \varepsilon), \ldots, B(x_s, \varepsilon)$ be a collection of disjoint balls in $B(p, R)$. Suppose that $l\delta/2 < R \leq (l+1)\delta/2$; then the volume of $B(p, R)$ is at most $N^{(l+1)}$ times the maximal volume of $(\delta/4)$-ball, therefore at most $N^{(l+1)}$ times the maximal volume of a chart, therefore at most $N^{(l+1)} e^{nK} \operatorname{vol} B(0, r) \leq F(R) = F(R, n, K, r)$. Conversely, each $B(x_i, \varepsilon)$ lies in some chart $\varphi : B(0, r) \to U \subset M$ whose preimage in $B(0, r)$ contains an $(e^{-K}\varepsilon)$-ball. Thus $\operatorname{vol} B(p_i, \varepsilon) \geq e^{-2nK} \operatorname{vol} B(0, \varepsilon)$. All in all we get $F(R) \geq \operatorname{vol} B(p, R) \geq \sum \operatorname{vol} B(p_i, \varepsilon) \geq s \, e^{-2nK} \operatorname{vol} B(0, \varepsilon)$. Thus $s \leq N(\varepsilon) = F(R) e^{2nK} (\operatorname{vol} B(0, \varepsilon))^{-1}$.

Now select a sequence (M_i, g_i, p_i) in $\mathcal{H}$. From the previous considerations we can assume that $(M_i, g_i, p_i) \to (X, d, p)$ converge to some metric space in the Gromov–Hausdorff topology. It will be necessary in many places to pass to subsequences of (M_i, g_i, p_i) using various diagonal processes. Whenever this happens, we shall not reindex the family, but merely assume that the sequence was chosen to have the desired properties from the beginning. For each (M_i, g_i, p_i) choose charts $\varphi_{is} : B(0, r) \to U_{is} \subset M_i$ satisfying (n1)–(n4). We can furthermore assume that the index set $\{s\} = \{1, 2, 3, 4, \cdots\}$ is the same for all M_i, that $p_i \in U_{i1}$, and that the balls $B(p_i, (l/2)\delta)$ are covered by the first N^l charts. Note that these N^l charts will then be contained in $D(p_i, (l/2)\delta + \lfloor e^K + 1 \rfloor\delta)$. Finally, for each l, the sequence $D(p_i, (l/2)\delta) = \{x_i : d(x_i, p_i) \leq (l/2)\delta\}$ converges to $D(p, (l/2)\delta) \subset X$, so we can choose a metric on the disjoint union $Y_l = (D(p, (l/2)\delta) \amalg \coprod_{i=1}^{\infty} D(p_i, (l/2)\delta))$ such that $p_i \to p$ and $D(p_i, (l/2)\delta) \to D(p, (l/2)\delta)$ in the Hausdorff distance inside this metric space.

FACT 5. (X, d, p) *is a riemannian manifold of class* $C^{k+1+\alpha}$ *with norm* $\leq K$. Obviously we need to find bijections $\varphi_s : B(0, r) \to U_s \subset X$ satisfying (n1)–(n4). For each s consider the maps $\varphi_{is} : B(0, r) \to U_{is} \subset Y_{l+2\lfloor e^K + 1 \rfloor}$. Inequality (a) of Fact 1 implies that this is a family of equicontinuous maps into the compact space $Y_{l+2\lfloor e^K + 1 \rfloor}$. The Arzela–Ascoli Theorem shows that this sequence must subconverge (in C^0-topology) to a map $\varphi_s : B(0, r) \subset Y_{l+2\lfloor e^K + 1 \rfloor}$ which also has Lipschitz constant e^K. Furthermore, inequality (b) will also hold for this map as it holds for all the φ_{is} maps. In particular, φ_s is one-to-one. Finally, since $U_{is} \subset D(p_i, (l/2)\delta + \lfloor e^K + 1 \rfloor)$ and $D(p_i, (l/2)\delta + \lfloor e^K + 1 \rfloor)$ Hausdorff converges to $D(p, (l/2)\delta + \lfloor e^K + 1 \rfloor) \subset X$, we see that $\varphi_x(B(0, r)) = U_s \subset X$. A simple diagonal argument says that we can pass to a subsequence of (M_i, g_i, p_i) having the property that $\varphi_{is} \to \varphi_s$ for all s.

In this way, we have constructed (topological) charts $\varphi_s : B(0,r) \to U \subset X$, and we can easily check that they satisfy (n1). Since φ_s also satisfy inequalities (a) and (b), they would also satisfy (n2) if they were differentiable (this being equivalent to the transition functions being C^1). Now the transition functions $\varphi_{is}^{-1} \circ \varphi_{it}$ converge to $\varphi_s^{-1} \circ \varphi_t$, because the φ_{is} converge to φ_s. Note that these transition functions are not defined on the same domains; but we do know that the domain for $\varphi_s^{-1} \circ \varphi_t$ is the limit of the domains for $\varphi_{is}^{-1} \circ \varphi_{it}$, so the convergence makes sense on all compact subsets of the domain of $\varphi_s^{-1} \circ \varphi_t$. Now $\|\varphi_{is}^{-1} \circ \varphi_{it}\|_{C^{k+1+\alpha}} \leq K$, so a further application of Arzela–Ascoli, followed by passage to subsequences, tells us that $\|\varphi_s^{-1} \circ \varphi_t\|_{C^{k+1+\alpha}} \leq K$, and that we can assume that the $\varphi_{is}^{-1} \circ \varphi_{it}$ converge to $\varphi_s^{-1} \circ \varphi_t$ in the $C^{k+1+\beta}$-topology. This then establishes (n2) and (n4). We now construct a compatible riemannian metric on X that satisfies (n3). For each s, consider the metric $g_{is} = g_{is\cdot\cdot}$ written out in its components on $B(0,r)$ with respect to the chart φ_{is}. Since all of the $g_{is\cdot\cdot}$ satisfy (n3), we can again use Arzela–Ascoli to make these functions converge on $B(0,r)$ in the $C^{k+\beta}$-topology to functions $g_{s\cdot\cdot}$, which also satisfy (n3).

The local "tensors" $g_{s\cdot\cdot}$ thus obtained satisfy the change of variables formulae required to make them into global tensors on X. This is because all the $g_{is\cdot\cdot}$ satisfy these properties, and everything that needs to converge in order for these properties to be carried over to the limit also converges. Recall that the rephrasing of (n2) gives the necessary C^0 bounds, and also shows that $g_{s\cdot\cdot}$ is positive definite. We now have exhibited a riemannian structure on X such that $\varphi_s : B(0,r) \to U_s \subset X$ satisfy (n1)–(n4) with respect to this structure. This, however, does not guarantee that the metric generated by this structure is identical to the metric we got from X being the pointed Gromov–Hausdorff limit of (M_i, g_i, p_i). However, since Gromov–Hausdorff convergence implies that distances converge, and since we know at the same time that the riemannian metric converges locally in coordinates, it follows that the limit riemannian structure must generate the "correct" metric, at least locally, and therefore also globally.

FACT 6. $(M_i, g_i, p_i) \to (X, d, p) = (X, g, p)$ *in the pointed $C^{k+\beta}$-topology.* We assume the setup is as in Fact 5, where charts φ_{is}, transitions $\varphi_{is}^{-1} \circ \varphi_{it}$, and metrics $g_{is\cdot\cdot}$, converge to the same items in the limit space. Let's agree that two maps f, g between subsets in M_i and X are $C^{k+1+\beta}$ close if all the coordinate compositions $\varphi_s^{-1} \circ g \circ \varphi_{it}$ and $\varphi_s^{-1} \circ f \circ \varphi_{it}$ are $C^{k+1+\beta}$ close. Thus we have a well defined $C^{k+1+\beta}$ topology on maps from M_i to X. Our first observation is that $f_{is} = \varphi_{is} \circ \varphi_s^{-1} : U_s \to U_{is}$ and $f_{it} = \varphi_{it} \circ \varphi_t^{-1} : U_t \to U_{it}$ "converge to each other" in the $C^{k+1+\beta}$ topology. Furthermore, $(f_{is})^* g_i|_{U_{is}}$ converges to $g|_{U_s}$ in the $C^{k+\beta}$ topology. These are just restatements of what we already know. In order to finish the proof, we therefore only need to construct diffeomorphisms $F_{il} : \Omega_l = \bigcup_{s=1}^{l} U_s \to \Omega_{il} = \bigcup_{s=1}^{l} U_{is}$, which are closer and closer to the f_{is}, for $s = 1, \ldots, l$ maps (and therefore all f_{is}) as $i \to \infty$. We will construct F_{il} by induction on l and large i depending on l. For this purpose we shall need a partition of unity $\{\lambda_s\}$ on X subordinate to $\{U_s\}$. We can find such a partition since the covering $\{U_s\}$ is locally finite by choice, and we can furthermore assume that λ_s is $C^{k+1+\alpha}$.

For $l = 1$ simply define $F_{i1} = f_{i1}$.

Suppose we have maps $F_{il} : \Omega_l \to \Omega_{il}$, for large i, that get arbitrarily close to f_{is}, for $s = 1, \ldots, l$, as $i \to \infty$. If $U_{l+1} \cap \Omega_l = \varnothing$, we just define $F_{il+1} = F_{il}$ on Ω_{il} and $F_{il+1} = f_{il+1}$ on U_{l+1}. In case $U_{l+1} \subset \Omega_l$, we simply let $F_{il+1} = F_{il}$. Otherwise we know that F_{il} and f_{il+1} are as close as we like in the $C^{k+1+\beta}$-topology when $i \to \infty$. So the natural thing to do is to average them on U_{l+1}. Define F_{il+1} on U_{l+1} as

$$
F_{il+1}(x) = \varphi_{il+1} \circ \left(\sum_{s=l+1}^{\infty} \lambda_s(x)\, \varphi_{il+1}^{-1} \circ f_{il+1}(x) + \sum_{s=1}^{l} \lambda_s(x)\, \varphi_{il+1}^{-1} \circ F_{il}(x) \right)
$$

$$
= \varphi_{il+1} \circ (\mu_1(x)\varphi_{l+1}^{-1}(x) + \mu_2(x)\varphi_{il+1}^{-1} \circ F_{il}(x)).
$$

This map is clearly well defined on U_{l+1}, since $\mu_2(x) = 0$ on $U_{l+1} - \Omega_l$; since $\mu_1(x) = 0$ on Ω_l, the map is a smooth ($C^{k+1+\alpha}$) extension of F_{il}. Now consider this map in coordinates:

$$
\varphi_{il+1}^{-1} \circ F_{il+1} \circ \varphi_{l+1}(y) = \mu_1 \circ \varphi_{l+1}(y)\, y + \mu_2 \circ \varphi_{l+1}(y)\, \varphi_{il+1}^{-1} \circ F_{il} \circ \varphi_{l+1}(y)
$$

$$
= \tilde{\mu}_1(y)F_1(y) + \tilde{\mu}_2(y)F_2(y).
$$

Now

$$
\|\tilde{\mu}_1 F_1 + \tilde{\mu}_2 F_2 - F_1\|_{k+1+\beta} = \|\tilde{\mu}_1(F_1 - F_1) + \tilde{\mu}_2(F_2 - F_1)\|_{k+1+\beta}
$$

$$
\leq \|\tilde{\mu}_2\|_{k+1+\beta}\, \|F_2 - F_1\|_{k+1+\beta}.
$$

This inequality is valid on all of $B(0, r)$, despite the fact that F_2 is not defined on all of $B(0, r)$, because $\tilde{\mu}_1 F_1 + \tilde{\mu}_2 F_2 = F_1$ on the region where F_2 is undefined. By assumption, $\|F_2 - F_1\|_{k+1+\beta}$ converges to 0 as $i \to \infty$, so F_{il+1} is $C^{k+1+\beta}$-close to f_{is}, for $s = 1, \ldots, l+1$, as $i \to \infty$. It remains to be seen that F_{il+1} is a diffeomorphism. But we know that $\tilde{\mu}_1 F_1 + \tilde{\mu}_2 F_2$ is an embedding since F_1 is and the space of embeddings is open in the C^1-topology and $k + 1 + \beta > 1$. Also the map is one-to-one since the images $f_{il+1}/(U_{l+1} - \Omega_l)$ and $F_{il}(\Omega_l)$ don't intersect.   □

COROLLARY 2.3. *The subclasses $\mathcal{H}(D) \subset \mathcal{H} = \mathcal{H}^{k+\alpha}(n, Q, r)$ and $\mathcal{H}(V) \subset \mathcal{H}$, where the elements in addition satisfy* diam $\leq D$ *and* vol $\leq V$, *respectively, are compact in the $C^{k+\beta}$-topology. In particular, $\mathcal{H}(D)$ and $\mathcal{H}(V)$ contain only finitely many diffeomorphism types.*

PROOF. Use notation as in the Fundamental Theorem. If $\mathrm{diam}(M, g, p) \leq D$, then clearly $M \subset B\left(p, \frac{1}{2}k\delta\right)$ for $k > (2/\delta)D$. Hence each element in $\mathcal{H}(D)$ can be covered by $\leq N^k$ charts. Thus $C^{k+\beta}$-convergence is actually in the unpointed topology, as desired.

If instead $\mathrm{vol}\, M \leq V$, we can use Fact 4 in the proof to see that we can never have more than $k = Ve^{2nk}(\mathrm{vol}\, B(0,\varepsilon))^{-1}$ disjoint ε-balls. In particular, $\mathrm{diam} \leq 2\varepsilon k$, and we can use the above argument.

Clearly compactness in any $C^{k+\beta}$-topology implies that the class cannot contain infinitely many diffeomorphism types. $\square$

This corollary is essentially contained in [Cheeger 1970]. Technically speaking, our proof comes pretty close to Cheeger's original proof, but there are some differences: notably, the use of Gromov–Hausdorff convergence which was not available to Cheeger at the time. Another detail is that our proof centers on the convergence of the charts themselves rather than the transition functions. This makes it possible for us to start out with the apparently weaker assumption that the covering $\{U_s\}$ has δ as a Lebesgue number, rather than assuming that M can be covered by chart images of euclidean $(r/2)$-balls that lie in transition domains $U_s \cap U_t$.

COROLLARY 2.4. *The norm* $\|(M,g)\|_{k+\alpha,r}$ *is always realized by some charts* $\varphi_s : B(0,r) \to U_s$ *satisfying* $(n1)$–$(n4)$ *with* $\|(M,g)\|_{k+\alpha,r}$ *in place of Q.*

PROOF. Choose appropriate charts $\varphi_s^Q : B(0,r) \to U_s^Q \subset M$ for each $Q > \|(M,g)\|_{k+\alpha,r}$, and let $Q \to \|(M,g)\|_{k+\alpha,r}$. If the charts are chosen to conform with the proof of the fundamental theorem, we will obviously get some limit charts with the wanted properties. $\square$

COROLLARY 2.5. M *is a flat manifold if* $\|(M,g)\|_{k+\alpha,r} = 0$ *for some* r, *and* M *is euclidean space with the canonical metric if* $\|(M,g)\|_{k+\alpha,r} = 0$ *for all* $r > 0$.

PROOF. Using the previous corollary, M can be covered by charts $\varphi_s : B(0,r) \to U_s \subset M$ satisfying $|d\varphi_s| \equiv 1$. This clearly makes M locally euclidean, and hence flat. If M is not euclidean space, the same reasoning clearly shows that $\|(M,g)\|_{k+\alpha,r} > 0$ for $r > \mathrm{inj}\,\mathrm{rad}(M,g)$. $\square$

Finally we should mention that all properties of this norm concept would not change if we changed (n1)–(n4) to, say,

(n1$'$) U_s has Lebesgue number $f_1(n,Q,r)$,
(n2$'$) $(f_2(n,Q))^{-1} \leq |d\varphi_s| \leq f_2(n,Q)$,
(n3$'$) $r^{|j|+\alpha}\|\partial^j g_s..\|_\alpha \leq f_3(n,Q)$ for $0 \leq |j| \leq k$,
(n4$'$) $\|\varphi_s^{-1} \circ \varphi_t\|_{k+\alpha} \leq f_4(n,Q,r)$,

for continuous functions $f_1, \ldots, f_4$ with $f_1(n,0,r) = 0$ and $f_2(n,0) = 1$. The key properties we want to preserve are the continuity of $\|(M,g)\|_{k+\alpha,r}$ with respect to r, the Fundamental Theorem, and the characterization of flat manifolds and euclidean space.

We should also mention that we could develop an L^p-norm concept, but we shall delay this till Section 4, in order to do it in the context of harmonic coordinates.

Another interesting thing happens if in the definition of $\|(M,g)\|_{k+\alpha,r}$ we let $k = \alpha = 0$. Then (n3) no longer makes sense, because $\alpha = 0$, but aside from that we still have a C^0-norm concept. The class $\mathcal{H}^0(n, Q, r)$ is now only compact in the pointed Gromov–Hausdorff topology, but the characterization of flat manifolds is still valid. The subclasses $\mathcal{H}(D)$ and $\mathcal{H}(V)$ are also only compact with respect to the Gromov–Hausdorff topology, and the finiteness of diffeomorphism types apparently fails.

It is, however, possible to say more. If we investigate the proof of the Fundamental Theorem we see that the problem lies in constructing the maps $F_{ik} : \Omega_k \to \Omega_{ik}$, because we now only have convergence of the coordinates in the C^0 (or C^α, for $\alpha < 1$) topology, so the averaging process fails as it is described. We can, however, use a deep theorem from topology about local contractibility of homeomorphism groups [Edwards and Kirby 1971] to conclude that two C^0-close topological embeddings can be "glued" together in some way without altering them too much in the C^0-topology. This makes it possible to exhibit topological embeddings $F_{ik} : \Omega \hookrightarrow M_i$ such that the pullback metrics (not riemannian metrics) converge. As a consequence, we see that $\mathcal{H}(D)$ and $\mathcal{H}(V)$ contain only finitely many homeomorphism types. This is exactly the content of the original version of Cheeger's Finiteness Theorem, including the proof as we have outlined it. But, as we have pointed out earlier, he also considered the easier to prove finiteness theorem for diffeomorphism types given better bounds on the coordinates.

3. The Cheeger–Gromov Compactness Results

The focus of this section is the relationship between volume, injectivity radius, sectional curvature and the norm concept from Section 2.

First let's see what exponential coordinates can do for us. If (M, g) is a riemannian manifold with $|\sec M| \leq K$ and $\operatorname{inj rad} M \geq i_0$, we know from the Rauch comparison theorems that there is a continuous function $f(r, i_0, K)$, for $r < i_0$, such that $f(0, i_0, K) = 1$, $f(r, i_0, 0) = 1$, and $\exp_p : B(0, r) \subset T_p M \to B(p, r) \subset M$ satisfies $(f(r, i_0, K))^{-1} \leq |d\exp_p| \leq f(r, i_0, K)$. In particular we have:

THEOREM 3.1. *For every $Q > 0$ there exists $r > 0$ depending only on i_0, K such that any complete (M, g) with $|\sec M| \leq K$ and $\operatorname{inj rad} M \geq i_0$ satisfies $\|(M, g)\|_{0,r} \leq Q$. Furthermore, if (M_i, g_i, p_i) satisfy $\operatorname{inj rad} M_i \geq i_0$ and $|\sec M_i| \leq K_i \to 0$, then a subsequence will converge in the pointed Gromov–Hausdorff topology to a flat manifold with $\operatorname{inj rad} \geq i_0$.*

The proof follows immediately from our constructions in Section 2.

This theorem does not seem very satisfactory, because even though we have assumed a C^2 bound on the riemannian metric, locally we only get a C^0 bound.

To get better bounds under the same circumstances, we must look for different coordinates. Our first choice for alternate coordinates is distance coordinates.

LEMMA 3.2. *Given a riemannian manifold (M, g) with* inj rad $\geq i_0$ *and* $|\sec| \leq K$, *and given* $p \in M$, *the Hessians* $H(v) = \operatorname{Hess} r(x)(v) = D_v \operatorname{grad} r$ *of the distance function* $r(x) = d(x, p)$ *has* grad r *as an eigenvector with eigenvalue 0, all other eigenvalues lie in the interval*

$$[\sqrt{K} \cot(r(x)\sqrt{K}), \ \sqrt{K} \coth(r(x)\sqrt{K})].$$

PROOF. We know that the operator H satisfies the Riccati equation $\partial H/\partial r + H^2 = -R$, where $R(v) = R(v, \partial/\partial r)(\partial/\partial r) = R(v, \operatorname{grad} r) \operatorname{grad} r$. The eigenvalues for $R(v)$ are assumed to lie in the interval $[-K, K]$, so elementary differential equation theory implies our estimate. □

Now fix $(M, g), p \in M$ as in the lemma, and choose an orthonormal basis $e_1, \ldots, e_n$ for $T_p M$. Then consider the geodesics $\gamma_i(t)$ with $\gamma_i(0) = p$, $\dot{\gamma}_i(0) = e_i$ and, together with them, the distance functions $r_i(x) = d(x, \gamma_i(i_0/(4\sqrt{K})))$. These distance functions will then have uniformly bounded Hessians on $B(p, \delta)$, for $\delta = i_0/(8\sqrt{K})$. Set $\varphi_p(x) = (r_1(x), \ldots, r_n(x))$ and $g_{pij} = \langle \partial/\partial r_i, \partial/\partial r_j \rangle$. Note that the inverse of g_{pij} is $g_p^{ij} = \langle \operatorname{grad} r_i, \operatorname{grad} r_j \rangle$.

THEOREM 3.3. *Given i_0 and $K > 0$ there exist $Q, r > 0$ such that any (M, g) with* inj rad $\geq i_0$ *and* $|\sec| \leq K$ *satisfies* $\|(M, g)\|_{0+1,r} \leq Q$.

PROOF. The inverses of the φ_p are our potential charts. First observe that $g_{pij}(p) = \delta_{ij}$, so the uniform Hessian estimate shows that $|d\varphi_p| \leq e^Q$ and $|d\varphi_p^{-1}| \leq e^Q$ on $B(p, \varepsilon)$, where Q, ε depend only on i_0, K. The proof of the inverse function theorem then tells us that there is $\hat{\varepsilon} > 0$ depending only on Q, n such that $\varphi_p : B(0, \hat{\varepsilon}) \to \mathbb{R}^n$ is one-to-one. We can then easily find r such that $\varphi_p^{-1} : B(0, r) \to U_p \subset B(p, \varepsilon)$ satisfies (n2). The conditions (n3), (n4) now immediately follow from the Hessian estimates, except we might have to increase Q somewhat. Finally (n1) holds since we have coordinates centered at every $p \in M$. □

Notice that Q cannot be chosen arbitrarily small, because our Hessian estimates cannot be improved by going to smaller balls. This will be taken care of in the next section, by using even better coordinates. Theorem 3 was first proved in [Gromov 1981b] as stated. The reader should be aware that what Gromov refers to as a $C^{1,1}$-manifold is in our terminology a manifold with $\|(M, h)\|_{0+1,r} < \infty$, i.e., C^{0+1}-bounds on the riemannian metric.

Without going much deeper into the theory we can easily prove:

LEMMA 3.4 (CHEEGER'S LEMMA). *Given a compact n-manifold (M, g) with* $|\sec| \leq K$ *and* vol $B(p, 1) \geq v > 0$ *for all $p \in M$, we have* inj rad $M \geq i_0$, *where i_0 depends only on n, K and v.*

PROOF. The proof goes by contradiction, using Theorem 3.3. Assume we have (M_i, g_i) with inj rad $M_i \to 0$ and satisfying the assumptions of the lemma. Find $p_i \in M_i$ such that inj $\mathrm{rad}_{p_i} = $ inj rad M_i and consider the pointed sequence (M_i, h_i, p_i), where $h_i = ($inj rad $M_i)^{-2} g_i$ is rescaled so that inj $\mathrm{rad}(M_i, h_i) = 1$ and $|\sec(M_i, h_i)| \le $ inj $\mathrm{rad}(M_i, g_i) K = K_i \to 0$. Theorem 3.3, together with the Fundamental Theorem 2.2, then implies that some subsequence (M_i, h_i, p_i) will converge in the pointed $C^{0+\alpha}$ topology $(\alpha < 1)$ to a flat manifold (M, g, p).

The first observation about (M, g, p) is that inj $\mathrm{rad}(p) \le 1$. This follows because the conjugate radius for (M_i, h_i) is at least $\pi/\sqrt{K_i}$, which goes to infinity, so Klingenberg's estimate for the injectivity radius implies that there must be a geodesic loop of length 2 at $p_i \in M_i$. Since (M_i, h_i, p_i) converges to (M, g, p) in the pointed $C^{0+\alpha}$-topology, the geodesic loops must converge to a geodesic loop in M based at p, having length 2. Hence inj $\mathrm{rad}(M) \le 1$.

The other contradictory observation is that (M, g) is $\mathbb{R}^n$ with the canonical metric. Recall that vol $B(p_i, 1) \ge v$ in (M_i, g_i), so relative volume comparison shows that there is a $v'(n, K, v)$ such that vol $B(p_i, r) \ge v' r^n$, for $r \le 1$. The rescaled manifold (M_i, h_i) therefore satisfies vol $B(p_i, r) \ge v' r^n$, for $r \le$ (inj $\mathrm{rad}(M_i, g_i))^{-1}$. Using again the convergence of (M_i, h_i, p_i) to (M, g, p) in the pointed $C^{0+\alpha}$-topology, we get vol $B(p, r) \ge v' r^n$ for all r. Since (M, g) is flat, this shows that it must be euclidean space. $\qquad \square$

This lemma was proved by a more direct method in [Cheeger 1970], but we have included this perhaps more convoluted proof in order to show how our convergence theory can be used. The lemma also shows that Theorem 3.3 remains true if the injectivity radius bound is replaced by a lower bound on the volume of balls of radius 1.

Theorem 3.3 can be generalized in another interesting direction.

THEOREM 3.5. *Given $i_0, k > 0$ there exist Q, r depending on i_0, k such that any manifold (M, g) with inj rad $\ge i_0$ and sec $\ge -k^2$ satisfies $\|(M, g)\|_{0+1, r} \le Q$.*

PROOF. It suffices to get some Hessian estimate for distance functions $r(x) = d(x, p)$. As before, Hess $r(x)$ has eigenvalues $\le k \coth(kr(x))$. Conversely, if $r(x_0) < i_0$, then $r(x)$ is supported from below by $f(x) = i_0 - d(x, y_0)$, where $y_0 = \gamma(i_0)$, and γ is the unique unit speed geodesic which minimizes the distance from p to x_0. Thus Hess $r(x) \ge$ Hess $f(x)$ at x_0. But Hess $f(x)$ has eigenvalues at least $-k \coth(d(x_0, y_0)k) = -k \coth(k(i_0 - r(x_0)))$ at x_0. Hence we have two-sided bounds for Hess $r(x)$ on appropriate sets. The proof can then be finished as for Theorem 3.3. $\qquad \square$

Interestingly enough, this theorem is optimal in two ways. Consider rotationally symmetric metrics $dr^2 + f_\varepsilon^2(r) \, d\theta^2$, where f_ε is concave and satisfies $f_\varepsilon(r) = r$ for $0 \le r \le 1 - \varepsilon$ and $f_\varepsilon(r) = \frac{3}{4} r$, with $1 + \varepsilon \le r$. These metrics have sec ≥ 0 and inj rad $= \infty$. As $\varepsilon \to 0$ we get a C^{1+1} manifold with a C^{0+1} riemannian metric (M, g). In particular, $\|(M, g)\|_{0+1, r} < \infty$ for all r. Limit spaces of sequences with inj rad $\geq i_0$ and sec $\geq k$ can therefore not in general be assumed to be smoother than the above example.

With a more careful construction, we can also find g_ε with $g_\varepsilon(r) = \sin(r)$ for $0 \leq r \leq \pi/2 - \varepsilon$ and $g_\varepsilon(r) = 1$ for $r \geq 1 + \varepsilon$, having the property that the metric $dr^2 + g_\varepsilon^2(r)\, d\theta^2$ satisfies $|\sec| \leq 4$ and inj rad $\geq \frac{1}{4}$. As $\varepsilon \to 0$ we get a limit metric of class C^{1+1}. So while we may suspect (this is still unknown) that limit metrics from Theorem 3.3 are C^{1+1}, we only prove that $\|(M, g)\|_{1+\alpha, r} \leq f(r)$ with $f(r) \to 0$ as $r \to 0$, where $f(r)$ depends only on $n = \dim M$, i_0 ($\leq$ inj rad M), and K ($\geq |\sec M|$); see Theorems 4.1 and 4.2.

4. The Best Possible of All Possible . . .

We are now ready to introduce and use harmonic coordinates in our convergence theory. We will use Einstein summation convention whenever convenient.

The Laplace–Beltrami operator Δ on (M, g) is defined as $\Delta = \text{trace}(\text{Hess})$. In coordinates one can compute

$$\Delta = \frac{1}{g} \sum \frac{\partial}{\partial x^i} \left(g\, g^{ij} \frac{\partial}{\partial x^j} \right) = \sum g^{ij} \frac{\partial^2}{\partial x^i \partial x^j} + \sum \frac{1}{g} \frac{\partial(g\, g^{ij})}{\partial x^i} \frac{\partial}{\partial x^j}$$

$$= \frac{1}{g} \partial_i (g\, g^{ij} \partial_j) = g^{ij} \partial_i \partial_j + \frac{1}{g} \partial_i (g\, g^{ij}) \partial_j,$$

where g_{ij} are the metric components, (g^{ij}) is the inverse of (g_{ij}), and $g = \sqrt{\det g_{ij}}$. A function u is said to be harmonic (on (M, g)) if $\Delta u = 0$. Notice that if we scale g to $\lambda^2 g$ then Δ changes to $\lambda^{-2}\Delta$; hence the concept of harmonicity doesn't change. A coordinate system $\varphi : U \to \mathbb{R}^n$ is said to be harmonic if each of the coordinate functions are harmonic. In harmonic coordinates, the Laplace operator has the form $\Delta = g^{ij} \partial_i \partial_j$, since

$$0 = \Delta x^k = g^{ij} \partial_i \partial_j x^k + \frac{1}{g} \partial_i (g\, g^{ij}) \partial_j x^k = 0 + \frac{1}{g} \partial_i (g\, g^{ij} \delta_j^k) = \frac{1}{g} \partial_i (g\, g^{ij}).$$

The nicest possible coordinates that one could ask for would be linear coordinates (Hess $\equiv 0$). But such coordinates clearly only exist on flat manifolds. In contrast, any riemannian manifold admits harmonic coordinates around every point. It turns out that these coordinates, while harder to construct, have much nicer properties than both exponential and distance coordinates. First of all we have the important equation $\Delta g_{ij} + Q(g_{ij}, \partial g_{ij}) = \text{Ric}_{ij}$ for the Ricci tensor in harmonic coordinates, where Q is a term quadratic in ∂g_{ij}. We shall often write this equation in symbolic form as $\Delta g + Q(g, \partial g) = \text{Ric}$, and merely imagine the appropriate indices.

This equation was used in [DeTurck and Kazdan 1981] to prove that the metric always has maximal regularity in harmonic coordinates (if it is not C^∞). And in a very important paper [Jost and Karcher 1982], Theorem 3.3 was improved as follows:

THEOREM 4.1. *Given (M, g) as in Theorem 3.3, then for every $\alpha < 1$ and $Q > 0$ there exists r depending on i_0, K, n, α, Q such that $\|(M, g)\|_{1+\alpha, r} \leq Q$.*

The coordinates used to give this improved bound were harmonic coordinates. We should point out their original theorem has been rephrased into our terminology. With these improved coordinates, the following result was then proved in [Peters 1987; Greene and Wu 1988]:

THEOREM 4.2. *The class of riemannian n-manifolds with $|\sec| \leq K$, diam $\leq D$, vol $\geq v$, for fixed but arbitrary $K, D, v > 0$, is precompact in the $C^{1,\alpha}$-topology for any $\alpha < 1$, and contains only finitely many diffeomorphism types.*

This is, of course, a direct consequence of Corollary 2.3, Lemma 3.4, and Theorem 4.1.

It is the purpose of this section to set up the theory of harmonic coordinates so that we can prove Anderson's generalization of these two results. For this purpose, let us define for a riemannian manifold (M, g) the (harmonic) $L^{p,k}$ norm at the scale of r, denoted $\|(M, g)\|_{p,k,r}$, as the infimum of all $Q \geq 0$ such that there are charts $\varphi_s : B(0, r) \to U_s \subset M$ satisfying:

(h1) $\delta = \frac{1}{10} e^{-Q} r$ is a Lebesgue number for the covering $\{U_s\}$.

(h2) $|d\varphi_s^{-1}| \leq e^Q$ and $|d\varphi_s| \leq e^Q$ on $B(0, r)$.

(h3) $r^{|j|-n/p} \|\partial^j g_{s..}\|_p \leq Q$ for $1 \leq |j| \leq k$.

(h4) $\varphi_s^{-1} : U_s \to B(0, r)$ are harmonic coordinates.

Here $\|f\|_p = (\int f^p)^{1/p}$ in the usual L^p-norm on euclidean space.

Several comments are in order. All riemannian manifolds in this section are C^∞, so our harmonic coordinates will automatically also be C^∞. Thus we are only concerned with how the metric is bounded, and not with how smooth it might be. When $k = 0$ condition (h3) is obviously vacuous, so we will always assume $k \geq 1$; for analytical reasons we will in addition assume that $p > n$ if $k = 1$ and $p > n/2$ if $k \geq 2$. The transition condition (n4) has been replaced with the somewhat different harmonicity condition. To recapture (n4), we need to use the L^p elliptic estimates from Appendix A together with Appendix B. In harmonic coordinates we have $\Delta = g^{ij} \partial_i \partial_j$, and (h2), (h3) imply that the eigenvalues for g^{ij} are in $[e^{-Q}, e^Q]$ and that $\|g^{ij}\|_{p,k} \leq \tilde{Q}$, where $\tilde{Q}$ depends on n, Q. Thus the L^p estimates say that $\|u\|_{p,k+1} \leq C(\|\Delta u\|_{p,k-1} + \|u\|_{L^p})$. In particular we get $L^{p,k+1}$ bounds on transition functions on compact subsets of domain of definition, since they satisfy $\Delta = 0$.

In the case where $k - n/p > 0$ is not an integer, we have a continuous embedding $L^{p,k} \subset C^{k-n/p}$; we can therefore bound $\|(M, g)\|_{k-n/p, \tilde{r}}$, for $\tilde{r} < r$, in terms of $\|(M, g)\|_{p,k,r}$. This, together with the Fundamental Theorem 2.2, immediately implies:

THEOREM 4.3. *Let $Q, r > 0$, $p > 1$ and $k \in \mathbb{N}$ be given. The class $\mathcal{H}^{p,k}(n, Q, r)$ of n-dimensional riemannian manifolds with $\|(M, g)\|_{p,k,r} \leq Q$ is precompact in the pointed $C^{l+\alpha}$ topology for $l + \alpha < k - n/p$.*

The concept of $L^{p,k}$-convergence for riemannian manifolds is defined as for $C^{l+\alpha}$-convergence, using convergence in appropriate fixed coordinates for the pullback metrics.

Before discussing how to achieve $L^{p,k}$-convergence, we warm up with some elementary properties of our new norm concept. Note that if $A \subset M$, then $\|(A,g)\|_{p,k,r}$ is defined just like $\|(A,g)\|_{l+\alpha,r}$.

PROPOSITION 4.4. *Let* (M,g) *be a* C^∞ *riemannian manifold.*

(i) $\|(A, \lambda^2 g)\|_{p,k,\lambda r} = \|(A,g)\|_{p,k,r}.$

(ii) $\|(A,g)\|_{p,k,r}$ *is finite for some* r, *then it is finite for all* r, *and in this case it will be a continuous function of* r *that approaches* 0 *as* $r \to 0$.

(iii) *For any* $D > 0$, *we have*

$$\|(M,g)\|_{p,k,r} = \sup\{\|(A,g)\|_{p,k,r} : A \subset M, \operatorname{diam} A \leq D\}.$$

(iv) *If the sequence* (M_i, g_i, p_i) *converges to* (M, g, p) *in the pointed* $L^{p,k}$-*topology, and all spaces are* C^∞ *riemannian manifolds, then for each bounded* $B \subset M$ *there are bounded sets* $B_i \subset M_i$ *such that*

$$\|(B_i, g_i)\|_{p,k,r} \to \|(B,g)\|_{p,k,r}.$$

PROOF. Parts (i) and (ii) are proved as before, except for the statement that $\|(M,g)\|_{p,k,r}$ converges to 0 as $r \to 0$. To see this, observe that we know that $\|(M,g)\|_{2k+\alpha,r} \to 0$ as $r \to 0$. Then approximate these coordinates by harmonic maps by solving Dirichlet boundary value problems. This will then give the right type of harmonic coordinates.

Property (iii) is unique to norms that come from harmonic coordinates. What fails in the general case is the condition (n4). Clearly $K = \sup\{\|(A,g)\|_{p,k,r} : A \subset M, \operatorname{diam} A \leq D\} \leq \|(M,g)\|_{p,k,r}$. Conversely, choose $Q > K$. If every set of diameter $\leq D$ can be covered with coordinates satisfying (h1)–(h4), then M can obviously be covered by similar coordinates.

To prove (iv), suppose that (M_i, g_i, p_i) converges to (M, g, p) in the pointed $L^{k,p}$-topology and that B is a bounded subset of M. If $B \subset B(p, R) \subset \Omega$ and we have $F_i : \Omega \to \Omega_i \supset B(p_i, R)$ such that $F_i^* g_i \to g$ in the $L^{k,p}$-topology, we can just let $B_i = F_i(B)$.

Let's first prove the inequality $\limsup \|(B_i, g)\|_{p,k,r} \leq \|(B,g)\|_{p,k,r}$. For this purpose, choose $Q > \|(B,g)\|_{p,k,r}$; and then using (i) (see also the proof part (i) of Proposition 2.1), choose $\varepsilon > 0$ such that $\|(B,g)\|_{p,k,(r+\varepsilon)} < Q$.

Then select a finite collection of charts $\varphi_s : B(0, r + \varepsilon) \to U_s \subset M$ that realizes this inequality. Define $U_{is} = F_i(\varphi_s(D(0, r + \varepsilon/2)))$; then U_{is} is a closed disk with boundary $\partial U_{is} = F_i(\varphi_s(S(0, r+\varepsilon/2)))$. On each U_{is}, solve the Dirichlet boundary value problem

$$\psi_{is} : U_{is} \to \mathbb{R}^n, \text{ with } \Delta_i \psi_{is} = 0 \text{ and } \psi_{is}|_{\partial U_{is}} = \varphi_s^{-1} \circ F_i^{-1}|_{\partial U_{is}}.$$

Then on each $\varphi_s\left(D\left(0,\, r+\varepsilon/2\right)\right)$ we have two maps: φ_s^{-1}, which satisfies $\Delta\varphi_s^{-1} \equiv 0$, and $\psi_{is}\circ F_i$, which satisfies $\Delta_i\psi_{is}\circ F_i \equiv 0$, where Δ_i also denotes the Laplacian of the pullback metric $F_i^* g_i$. We know that $F_i^* g_i \to g$ in $L^{p,k}$ with respect to the fixed coordinate system φ_s on $B(0,\, r+\varepsilon)$. What we want to show is that the metric $F_i^* g_i$ in the harmonic coordinates ψ_{is} also converges to g in the $L^{p,k}$ topology on $B(0,r)$, or equivalently on $\varphi_s(B(0,r))$. This will clearly be true if we can prove that $\|\varphi_s^{-1} - \psi_{is}\circ F_i\|_{p,k+1}$ converges to as $i \to \infty$. If we write Δ_i in the fixed coordinate system φ_s, then the elliptic estimates for divergence operators (see Theorem A.3) tell us that

$$\|\varphi_s^{-1} - \psi_{is}\circ F_i\|_{p,2,B(0,r+\varepsilon/2)} \le C\,\|\Delta_i(\varphi_s^{-1} - \psi_{is}\circ F_i)\|_{p,B(0,r+\varepsilon/2)}$$
$$= C\,\|\Delta_i\varphi_s^{-1}\|_{p,B(0,r+\varepsilon/2)} \qquad (4.1)$$

when $k = 1$ and $p > n$, while

$$\|\varphi_s^{-1} - \psi_{is}\circ F_i\|_{p,k+1,B(0,r)}$$
$$\le C\left(\|\Delta_i\varphi_s^{-1}\|_{p,k-1,B(0,r+\varepsilon/2)} + \|\varphi_s^{-1} - \psi_{is}\circ F_i\|_{p,B(0,r+\varepsilon/2)}\right) \qquad (4.2)$$

when $k \ge 2$ and $p > n/2$.

To use these inequalities, observe that $\|\Delta_i\varphi_s^{-1}\|_{p,k-1,B(0,r+\varepsilon/2)}$ approaches 0 as $i \to \infty$ since the coefficients of $\Delta_i = a^{ij}\partial_i\partial_j + b^j\partial_j$ converge to those of Δ in $L^{p,k-1}$, and $\Delta\varphi_s^{-1} = 0$ (see also Appendix B). Inequality 4.1 therefore takes care of the cases when $k = 1$ and $p > n$.

For the remaining cases, when $k \ge 2$ and $p > n/2$, recall that we have a Sobolev embedding $L^{2p,1} \supset L^{p,k}$. Thus we can again use 4.1 to see that $\|\varphi_s^{-1} - \psi_{is}\circ F_i\|_{p,B(0,r+\varepsilon/2)}$ approaches 0 as $i \to \infty$. Inequality 4.2 then shows that $\|\varphi_s^{-1} - \psi_{is}\circ F_i\|_{p,k+1,B(0,r)} \to 0$ as $i \to \infty$.

To check the reverse inequality $\liminf \|(B_i,g)\|_{p,k,r} \ge \|(B,g)\|_{p,k,r}$, fix some $Q > \liminf \|(B_i,g)\|_{p,k,r}$. For some subsequence of (M_i,g_i,p_i) we can then choose charts $\varphi_{is} : B(0,r) \to U_{is} \subset M_i$ satisfying (h1)–(h4) that converge to limit charts $\varphi_s : B(0,r) \to U_s \subset M$ satisfying (h1)–(h3). Since the metrics converge in the $L^{k,p}$-topology, $k \ge 1$, the Laplacians must also converge, and so we can easily check that the charts φ_s^{-1} are harmonic and therefore give charts showing that $\|(B,g)\|_{p,k,r} \le Q$. $\qquad\square$

One of the important consequences of the Fundamental Theorem is that convergence of the metric components in appropriate charts implies global convergence of the manifolds. The key step in getting these global diffeomorphisms is to glue together compositions of charts using a center of mass construction. It is clearly desirable to have a similar local-to-global convergence construction for $L^{p,k}$ convergence, but the method as we described fails. This is because products of $L^{p,k}$ functions are not necessarily $L^{p,k}$. If, however, one of the functions in the product or composition has universal C^∞ bounds, there won't be any problems. So we arrive at the following result:

LEMMA 4.5. *Suppose (M_i, g_i, p_i) and (M, g, p) are pointed C^∞ riemannian man-ifolds and that we have charts $\varphi_{is} : B(0, r) \to U_{is} \subset M_i$ as in the proof of the Fundamental Theorem 2.2 satisfying (h1), (h2), and (h4), and such that the metric components g_{is} converge in the $L^{p,k}$-topology, to the limit metric (M, g). Then a subsequence of (M_i, g_i, p_i) will converge to (M, g, p) in the pointed $L^{p,k}$ topology.*

PROOF. The proof is almost word for word the same as that of the Fundamental Theorem. The important observation is that the limit coordinates $\varphi_s : B(0, r) \to U_s \subset M$ satisfy (h4) and are therefore C^∞, since (M, g) was assumed to be C^∞. The partition of unity on (M, g) is therefore also C^∞. Thus the compositions $\varphi_{is} \circ \varphi_s^{-1}$ are all close in the $L^{p,k+1}$ topology, and we also have

$$\|\mu_1 F_1 + \mu_2 F_2 - F_1\|_{p,k+1} \leq \|\mu_2\|_{C^\infty} \|F_2 - F_1\|_{p,k+1}.$$

Finally, $\mu_1 F_1 + \mu_2 F_2$ is also C^1 close to F_1, since $k + 1 - n/p > 1$ and $L^{p,k+1} \subset C^1$ is continuous. □

5. Generalized Convergence Results

Many of the results in this section have also been considered by Gao and D. Yang, but we will follow the approach taken by Anderson. The survey paper [Anderson 1993] has many good references for further applications and to the papers where many of these things were first considered.

The machinery developed in the previous section makes it possible to state and prove the most general results immediately. Rather than assuming point-wise bounds on curvature, we will use L^p bounds. Our notation is $\|\operatorname{Ric}\|_p = \left(\int_M |\operatorname{Ric}|^p d\operatorname{vol}\right)^{1/p}$ and $\|R\|_p = \left(\int_M |R|^p d\operatorname{vol}\right)^{1/p}$, where $|\operatorname{Ric}|$ and $|R|$ are the pointwise bound on the Ricci tensor and curvature tensor. Note that if we scale the metric g to $\lambda^2 g$, the curvature tensors are multiplied by λ^{-2} and the volume form by λ^n, so the new L^p norms on curvature are $\|\operatorname{Ric}_{new}\|_p = \lambda^{(n/p)-2}\|\operatorname{Ric}_{old}\|_p$ and $\|R_{new}\|_p = \lambda^{(n/p)-2}\|R_{old}\|_p$. So if $p > n/2$ the L^p bounds scale just as the $C^0 = L^\infty$ bounds, while if $p = n/2$ they remain unchanged. We shall, therefore, mostly be concerned with L^p bounds on curvature, for $p > n/2$. Another heuristic reason for doing so is that if $f \in L^{p,2}$ for $p \in (n/2, n)$, then $f \in C^{2-n/p} \cap L^{1,q}$, where $1/q = 1/p - 1/n$. In particular we have C^0 control on f. So L^p bounds $p > n/2$ on curvature should somehow control the geometry.

With this philosophy behind us, we can state and prove Anderson's conver-gence results.

THEOREM 5.1. *Let $p > n/2$, $\Lambda \geq 0$, and $i_0 > 0$ be given. For every $Q > 0$ there is an $r = r(n, p, \Lambda, i_0)$ such that any riemannian n-manifold (M, g) with $\|\operatorname{Ric}\|_p \leq \Lambda$ and inj rad $\geq i_0$ satisfies $\|(M, g)\|_{p,2,r} \leq Q$.*

PROOF. Suppose this were not true. Then for some $Q > 0$ we could find (M_i, g_i) with $\|(M_i, g_i)\|_{p,2,r_i} \geq Q$ for some sequence $r_i \to 0$. By further decreasing r_i we can actually assume $\|(M_i, g_i)\|_{p,2,r_i} = Q$ for $r_i \to 0$. Now define $h_i = r_i^{-2} g_i$; then $\|(M_i, h_i)\|_{p,2,1} = Q$. Next select $p_i \in M_i$ such that $\|(A_i, h_i)\|_{p,2,1} \geq Q/2$, for all $A_i \ni p_i$. After possibly passing to a subsequence, we can assume that (M_i, h_i, p_i) converges to (M, g, p) in the pointed $C^{l+\alpha}$, topology, where $l + \alpha < 2 - n/p$. Together with this convergence we will select charts $\varphi_{is} : B(0, r) \to U_{is} \subset M_i$ and $\varphi_s : B(0, r) \to U_s \subset M$ such that the φ_{is} satisfy (h1)–(h4) for some $\tilde{Q} > Q$, $r > 1$, and converging to φ_s in $C^{l+1+\alpha}$. Denote the metric components of h_i in φ_{is} by $h_{is\cdots}$, and similarly denote by $g_{s\cdots}$ the components for g in φ_s. By assumption the $h_{is\cdots}$ are bounded in $L^{p,2}$. If $p > n$ then $2 - n/p > 1$, so $l = 1$; hence we can assume that the $h_{is\cdots}$ converge to $g_{s\cdots}$ in $C^{1+\alpha}$, with $\alpha = 1 - n/p$. If $p < n$ then $1/(2p) > 1/p - 1/n$, and we can assume that the $h_{is\cdots}$ converge to $g_{s\cdots}$ in $L^{2p,1}$. In particular, g has a well defined Laplacian Δ on M, which maps $L^{p,2}$ into L^p, and the Laplacians Δ_i for h_i on M_i converge to Δ in $L^{2p,1}$. So if $u \in L^{p,2}$ then $\Delta_i u \to \Delta u$ in L^p.

We will now establish simultaneously two contradictory statements, namely that (M, g) is $\mathbb{R}^n$ with the canonical metric, and that the (M_i, h_i, p_i) converge to (M, g, p) in the pointed $L^{p,2}$ topology. These statements are contradictory since we know that $\|(\mathbb{R}^n, \text{can})\|_{p,2,1} = 0$, and therefore Proposition 4.4(iv) implies that $\|(B(p_i, \delta), h_i)\|_{p,2,1} \to 0$ as $i \to \infty$, which goes against our assumption that

$$\|(B(p_i, \delta), h_i)\|_{p,2,1} \geq Q/2.$$

Since we are using harmonic coordinates, $h_{is\cdots}$ satisfies

$$\Delta_i h_{is\cdots} + Q(h_{is\cdots}, \partial h_{is\cdots}) = \text{Ric}_{is\cdots},$$

where $\text{Ric}_{is\cdots}$ represents the components of the Ricci tensor for h_i in the φ_{is} coordinates. Since $h_i = r_i^{-2} g_i$ and $r_i \to 0$ we know that $\|\text{Ric}_{is\cdots}\|_p \to 0$ as $i \to 0$. Also, $h_{is\cdots} \to g_{s\cdots}$ in C^0 and $\partial h_{is\cdots} \to \partial g_{s\cdots}$ in L^{2p}. Therefore $Q(h_{is\cdots}, \partial h_{is\cdots}) \to Q(g_{s\cdots}, \partial g_{s\cdots})$ in L^p, since Q is a universal function of class C^∞ in its arguments and quadratic in ∂g. Thus we conclude that $\Delta_i h_{is\cdots} \to \Delta g_{s\cdots}$ in L^p. We can then use the L^p estimates from Appendix A (Theorem A.2) to see that

$$\|h_{is\cdots} - g_{s\cdots}\|_{p,2}$$
$$\leq C_i \|\Delta_i(h_{is\cdots} - g_{s\cdots})\|_{L^p} + C_i \|h_{is\cdots} - g_{s\cdots}\|_{L^p}$$
$$\leq C_i \|\Delta_i(h_{is\cdots}) - \Delta(g_{s\cdots})\|_{L^p} + C_i \|(\Delta - \Delta_i) g_{s\cdots}\|_{L^p} + C_i \|h_{is\cdots} - g_{s\cdots}\|_{L^p},$$

which converges to 0 as $i \to \infty$. As usual, C_i is bounded by the coefficients of Δ_i, which we know are universally bounded. There is also the slight detail about the left-hand side being measured on a compact subset of $B(0, r)$. But this can be finessed because $\tilde{Q} > Q = \|(M_i, h_i, p_i)\|_{p,2,1}$, so we can find a universal $r = 1 + \varepsilon$ by the proof of Proposition 4.4(i) such that $\|(M_i, h_i)\|_{p,2,r} < \tilde{Q}$. Thus we have the convergence taking place on $B(0, 1)$, and we conclude using Proposition 4.4(iv) that $(M_i, h_i, p_i) \to (M, g, p)$ in the pointed $L^{p,2}$ topology. (We will show below that (M, g) is C^∞.)

The limit metric (M, g) now satisfies $\Delta g = -Q(g, \partial g)$ in our weak $L^{p,2}$ harmonic coordinates. If $p > n$, we have $g \in C^{1+\alpha}$ as already observed; thus the leading coefficients for Δ are in C^1, and $Q(g, \partial g)$ is in C^α. Then standard regularity theory implies $g \in C^{2+1+\alpha}$. A bootstrap argument then shows $g \in C^\infty$. The equation $\Delta g + Q = 0$ then shows that M is Ricci-flat. We must now recall that $\operatorname{inj rad}(M_i, h_i) = r_i^{-1} i_0$, which goes to ∞, so $\operatorname{inj rad}(M, g) = \infty$. The Cheeger–Gromoll splitting theorem then implies that $(M, g) = (\mathbb{R}^n, \operatorname{can})$.

We must now deal with the case where $p \in (n/2, n)$. Define $\tilde{p} = (2/p - 2/n)^{-1}$, so $f \in L^{p,2}$ implies $f \in L^{2\tilde{p},1} \cap C^{2-n/p}$. Since $g \in L^{p,2}$, we must therefore have $Q(g, \partial g) \in L^{\tilde{p}}$. But then $\Delta g \in L^{\tilde{p}}$, and $g \in L^{\tilde{p},2}$ since the coefficients for Δ are C^0. Now $\tilde{p} > n/2 + 2(p - n/2) = 2p - n/2 = p + (p - n/2)$. Iterating this procedure k times, we get $g \in L^{2,q}$ for $q > p + k(p - n/2)$. If k is big enough, we clearly have $q > n$, and we get $g \in C^{1+\beta}$, where $\beta = 1 - n/q > 0$. We can then proceed as above. $\qquad \square$

This theorem has some immediate corollaries.

COROLLARY 5.2.   *Given $\Lambda \geq 0$, $i_0, Q > 0$, and $p > 1$, there is an $r = r(n, p, Q, \Lambda, i_0)$ such that any riemannian n-manifold (M, g) with $|\mathrm{Ric}| \leq \Lambda$ and $\operatorname{inj rad} \geq i_0$ satisfies $\|(M, g)\|_{p,2,r} \leq Q$.*

This corollary clearly implies that we have $C^{1+\alpha}$ control ($\alpha < 1$) on the metric, given C^0 bounds on $|\mathrm{Ric}|$ and lower bounds on injectivity radius.

COROLLARY 5.3.   *Given Λ, $i_0, V > 0$, and $p > n/2$, the class of n-manifolds with: $\|\mathrm{Ric}\|_p \leq \Lambda$, $\operatorname{inj rad} \geq i_0$, and $\mathrm{vol} \leq V$ is precompact in the C^α-topology, where $\alpha < 2 - n/p$.*

Recall that, when we had bounded sectional curvature, we could replace the injectivity radius bound by a lower volume bound on balls of radius 1. This is no longer possible when we only bound the Ricci curvature, because there are many nonflat Ricci-flat metrics with volume growth of order r^n—for example, the Eguchi–Hanson metric described in Appendix C. Also, if we assume $\|R\|_p \leq \Lambda$, we can construct examples with lower volume bounds on balls of radius 1, but without any control on the metric. This is partly due to the fact that we have no way of showing that these assumptions imply the volume growth condition $\mathrm{vol} B(p, r) \geq v' r^n$ for $r \leq 1$. To get such a condition, one must have at least a lower bound on Ricci curvature. The best we can do is therefore this:

THEOREM 5.4.   *Given $p > n/2$, $\Lambda \geq 0$, $v > 0$, and $Q > 0$, there is an $r = r(n, p, Q, \Lambda, v)$ such that any n-manifold (M, g) with $\|R\|_p \leq \Lambda$ and $\mathrm{vol} B(p, s) \geq v s^n$ for all $p \in M$ and $s \leq 1$ satisfies $\|(M, g)\|_{p,2,r} \leq Q$.*

PROOF.   The setup is identical to the proof of Theorem 5.1. This time the limit space will be flat, since $\|R\|_p$ varies continuously in the $L^{p,2}$ topology, and have $\mathrm{vol} B(p, r) \geq v r^n$ for all r. Thus $(M, g) = (\mathbb{R}^n, \operatorname{can})$. $\qquad \square$

Considering that the proofs of Theorems 5.1 and 5.4 only differ in how we characterize euclidean space, it would seem reasonable to conjecture that any appropriate characterization of euclidean space should yield some kind of theorem of this type. We give some examples of this now:

EXAMPLE 5.5. $(\mathbb{R}^n, \text{can})$ is the only space with Ric ≥ 0 and inj rad $= \infty$, by the Cheeger–Gromoll splitting theorem. In [Anderson and Cheeger 1992], it is proved by a slightly different method from what we have described above that, for any $p > n$, the $L^{1,p}$ norm at some scale r is controlled by dimension, lower Ricci curvature bounds, and lower injectivity radius bounds.

EXAMPLE 5.6. If in Theorems 5.1 and 5.4 we also assume that the covariant derivatives of Ric up to order k are bounded in L^p, where $p > n/2$, we can control the $L^{p,k+2}$ norm. The proof of this is exactly like the proof of Theorems 5.1 and 5.4: we just use the stronger assumption that Ric goes to zero in the $L^{p,k}$ topology.

EXAMPLE 5.7. Let ω_n be the volume of the unit ball in euclidean space. Then $(\mathbb{R}^n, \text{can})$ is characterized as the only Ricci-flat manifold with vol $B(p,r) \geq (\omega_n - \varepsilon_n)r^k$ for all r, where $\varepsilon_n > 0$ is a universal constant only depending on dimension. The existence of such a constant was established in [Anderson 1990b]. Note that relative volume comparison shows that the volume growth need only be checked for one $p \in M$. Also if (M, g, p) satisfies these conditions, then $(M, \lambda^2 g, p)$ satisfies these conditions for any $\lambda > 0$.

It is actually an interesting application of Theorem 5.1 to prove that this really gives a characterization of $(\mathbb{R}^n, \text{can})$. We proceed by contradiction on the existence of such an $\varepsilon_n > 0$. Using scale invariance of norm and the conditions, we can then find a sequence (M_i, g_i, p_i) of Ricci-flat manifolds such that vol $B(p_i, r) \geq (\omega_n - i^{-1})r^n$ for all r, and a sequence $R_i \to \infty$ such that, say, $Q \geq \|(A_i, g_i)\|_{2n,2,1} \geq Q/2$, where $p_i \in A_i \subset B(p_i, R_i)$, for some $Q > 0$. The proof of Theorem 5.1 then shows that $(B(p_i, R_i), p_i) \to (M, g, p)$ in the $L^{2n,2}$-topology. But this limit space is now Ricci-flat and satisfies vol $B(p, r) = \omega_n r^n$ for all r, and must, therefore, be $(\mathbb{R}^n, \text{can})$. This contradicts continuity of the $L^{2n,2}$ norm in the $L^{2n,2}$ topology.

We can then control the $L^{p,2}$ norm in terms of r_0 and Λ, where $\| \text{Ric} \|_p \leq \Lambda$ and vol $B(p,r) \geq (\omega_n - \varepsilon_n)r^n$, for $r \leq r_0$. If we impose the stronger curvature bound $|\text{Ric}| \leq \Lambda$, then relative volume comparison shows that we need only check that vol $B(p, r_0) \geq (\omega_n - \varepsilon_n)(v(n, -\Lambda, 1))^{-1}v(n, -\Lambda, r_0)$, where $v(n, -\Lambda, r)$ is the volume of an r-ball in constant curvature $-\Lambda$ and dimension n.

As already pointed out, the Eguchi–Hanson metric shows that one cannot expect to get as nice control on the metric with Ricci curvature and general volume bounds as one gets in the presence of sectional curvature bounds. But if we also assume $L^{n/2}$ bounds on R, then we can do better than in Example 5.7.

EXAMPLE 5.8. In [Bando et al. 1989] it is proved that any complete n-manifold with Ric $\equiv 0$, $\|R\|_{n/2} < \infty$ and vol $B(p, r) \geq vr^n$, for all r and some $v > 0$, is an asymptotically locally euclidean space. This implies, in particular, that lim vol $B(p, r)/r^n = \omega_n/k$ for some integer k (in fact, k is the order of the fundamental group at ∞). If, therefore, $v > \frac{1}{2}\omega_n$, then $k = 1$. But then relative volume comparison and nonnegative Ricci curvature implies that the space is euclidean space.

This again leads to two results. First: The $L^{2,p}$ norm for any p is controlled by Λ, r_0, and ε, where $|\text{Ric}| \leq \Lambda$, $\|R\|_{n/2} \leq \Lambda$, and

$$\text{vol} \, B(p, r_0) \geq \left(\tfrac{1}{2}\omega_n + \varepsilon\right) \left(v(n, -\Lambda, 1)\right)^{-1} v(n, -\Lambda, r_0).$$

Second: The $L^{2,p}$ norm is controlled by Λ, r_0 and ε where $\|\text{Ric}\|_p \leq \Lambda$, $\|R\|_{n/2} \leq \Lambda$ and vol $B(p, r) \geq \left(\tfrac{1}{2}\omega_n + \varepsilon\right) r^n$, for $r \leq r_0$.

For even more general volume bounds, we have:

EXAMPLE 5.9. With a contradiction argument similar to the one used in Example 5.7, one can easily see that there is an $\varepsilon(n, v) > 0$ such that any complete n-manifold with Ric $\equiv 0$, $\|R\|_{n/2} \leq \varepsilon(n, v)$ and vol $B(p, r) \geq vr^n$ for all r is $(\mathbb{R}^n, \text{can})$. Hence given $v, r_0, \Lambda > 0$ with $|\text{Ric}| \leq \Lambda$, vol $B(p, r_0) \geq v$, and the $(n/2)$-norm of R on balls of radius r_0 satisfying $\|R|_{B(p, r_0)}\|_{n/2} \leq \varepsilon(n, v)$ we can control the $L^{p,2}$-norm for any p. We can, as before, modify this argument if we only wish to assume L^p bounds on Ric (see also [Yang 1992] for a different approach to this problem).

This result seems quite promising for getting our hands on manifolds with given upper bounds on $|\text{Ric}|$, $\|R\|_{n/2}$, and lower bounds on vol $B(p, 1)$. The Eguchi–Hanson metric will, however, still give us trouble. This metric has Ric $\equiv 0$, $\|R\|_{n/2} < \infty$, and vol $B(p, r) \geq \frac{1}{2}\omega_n r^n$ for all p. All these quantities are scale-invariant, but if we multiply by larger and larger constants, the space will converge in the Gromov–Hausdorff topology to the euclidean cone over $\mathbb{RP}^3$, which is not even a manifold. Away from the vertex of the cone, the convergence is actually in any topology we like, since Ric $\equiv 0$. What happens is that $\|R\|_{n/2}$, while staying finite, will concentrate around the singularity that develops and go to zero elsewhere. Questions related to this are investigated in [Anderson and Cheeger 1991], where the authors prove that the class of n-manifolds with $|\text{Ric}| \leq \Lambda$, $\|R\|_{n/2} \leq \Lambda$, diam $\leq D$, and vol $\geq v$ contains only finitely many diffeomorphism types. The idea is that these bounds give $L^{p,2}$-control on each of the spaces away from at most $N(n, \Lambda, D, v)$ points, and that each of these bad points can only degenerate in a very special fashion that looks like the degeneration of the Eguchi–Hanson metric.

Recall that the injectivity radius of a closed manifold is computed as either half the length of the shortest closed geodesic (abbreviated scg(M, g)) or the distance to the first conjugate point. Thus the injectivity radius of a manifold with bounds on $|R|$ is completely controlled by $\mathrm{scg}(M)$. In the next example, we shall see how at least the $L^{p,2}$ norm is controlled by $\|R\|_p$ and $\mathrm{scg}(M)$. Some auxiliary comments are needed before we proceed. Let us denote by $\mathrm{sgl}(M,g)$ the shortest geodesic loop. (A closed geodesic is a smooth map $\gamma : S^1 \to M$ with $\ddot{\gamma} \equiv 0$, while a geodesic loop is a smooth map $\gamma : [0,l] \to M$ with $\gamma(0) = \gamma(l)$ and $\ddot{\gamma} \equiv 0$. The base point for a loop is $\gamma(0)$.) On a closed manifold M we have $\mathrm{scg}(M,g) \geq \mathrm{sgl}(M,g)$. On a complete flat manifold, $\mathrm{scg}(M,g) = \mathrm{sgl}(M,g)$, and each geodesic loop realizing $\mathrm{sgl}(M,g)$ is a noncontractible closed geodesic.

EXAMPLE 5.10. (See also [Anderson 1991]). Our characterization of $(\mathbb{R}^n, \mathrm{can})$ is that it is flat and contains no closed geodesics. First we claim that given $p > n/2$ and $l, \Lambda, Q > 0$ there is an $r(n,p,l,\Lambda,Q)$ such that any closed manifold with $\|R\|_p \leq \Lambda$ and $\mathrm{sgl} \geq l$ satisfies $\|(M,g)\|_{p,2,r} \leq Q$. This is established by our usual contradiction argument. If the statement were not true, we could find a sequence (M_i, g_i) with $\|R(g_i)\|_p \to 0$, $\mathrm{sgl}(M_i, g_i) \to \infty$, and $\|(M_i, g_i)\|_{p,2,1} = Q > 0$ for all i. For each (M_i, g_i) select $p_i \in M_i$ such that $\|(A_i, g_i)\|_{p,2,1} = Q$ for any $A_i \ni p_i$. Then the pointed sequence (M_i, g_i, p_i) will subconverge to a flat manifold (M, g, p) in the pointed $L^{p,2}$ topology. Thus (M, g, p) also has the property that $\|(A, g)\|_{p,2,1} = Q$ for all $A \ni p$. This implies that $\mathrm{inj\,rad}(M,g) < 1$ at p; thus we have a geodesic loop $\gamma : [0,l] \to M$ of length $l < 2$ and $\gamma(0) = \gamma(l) = p$. Using the embeddings $F_{ik} : \Omega_k \to M_i$, we can then construct a loop $c_i : [0, l_i] \to M_i$ of length $l_i \to l$ with $c_i(0) = c_i(l_i) = p_i$. Since γ is not contractible, the loops c_i will not be contractible in $B(p_i, 1)$ for large i. We can, therefore, shorten c_i through loops based at p_i until we get a nontrivial geodesic loop $\gamma_i : [0, \tilde{l}_i] \to M_i$ of length $\tilde{l}_i \leq l_i$ based at p_i. This, however, contradicts our assumption that $\mathrm{sgl}(M_i, g_i) \to \infty$.

We can now use this to establish the following statement: Given $p > n/2$ and $\Lambda, l, Q > 0$, there is an $r = r(n, p, Q, \Lambda, l)$ such that any closed manifold with $\|R\|_p \leq \Lambda$ and $\mathrm{scg} \geq l$ satisfies $\|(M,g)\|_{p,2,r} \leq Q$. We contend that there is an $\tilde{l} = \tilde{l}(n, p, \Lambda, l)$ such that any such manifold also satisfies $\mathrm{sgl} \geq \tilde{l}$. Otherwise, we could find a pointed sequence (M_i, g_i, p_i) with $\|R(g_i)\|_p \to 0$, $\mathrm{scg}(M_i, g_i) \to \infty$, and $\mathrm{sgl}(M_i, g_i) = 2$. Let $\gamma_i : [0, 2] \to M_i$ be a geodesic loop at p_i of length 2. From our previous considerations, it follows that (M_i, g_i, p_i) subconverges to a flat manifold (M, g, p) in the pointed $L^{p,2}$ topology. The loops γ_i will converge to a geodesic loop $\gamma : [0, 2] \to M$ of length 2 and $\gamma(0) = \gamma(2) = p$. If $\mathrm{sgl}(M,g) = \mathrm{scg}(M,g) < 2$ we can, as before, find geodesic loops in M_i of length less than 2 for large i. Hence γ is actually a closed geodesic of length 2 which is noncontractible. This implies that the geodesic loops γ_i are noncontractible inside $B(p_i, 2)$ for large i. By assumption γ_i cannot be a closed geodesic, and is therefore not smooth at p_i. Thus the closed curve γ_i, when based at $\gamma_i(\varepsilon)$ for some fixed but small $\varepsilon > 0$, can be shortened through curves based at $\gamma_i(\varepsilon)$ to a nontrivial geodesic loop of length less than 2. This again contradicts the assumption that $\mathrm{sgl}(M_i, g_i) = 2$.

Our proof of this last result deviates somewhat from that of [Anderson 1991]. This is required by our insisting on using norms rather than radii to measure smoothness properties of manifolds.

EXAMPLE 5.11. If a complete manifold satisfies $\mathrm{Ric} \geq -\Lambda$, $\mathrm{vol}\, B(p,1) \geq v$ for all p, and the conjugate radius is at least r_0, then the injectivity radius has a lower bound, $\mathrm{inj\,rad} \geq i_0(n, \Lambda, v, r_0)$. This result was observed by Zhu and the author, and independently by Dai and Wei. It follows from the proof of the injectivity radius estimate in [Cheeger et al. 1982]: the above bounds are all one needs in order for the proof to work. Thus we conclude that the class of n-manifolds with $|\mathrm{Ric}| \leq \Lambda$, $\mathrm{vol} \geq v$, $\mathrm{diam} \leq D$, and $\mathrm{conj\,rad} \geq r_0$ is precompact in the $C^{1+\alpha}$-topology for all $\alpha < 1$.

The local model for a class of manifolds is the type of space characterized by the inequalities and equations one gets by multiplying the metrics in the class by constants going to infinity and observing how the quantities defining the class changes.

Some examples are: The class of manifolds with $\|\mathrm{Ric}\|_p \leq \Lambda$ and $\mathrm{inj\,rad} \geq i_0$ has local model $\mathrm{Ric} \equiv 0$, $\mathrm{inj\,rad} \equiv \infty$, which, as we know, is $(\mathbb{R}^n, \mathrm{can})$. The class of manifolds with $\|R\|_p \leq \Lambda$ has local model $R \equiv 0$, i.e., flat manifolds. The class with $|\mathrm{Ric}| \leq \Lambda$, $\|R\|_{n/2} \leq M$, and $\mathrm{vol}\, B(p, r_0) \geq v$ has local model $\mathrm{Ric} \equiv 0$, $\|R\|_{n/2} < \infty$, $\mathrm{vol}\, B(p, r) \geq v' r^n$ for all r; these are the nice asymptotically locally euclidean spaces.

In all our previous results, the local model has been $(\mathbb{R}^n, \mathrm{can})$. It is, therefore, reasonable to expect any class that has $(\mathbb{R}^n, \mathrm{can})$ as local model to satisfy some kind of precompactness condition. If the local model is not unique, the situation is obviously somewhat more complicated, but it is often possible to say something intelligent. It is beyond the scope of this article to go into this. Instead we refer the reader to [Anderson 1993; Fukaya 1990; Cheeger et al. 1992; Anderson and Cheeger 1991].

As a finale, Table 1 lists schematically some of the results we have proved.

6. Applications of Convergence Theory to Pinching Problems

In this section we will present some pinching results inspired by the work in [Gao 1990]. Our results are more general and the proofs rely only on the convergence results and their proofs given in previous sections, not on the work in [Gao 1990].

For $p > n/2 \geq 1$ and $Q, r > 0$, we have the class $\mathcal{H}(n, p, Q, r)$ of pointed C^∞ riemannian n-manifolds with $\| \cdot \|_{p,2,r} \leq Q$. We will be concerned with the subclass $\mathcal{H}(V) = \mathcal{H}(n, p, Q, r, V)$ of compact manifolds that in addition have volume at most V. If $p > n$ we know that $\mathcal{H}(V)$ is precompact in the $C^{1+\alpha}$ topology, for $\alpha < 1 - n/p$, while if $n/2 < p < n$ the class $\mathcal{H}(V)$ is precompact

	Class	Local model $= \mathbb{R}^n$	Precompact in C^β-topology
1	$\|R\| \leq \Lambda, (\|\nabla^k \mathrm{Ric}\| \leq M)$ $\mathrm{vol} \geq v$ $\mathrm{diam} \leq D$	$R \equiv 0$ $\mathrm{vol}\, B(p,r) \geq v' \cdot r^n$	$\beta < 2(+k)$
2	$\|\mathrm{Ric}\|_p \leq \Lambda, (\|\nabla^k \mathrm{Ric}\|_p \leq M)$ $\mathrm{inj} \geq i_0$ $\mathrm{vol} \leq V$ or $\mathrm{diam} \leq D$	$\mathrm{ric} \equiv 0$ $\mathrm{inj} \equiv \infty$	$\beta < 2(+k) - \frac{n}{p}$, where $p > \frac{n}{2}$
3	$\|R\|_p \leq \Lambda, (\|\nabla^k \mathrm{Ric}\|_p \leq M)$ $\mathrm{scg} \geq l_0$ $\mathrm{vol} \leq V$ or $\mathrm{diam} \leq D$	$R \equiv 0$ $\mathrm{scg} \equiv \infty$	$\beta < 2(+k) - \frac{n}{p}$, where $p > \frac{n}{2}$
4	$\|\mathrm{Ric}\| \leq \Lambda$ $\mathrm{vol} \geq v$ $\mathrm{diam} \leq D$ $\|R\|_{B(p,r_0)}\|_{\frac{n}{2}} \leq \varepsilon(n, v \cdot D^{-n})$	$\mathrm{Ric} \equiv 0$ $\mathrm{vol}\, B(p,r) \geq v' \cdot r^n$ $\|R\|_{\frac{n}{2}} \leq \varepsilon(n, v')$	$\beta < 2$
5	$\|\mathrm{Ric}\| \leq \Lambda$ $\mathrm{vol} \geq v$ $\mathrm{diam} \leq D$ $\mathrm{conj} \geq r_0$	$\mathrm{Ric} \equiv 0$ $\mathrm{vol}\, B(p,r) \geq v' \cdot r^n$ $\mathrm{conj} \equiv \infty$	$\beta < 2$

Table 1. Comparison of results. All classes consist of connected, closed riemannian n-manifolds.

in both the C^α and $L^{1,q}$, topologies, for $\alpha < 2 - n/p$ and $1/q > 1/p - 1/n$. Note that the volume bound tells us that any convergence takes place in the unpointed sense, so we are basically reducing ourselves to studying convergence of riemannian metrics on a fixed manifold.

Let $g \circ g$ be the Kulkarni–Nomizu product of g with itself.

THEOREM 6.1. *Fix* $l \in (1, p]$ *and* $c \in \mathbb{R}$. *There exists* $\varepsilon = \varepsilon(l, c, n, p, Q, r, v)$ *such that any* $(M, g) \in \mathcal{H}(V)$ *with* $\|\mathrm{Ric}_g - cg\|_l = \left(\int_M |\mathrm{Ric}_g - cg|^l\right)^{1/l} \leq \varepsilon$ *is* C^α, *close to an Einstein metric on* M, *where* $\alpha < 2 - n/p$. *If* $\|R_g - cg \circ g\|_l < \varepsilon$, *then* (M, g) *is* C^α *close to a constant curvature metric.*

PROOF. Note that we think of $|\mathrm{Ric}_g - cg|$ as a function on M whose value at $p \in M$ is $\sup\{|\mathrm{Ric}(v, v) - c| : p \in T_p M,\ g(v, v) = 1\}$. A similar definition is given for $\|R - cg \circ g\|_l$.

The proof is by contradiction on the existence of such an ε. For the sake of concreteness suppose $n/2 < p < n$. Using all of our previous work, we can therefore suppose that we we have a sequence of metrics g_i on a fixed manifold M such that $(M, g_i) \in \mathcal{H}(V)$ and $\|\mathrm{Ric}_{g_i} - c\, g_i\|_l$ converges to 0. We can furthermore assume that for each i, we have harmonic charts $\psi_{is} : B(0, r) \to M$, for $s = 1, \ldots, N$, so that $g_{is\cdot\cdot}$ converges on $B(0, r)$ to some limit riemannian metric $g_{s\cdot\cdot}$ in the $\psi_{is\cdot\cdot}$ coordinates. Note that the coordinates vary, but that they will themselves converge to limit charts ψ_s. As in the preliminaries of Theorem 5.1, we know that the limit coordinates are harmonic.

Now fix s and consider, in the ψ_{is} coordinates, the equation $\Delta_i g_i + Q(g_i, \partial g_i) = \mathrm{Ric}_{g_i}$. Here we know that $Q(g_i, \partial g_i)$ converges to $Q(g, \partial g)$ in $L^{q/2}$ where $1/q > 1/p - 1/n$, and g is the limit metric. Moreover, $\|\mathrm{Ric}_{g_i} - c\, g_i\|_l \to 0$. So we can conclude that, in the limit harmonic coordinates ψ_s, we have the equation $\Delta g + Q(g, \partial g) = c\, g$ or $\Delta g = -Q(g, \partial g) + c\, g$. Here the right-hand side is in $L^{q/2}$, with $q/2 > p$. So we can argue as in the proof of Theorem 5.1 that g is a C^∞ metric that satisfies the equation $\Delta g + Q(g, \partial g) = c\, g$ in harmonic coordinates. But this implies that the metric is Einstein, which contradicts our assumption that the g_i were not C^α close to an Einstein metric.

If we assume that $\|R - c\, g \circ g\|_l \to 0$, we have in particular

$$\|\mathrm{Ric}_g - (n-1)\, c\, g\|_l \to 0.$$

So the limit metric is again C^∞ and Einstein. We will, in addition, have $\|W\|_l \to 0$, where W is the Weyl tensor. Since $\|W\|_l$ varies semicontinuously in the C^α topology, the limit space must satisfy $\|W\|_l = 0$. But then the metric will also be conformally flat, and this makes it a constant curvature metric.   □

It is worthwhile pointing out that if we fix $c \in \mathbb{R}$ and assume that $\|\mathrm{Ric} - c\, g\|_p$ or $\|R - c\, g \circ g\|_p$ is small, where $p > n/2$, then we have L^p bounds on curvature. It is, therefore, possible to get some better pinching results in this case. Let's list some examples.

THEOREM 6.2. *Let $p > n/2 \geq 1$, $l, V > 0$, and $c \in \mathbb{R}$ be given. There exists $\varepsilon = \varepsilon(n, p, l, V, c)$ such that any (M^n, g) with $\mathrm{scg} \geq l$, $\mathrm{vol} \leq V$, and $\|R_g - c\, g \circ g\|_p \leq \varepsilon$ is $L^{p,2}$ close to a constant curvature metric on M.*

PROOF. The proof, of course, uses Example 5.9 and proceeds as Theorem 1 with the added finesse that, since $\|R - c\, g \circ g\|_p \to 0$, we can achieve convergence in $L^{p,2}$ rather than in the weaker C^α topology, $\alpha < 2 - n/p$.   □

THEOREM 6.3. *Let $p > n/2 \geq 1$, $i_0, D > 0$, and $c \in \mathbb{R}$ be given. There exists $\varepsilon = \varepsilon(n, p, i_0, D, c)$ such that any (M^n, g) with $\mathrm{inj\,rad} \geq i_0$, $\mathrm{diam} \leq D$ and $\|\mathrm{Ric} - c\, g\|_p \leq \varepsilon$ is $L^{p,2}$ close to an Einstein metric.*

PROOF. Same as before.   □

Finally, we can also improve Gao's result on $L^{n/2}$ curvature pinching.

THEOREM 6.4. *Let $n \geq 2$, $\Lambda, v, D > 0$, and $c \in \mathbb{R}$ be given. There exists $\varepsilon = \varepsilon(n, \Lambda, v, D)$ such that any (M^n, g) satisfying $|\operatorname{Ric}| \leq \Lambda$, $\operatorname{vol} \geq v$, $\operatorname{diam} \leq D$, and $\|R - c g \circ g\|_{n/2} \leq \varepsilon$ is $C^{1+\alpha}$ close to a constant curvature metric for any $\alpha < 1$.*

PROOF. Notice that the smallness of $\|R - c g \circ g\|_{n/2}$ implies that $\|R|_{B(p, r_0)}\|$ will be sufficiently small for some small r. Thus, the class is precompact in the $C^{1+\alpha}$ topology for any $\alpha < 1$. We can then use Theorem 6.1 to finish the proof. $\square$

When $\dim \geq 3$ we know from Schur's lemma and the identity $d\operatorname{Scal} = 2\operatorname{div}\operatorname{Ric}$ that pointwise constant sectional curvature implies constant curvature and that pointwise constant Ricci curvature implies constant Ricci curvature. This can be used to establish further pinching results, where instead of assuming that the curvature is close to a constant, we assume smallness of the tensors

$$R - (n(n-1))^{-1}\operatorname{Scal} g \circ g$$

or $\operatorname{Ric} - n^{-1}\operatorname{Scal} g$. With the background we have established so far, it is not hard to find the appropriate pinching theorems. Further details are left to the energetic reader.

Appendix A:
Useful Results from [Gilbarg and Trudinger 1983]

Fix a nice bounded domain $\Omega \subset \mathbb{R}^n$, $n \geq 2$. For $f \in C^\infty(\Omega)$ and $0 < \alpha \leq 1$, define

$$\|f\|_{k+\alpha, \Omega}$$
$$= \max\left\{\sup_{x \in \Omega} |\partial^j f(x)| : |j| \leq k\right\} + \max\left\{\sup_{x, y \in \Omega} \frac{|\partial^j f(x) - \partial^j f(y)|}{|x - y|^\alpha} : |j| = k\right\}.$$

Since we don't allow $\alpha = 0$, there should be no ambiguity about what $\|f\|_{k+1}$ is. Also define

$$\|f\|_{p, k, \Omega} = \sum_{0 \leq |j| \leq k} \left(\int_\Omega |\partial^j f(x)|^p dx\right)^{1/p}.$$

Then define $C^{k+\alpha}(\Omega)$ and $L^{p,k}(\Omega)$ as the completions of $C^\infty(\Omega)$ with respect to the norms $\|\cdot\|_{k+\alpha, \Omega}$ and $\|\cdot\|_{p, k, \Omega}$, respectively.

Sobolev embeddings. Any closed and bounded subset of $L^{p,k}$ is compact with respect to the weaker norms $\|\cdot\|_{l+\alpha}$ (or $\|\cdot\|_{q,l}$) provided $l + \alpha < k - n/p$ (or $l < k$ and $1/q > 1/p - (k-l)/n$). In other words, we have compact inclusions $C^{l+\alpha} \supset L^{p,k}$ for $l + \alpha < k - n/p$ and $L^{q,l} \supset L^{p,k}$ for $l < k$ and $1/q > 1/p - (k-l)/n$.

Elliptic estimates. Fix again a nice bounded $\Omega \subset \mathbb{R}^n$ and a compact $\Omega' \subset \overset{\circ}{\Omega}$. We shall work exclusively with functions $u \in C^{\infty}(\Omega)$ and operators whose coefficients are in $C^{\infty}(\Omega)$. The issues here are, therefore, quantitative rather than qualitative. Fix a second-order operator $L = a^{ij}\partial_i\partial_j$, where $\{a^{ij}\} : \Omega \to \mathbb{R}^n \times \mathbb{R}^n$ is a positive definite symmetric matrix with eigenvalues in $[e^{-Q}, e^{Q}]$ whose entries satisfy either $\|a^{ij}\|_{k+\alpha,\Omega} \leq Q$ or $\|a^{ij}\|_{p,l,\Omega} \leq Q$.

THEOREM A.1 (SCHAUDER ESTIMATES). *If* $\|a^{ij}\|_{k+\alpha} \leq Q$, *there exists* $C = C(k+\alpha, n, Q, \Omega' \subset \Omega)$ *such that*

$$\|u\|_{k+2+\alpha,\Omega'} \leq C \left(\|Lu\|_{k+\alpha,\Omega} + \|u\|_{C^0,\Omega}\right).$$

PROOF. For $k = 0$, this is [Gilbarg and Trudinger 1983, Theorem 6.2]. For $k \geq 1$, we can proceed by induction. Suppose we have established the inequality for some k. Fix ∂_j and use

$$
\begin{aligned}
\|\partial_j u\|_{k+2+\alpha} &\leq C(k)(\|L\partial_j u\|_{k+\alpha} + \|\partial_j u\|_{C^0}) \\
&\leq C(k)(\|\partial_j Lu\|_{k+\alpha} + \|[L,\partial_j]a\|_{k+\alpha} + \|\partial_j u\|_{C^0}) \\
&\leq C(k)(\|Lu\|_{k+1+\alpha} + \|a^{lm}\|_{k+\alpha}\|u\|_{k+2+\alpha} + \|\partial_j u\|_{C^0}) \\
&\leq C(k)(\|Lu\|_{k+1+\alpha} + (Q+1)\|u\|_{k+2+\alpha}) \\
&\leq C(k)(\|Lu\|_{k+1+\alpha} + (Q+1)C(k)(\|Lu\|_{k+\alpha} + \|u\|_{C^0})) \\
&\leq C(k+1)(\|Lu\|_{k+1+\alpha} + \|u\|_{C^0}).
\end{aligned}
$$

Since $\|u\|_{k+1+2+\alpha} \leq \|u\|_{k+2+\alpha} + \sum_j \|\partial_j u\|_{k+2+\alpha}$, we get the desired inequality. $\square$

THEOREM A.2 (L^p ESTIMATES). *If* $\mathrm{vol}\,\Omega \leq V$ *and* $\|a^{ij}\|_{p,k} \leq Q$, *where* $p > n$ *if* $k = 1$ *and* $p > n/2$ *if* $k \geq 2$, *there exists* $C = C(p, k, n, Q, \Omega' \subset \Omega, V)$ *such that*

$$\|u\|_{p,k+1,\Omega'} \leq C \left(\|Lu\|_{p,k-1,\Omega} + \|u\|_{L^p,\Omega}\right).$$

Furthermore, if $\Omega = B(0,r)$ *is a euclidean ball and* $u = 0$ *on* $\partial B(0,r) = S(0,r)$, *then we can find* $C = C(p, n, Q, r)$ *such that*

$$\|u\|_{p,2,B(0,r)} \leq C\left(\|Lu\|_{p,B(0,r)} + \|u\|_{p,B(0,r)}\right).$$

PROOF. The condition $p - n/k > 0$ implies by the Sobolev imbedding results that $\|a^{ij}\|_{l+\alpha,\Omega} \leq K(n,p,k,l,\alpha)\,Q$ for $l + \alpha < p - n/k$. In particular, the functions a^{ij} have moduli of continuity bounded on Ω in terms of Q, n, p, k. When $k = 1$, we can, therefore, use Theorem 9.11 (or 9.13 for the second inequality) in [Gilbarg and Trudinger 1983] to get the estimate. When $k \geq 2$, we use induction again. Assume the inequality holds for some k and that $\|a^{ij}\|_{p,k+1,\Omega} \leq Q$. Since $p > n/2$, we have $1/(2p) > 1/q - 1/n$, so $\|a^{ij}\|_{2p,k,\Omega} \leq K(n,p,k)\,Q$. Also, recall that $\|f\|_{p,\Omega} \leq C(p,q,\mathrm{vol}\,\Omega)\|f\|_{q,\Omega}$ for $q > p$ by Hölder's inequality. Now we compute as with the Schauder estimates:

$$\|\partial_l u\|_{p,k+1,\Omega'} \leq C(k,p)(\|L\partial_l u\|_{p,k-1} + \|\partial_l u\|_{L^p})$$
$$\leq C(k,p)(\|Lu\|_{p,k} + \|[L,\partial_l]u\|_{p,k-1} + \|\partial_l u\|_{L^p})$$
$$\leq C(k,p)(\|Lu\|_{p,k} + \|\partial_l a^{ij}\|_{2p,k-1}\|u\|_{2p,k-1} + \|\partial_l u\|_{L^p})$$
$$\leq C(k,p)(\|Lu\|_{p,k} + (KQ+1)\|u\|_{2p,k-1})$$
$$\leq C(k+1,p)(\|Lu\|_{p,k} + \|u\|_p).$$

By adding up, we therefore have

$$\|u\|_{p,k+2,\Omega'} \leq C(n,p)\left(\|Lu\|_{p,k,\Omega} + \|u\|_{p,\Omega}\right). \qquad \square$$

Now suppose our operator is given to us in divergence form, $L = \partial_i(a^{ij}\partial_j)$, where a^{ij} is a positive definite symmetric matrix with eigenvalues in $[e^{-Q}, e^Q]$. All Laplacians are written in this form unless we use harmonic coordinates, so it is obviously important for our theory to understand this case as well.

THEOREM A.3 (L^p ESTIMATES FOR DIVERGENCE OPERATORS). *Assume* $\Omega' \subset \subset \Omega$ *is compact,* Ω *is bounded with* $\mathrm{vol}\,\Omega \leq V$ *and* $\|a^{ij}\|_{p,k} \leq Q$, *where* $p > n$ *if* $k = 1$ *and* $p > n/2$ *if* $k \geq 2$. *There exists* $C = C(p,k,n,Q,\Omega' \subset \Omega, V)$ *such that*

$$\|u\|_{p,k+1,\Omega'} \leq C\left(\|Lu\|_{p,k-1,\Omega} + \|u\|_{p,\Omega}\right).$$

Furthermore, if $\Omega = B(0,r)$ *and* $u = 0$ *on* $\partial\Omega$, *there is* $C = C(p,n,r,Q)$ *such that*

$$\|u\|_{p,2,B(0,r)} \leq C\left(\|Lu\|_{p,B(0,r)}\right).$$

PROOF. We will concentrate on the second inequality. The proof of the first inequality is similar but simpler for the case where $k = 1$. For general k one can just use induction as before. The proof proceeds on two steps. First we show that there is a C such that $\|u\|_{p,2} \leq C\left(\|Lu\|_p + \|u\|_p\right)$. Then we show by a simple contradiction argument that there is a C such that $\|u\|_p \leq C\|Lu\|_p$. These two inequalities clearly establish our result.

Write $L = \partial_i(a^{ij}\partial_j) = a^{ij}\partial_i\partial_j + b^j\partial_j = \tilde{L} + b^j\partial_j$. We then have, from the previous L^p estimates,

$$\|u\|_{p,2,\Omega} \leq C_1\left(\|\tilde{L}u\|_{p,\Omega} + \|u\|_{p,\Omega}\right)$$
$$\leq C_1\left(\|Lu\|_{p,\Omega} + \|b^j\partial_j u\|_{p,\Omega} + \|u\|_{p,\Omega}\right)$$
$$\leq C_1\left(\|Lu\|_{p,\Omega} + \|b^j\|_{p,\Omega}\|\partial_j u\|_{C^0,\Omega} + \|u\|_{p,\Omega}\right)$$
$$\leq C_1\left(\|Lu\|_{p,\Omega} + Q\|\partial_j u\|_{C^0,\Omega} + \|u\|_{p,\Omega}\right).$$

Thus we need to bound $\|\partial_j u\|_{C^0}$ in terms of $\|Lu\|_p$ and $\|u\|_p$. This is done by a bootstrap method beginning with an L^2 estimate for $\partial_j u$. To get this L^2 estimate, we use

$$u\,Lu = u\,\partial_i(a^{ij}\partial_j u) = \partial_i u\,a^{ij}\,\partial_j u - \partial_i(u\,a^{ij}\partial_j u) \geq e^{-Q}\partial_i u\delta^{ij}\partial_j u - \partial_i(u\,a^{ij}\partial_j u).$$

Integration over $B(0,r)$, together with the fact that $u = 0$ on $\partial B(0,r)$, then yields

$$\int_{B(0,r)} u\,Lu \geq e^{-Q} \int_{B(0,r)} \sum_i (\partial_i u)^2 - \int_{B(0,r)}' \partial_i(u\,a^{ij}\partial_j u) = e^{-Q} \int_{B(0,r)} \sum_i (\partial_i u)^2.$$

Thus $\|\nabla u\|_2^2 \leq e^Q \|u\|_2 \|Lu\|_2 \leq e^Q c(n,r) \|u\|_p \|Lu\|_p$. In particular $\|\nabla u\|_2 \leq \max\{\|u\|_p, e^Q C(n,r)\|Lu\|_p\} \leq C_2(\|Lu\|_p + \|u\|_p)$. This implies that

$$\|u\|_{r_1,2} \leq C_1(\|\tilde{L}u\|_{r_1} + \|u\|_{r_1}) \leq C_1(\|Lu\|_{r_1} + \|b^j\|_p \|\partial_j u\|_2 + \|u\|_{r_2})$$
$$\leq C_3(\|Lu\|_p + \|u\|_p),$$

where $1/r_1 = 1/p + 1/2$. Now recall that we have a Sobolev inequality $\|\partial_j u\|_{q_1} \leq C(r_1, q_1, n)\|u\|_{r_1,2}$ where $1/q_1 = 1/r_1 - 1/n = 1/2 + (1/p - 1/n) > 0$; also $q_1 > 2$, so we now have a better bound on $\partial_j u$. Doing this over again we get $\|u\|_{r_2,2} \leq C_4 (\|Lu\|_p + \|u\|_p)$, where $1/r_2 = 1/p + 1/q_1 = (1/p - 1/n) + 1/p + 1/2$ and $\|\partial_j u\|_{q_2} \leq C \|u\|_{r_2,2}$, where $1/q_2 = 1/r_2 - 1/n = 1/2 + 2(1/p - 1/n)$. After l iterations of this type, we end up with

$$\|u\|_{r_l,2} \leq C_{2+l} (\|Lu\|_p + \|u\|_p), \|\partial_j u\|_{q_l} \leq C \|u\|_{r_l,2}$$

where $1/r_l = l(1/p - 1/n) + 1/p + 1/2$ and $1/q_l = 1/2 + (l + 1)(1/p - 1/n)$. This will end as soon as $1/2 + (l + 1)(1/p - 1/n) < 0$, in which case we can use the other type of Sobolev inequality: $\|\partial_j u\|_{C^0} \leq C \|u\|_{r_l,2}$ if $r_l > n$. Since the number of iterations we need to go through in order for this to happen only depends on p and n, we get the desired C^0 estimate on $\partial_j u$.

We now need to show that there is a constant $C(p,n,r)$ such that $\|u\|_p \leq C \|Lu\|_p$. If such a C does not exist we can find a sequence u_l with $\|u_l\|_p = 1$, $u_l = 0$ on $\partial\Omega$, and $L_l = \partial_i(a_l^{ij}\partial_j)$ with $\|L_l u_l\|_p \to 0$ as $l \to \infty$. Since all the operators L_l have the same bounds we get

$$\|u_l\|_{p,2} \leq C (\|L_l u_l\|_p + \|u_l\|_p) \leq 2C \quad \text{as } l \to \infty.$$

So we can, in addition, assume that u_l converges in the $C^{1+\alpha}$ topology, $\alpha < 1 - n/p$, to a function u with $\|u\|_p = 1$. On the other hand, we also know that

$$\|\nabla u_l\|_2^2 \leq e^Q \|u_l\|_2 \|L_l u_l\|_2 \to 0 \quad \text{as } l \to \infty.$$

The C^1 function u must therefore satisfy $\nabla u = 0$ and $u = 0$ on $\partial\Omega$, which implies that $u \equiv 0$ on Ω. This contradicts $\|u\|_p = 1$. $\qquad\square$

Elliptic regularity results. Let's return to the case where $L = a^{ij}\partial_j\partial_i$ and a^{ij} is a symmetric matrix whose eigenvalues lie in $[e^{-Q}, e^Q]$. Suppose $u \in L^{p,2}$ solves the equation $Lu = f, a^{ij} \in C^{l+\alpha}$ and $f \in L^{p,l}$ (or $f \in C^{l+\beta}$). Then $u \in L^{p,l+2}$ (or $C^{l+2+\beta}$).

Appendix B: On R and the coefficients of Δ

Suppose that we have a sequence of metrics on a fixed domain $B(0, r)$. How will the curvature operator R and the coefficients of Δ vary when these metrics are assumed to converge in the $L^{p,k}$ topology? We will show below that $\|R\|_p$ varies continuously in the $L^{p,2}$-topology if $p > n/2$ and that the coefficients of Δ vary continuously in the L^p-norm if the metrics converge in the $L^{p,1}$-topology, for $p > n$.

To make the formulation easier, suppose $G : \Omega \subset \mathbb{R}^n \to \mathbb{R}^n \times \mathbb{R}^n$ is a positive definite matrix whose entries g_{ij} are functions on some bounded domain $\Omega \subset \mathbb{R}^n$.

If we write $\Delta = g^{ij}\partial_{ij} + b^i\partial_i$, $b^i = \partial_j g^{ij} + (g^{ij}/g)\partial_j g$, then clearly $g^{ij} = G^{-1}$ and $b^i = \partial_j g^{ij} + \frac{1}{2}g^{ij}\partial_j(\log g)$. So we must check how G^{-1} and $\log G$ depend on G. Now $\partial_i G^{-1} = -G^{-1}(\partial_i G)G^{-1}$, so if we vary G in the $L^{p,1}$-topology, for $p > n$, then G, G^{-1}, $\det G$, and $\det G^{-1}$ vary continuously in the C^0 topology $\partial_i G$, $\partial_i G^{-1}$ vary continuously in the L^p topology. To see that b^i varies continuously in the L^p topology, it suffices to check that $\partial_j g = \partial_j \det G$ varies continuously in the L^p topology. But $\partial_j \det G = F(G, \partial_j G)$ for some algebraic function which is linear in the second matrix variable. This observation leads to the desired result.

To check the continuity of R, recall that $R^l_{ijk} = \partial_i \Gamma^l_{jk} - \partial_j \Gamma^l_{ik} + \sum_r \Gamma^r_{jk}\Gamma^l_{ir} - \Gamma^r_{ik}\Gamma^l_{jr}$, where

$$\Gamma^i_{jk} = \frac{1}{2}\sum_l g^{il}(\partial_j g_{kl} + \partial_k g_{lj} - \partial_l g_{jk}).$$

So if G varies continuously in $L^{p,2}$, for $p > n/2$, then ∂G varies continuously in L^{2p}; and from the above we conclude that Γ^i_{jk} varies continuously in $L^{1,p} \subset L^{2p}$, so R varies continuously in L^p.

If we vary G in $L^{p,k}$, for $k \geq 2$ and $p > n/2$, we can also easily check that R varies continuously in $L^{p,k-2}$, that G^{-1} varies continuously in $L^{p,k}$, and that b^i (first order term in Δ) varies continuously in $L^{p,k-1}$.

From the remarks about the coefficients of Δ, we see that conditions (h2) and (h3) imply that we can use the results from Appendix A without further ado.

Appendix C: The Eguchi–Hanson metric

(See also [Petersen and Zhu 1994].)

Let σ_1, σ_2 and σ_3 be the standard left- or right-invariant one-forms on $S^3 = \mathrm{SU}(2)$ with $d\sigma_i = -2\sigma_{i+1} \wedge \sigma_{i+2}$ (indices taken modulo 3). The canonical metric of curvature 1 can be written $ds^2 = \sigma_1^2 + \sigma_2^2 + \sigma_3^2$. We shall consider metrics on $I \times S^3$ of the form $dr^2 + \varphi^2(r)\sigma_1^2 + \psi^2(r)(\sigma_2^2 + \sigma_3^2)$, where $r \in I$ and I is an interval. We think of φ as being a function adjusting the length of the Hopf fiber in S^3. The volume form of this metric is $dr \wedge \varphi(r)\sigma_1 \wedge \psi(r)\sigma_2 \wedge \psi(r)\sigma_3$. Using this choice, we can define the Hodge $*$ operator, which, in particular, maps Λ^2 to Λ^2 and satisfies $*^2 = 1$. The two-forms $dr \wedge \varphi\sigma_1 \pm \psi^2\sigma_2 \wedge \sigma_3$, $dr \wedge \psi\sigma_2 \pm \psi\sigma_3 \wedge \varphi\sigma_1$, and $dr \wedge \psi\sigma_3 \pm \varphi\sigma_1 \wedge \psi\sigma_2$ have eigenvalues ± 1 for $*$, and form a basis for $\Lambda^2 T_p M$

for all $p \in M$. Normalizing these vectors by $1/\sqrt{2}$ we get an orthonormal basis for $\Lambda^2 T_p M$. In this basis, the curvature operator R has the form

$$\begin{pmatrix} A & B \\ B^* & D \end{pmatrix},$$

where A and D are the parts of the curvature tensor on the eigenspaces corresponding to the eigenvalues 1 and -1, respectively. It is a simple algebraic exercise to see that a four-dimensional manifold is Einstein if and only if $\sec(\pi) = \sec(\pi^\perp)$ for all planes π and their orthogonal components $\pi^\perp$. Using this, one can check that the metric is Einstein if and only if $B \equiv 0$. In our case, A, B, and D are diagonal matrices. If we set $S_{ij} = \sec(\sigma_i \wedge \sigma_j)$, where $\sigma_0 = dr$ etc., and further set $M = (\dot{\varphi}\psi - \dot{\psi}\varphi)/\psi^3$, the diagonal matrices bemoce

$$B = \tfrac{1}{2}\operatorname{diag}(-S_{01} + S_{23}, -S_{02} + S_{13}, -S_{03} + S_{12}),$$
$$A = \tfrac{1}{2}\operatorname{diag}(S_{01} + S_{23} - 4M, S_{02} + S_{13} + 2M, S_{03} + S_{12} + 2M),$$
$$D = \tfrac{1}{2}\operatorname{diag}(S_{01} + S_{23} + 4M, S_{02} + S_{13} - 2M, S_{03} + S_{12} - 2M).$$

The sectional curvatures can be computed using the tube formula, Gauss equations, and Codazzi–Maimardi equations, if we think of S^3 as sitting in $I \times S^3$ as a hypersurface for each r. We get

$$S_{23} = \frac{4\psi^2 - 3\varphi^2}{\psi^4} - \frac{\dot{\psi}^2}{\psi^2}, \qquad S_{j0} = -\frac{\ddot{\psi}}{\psi} \qquad \text{for } j = 2, 3,$$

$$S_{10} = -\frac{\ddot{\varphi}}{\varphi}, \qquad S_{1i} = \frac{\varphi^2}{\psi^4} - \frac{\dot{\varphi}\dot{\psi}}{\varphi\psi} \qquad \text{for } i = 2, 3.$$

Hence the metric is Einstein if

$$-\frac{\ddot{\varphi}}{\varphi} = \frac{4\psi^2 - 3\varphi^2}{\psi^4} - \frac{\dot{\psi}^2}{\psi^2}, \qquad -\frac{\ddot{\psi}}{\psi} = \frac{\varphi^2}{\psi^4} - \frac{\dot{\varphi}\dot{\psi}}{\varphi\psi}.$$

One can explicitly solve these equations, but we will only be interested in the special case where $\varphi = \psi\dot{\psi}$ and $\dot{\psi}^2 = 1 - (\alpha/6)\psi^2 + k\psi^{-4}$, where α is the Einstein constant and k is just some integration factor. When $\alpha = 6$ and $k = 0$, we see that $\psi = \sin r$ and $\varphi = \sin r \cos r$; this corresponds to the Fubini–Study metric on $\mathbb{C}P^2$. When $\alpha = 0$ and $k < 0$, we get a family of Ricci-flat metrics. These are the examples of Eguchi and Hanson. You may want to check that changing k in this case is the same as scaling the metric.

To get a smooth complete metric, we need the initial condition $\psi(0) = (-k)^{-1/4}$. This implies $\dot{\psi}(0) = 0$, $\varphi(0) = 0$, and $\dot{\varphi}(0) = 2$. Since $\dot{\varphi}(0)$ is twice as large as we would like, we divide out by the antipodal maps on S^3, which is still an isometry of $dr^2 + \varphi^2\sigma_1^2 + \varphi^2(\sigma_2^2 + \sigma_3^2)$. Then our metric lives on $I \times \mathbb{R}P^3 = I \times SO(3)$, and the Hopf fiber now has length π rather than 2π. As $r \to 0$, we can then check that the above metric defines a smooth metric on $TS^2 = ([0, \infty) \times \mathbb{R}P^3)/(0 \times \mathbb{R}P^3)$, where $0 \times \mathbb{R}P^3 \sim S^2$ by the Hopf fibration.

The level set $r = 0$ is a totally geodesic S^2 of constant curvature $4(-k)^{1/2}$, while
the level sets $r = r_0$ are $SO(3)$ with the metric $\varphi^2(r_0)\sigma_1^2 + \psi^2(r_0)(\sigma_2^2 + \sigma_3^2)$.

The differential equation $\dot{\psi}^2 = 1 + k\psi^{-4}$ shows that ψ is on the order of r as
$r \to \infty$. Thus $\dot{\psi} \simeq 1$, $\ddot{\psi} = -4k\psi^{-5} \simeq 0$ as $r \to \infty$. Using this, one can easily see
that the norm of the curvature operator, $|R|$, is of order r^{-6} at infinity.

The volume form satisfies

$$dr \wedge \varphi(r)\sigma_1 \wedge \psi^2(r)\sigma_2 \wedge \sigma_3 = \psi^3 (1 + k\psi^{-4})^{1/2} dr \wedge \sigma_1 \wedge \sigma_2 \wedge \sigma_3$$

$$\simeq r^3 dr \wedge \sigma_1 \wedge \sigma_2 \wedge \sigma_3.$$

Thus the metric at ∞ is very close to the cone over $\mathbb{RP}^3$, namely $dr^2 + r^2(\sigma_1^2 + \sigma_2^2 + \sigma_3)$ on $[0, \infty) \times SO(3)$. This implies that $\operatorname{vol} B(p, r) \geq \frac{1}{2}\omega_4 r^4$ for all r and p.

Also, $\|R\|_p = \int |R|^p d\operatorname{vol} \leq C \int_1^\infty r^{-3p} r^3 dr = C \int_1^\infty r^{3-3p} dr < \infty$ for $p > \frac{4}{3}$.
In particular, $\|R\|_{n/2} = \|R\|_2$ is finite.

As $k \to -\infty$, the level set $r = 0$ will degenerate to a point and the whole
metric converges to the cone metric $dr^2 + r^2(\sigma_1^2 + \sigma_2^2 + \sigma_3^2)$ on $[0, \infty) \times SO(3)$.
At the same time, $\|R\|_2$ stays bounded, but will go to zero on sets $r \geq \varepsilon > 0$.
What happends is that the collapse of S^2 absorbs all of the $\|R\|_2$.

We have now constructed some interesting complete Ricci-flat metrics. To
get compact examples, we can just use the fact that these examples as $k \to -\infty$
look like cones, and then glue them into compact examples which locally have
such cones. We illustrate this as done in [Anderson 1990a]: The graphs for φ
and ψ look approximately like r at ∞. The larger k is, the faster they get to
look linear. For large k, take φ, ψ on $[0, 5]$ and then bend them to have zero
derivatives at $r = 5$. This can be done keeping Ric ≥ 0 and without changing
the volume much. Next reflect φ and ψ in the line $r = 5$. This yields metrics on
$TS^2 \cup TS^2 = S^2 \times S^2$ that, near $r = 0$ and $r = 10$, look like the Eguchi–Hanson
metric, and otherwise don't change much. As $k \to -\infty$, these compact examples
degenerate at $r = 0, 10$ and the space in the limit becomes a suspension over $\mathbb{RP}^3$.
One can see that this process can happen through metrics with $0 \leq$ Ric ≤ 1000,
vol $\geq \frac{1}{5}$, diam ≤ 10, and $\|R\|_2 \leq 1000$.

Acknowledgements

The author would like to thank M. Cassorla, X. Dai, R. Greene, K. Grove,
S. Shteingold, G. Wei and S.-H. Zhu for many helpful discussions on the subject
matter presented here. In addition, I would like to extend special thanks to M.
Christ and S.-Y. Cheng for helping me getting a firm grip on elliptic regularity
theory and proving the results in Appendix A. Finally, many thanks are due
MSRI and its staff for their support and hospitality during my stay there.

References

[Anderson 1990a] M. T. Anderson, "Short geodesics and gravitational instantons", *J. Diff. Geom.* **31** (1990), 265–275.

[Anderson 1990b] M. T. Anderson, "Convergence and rigidity of manifolds under Ricci curvature bounds", *Invent. Math.* **102** (1990), 429–445.

[Anderson 1991] M. T. Anderson, "Remarks on the compactness of isospectral set in low dimensions", *Duke Math. J.* **63** (1991), 699–711.

[Anderson 1993] M. Anderson, "Degenerations of metrics with bounded curvature and applications to critical metrics of Riemannian functionals", pp. 53–79 in *Differential geometry: Riemannian geometry* (Los Angeles, 1990), edited by R. Greene and S. T. Yau, Proc. Symp. Pure Math. **54**, AMS, Providence, RI, 1993.

[Anderson and Cheeger 1991] M. Anderson and J. Cheeger, "Diffeomorphism finiteness for manifolds with Ricci curvature and $L^{n/2}$ norm of curvature bounded", *Geom. Funct. Anal.* **1** (1991), 231–252.

[Anderson and Cheeger 1992] M. Anderson and J. Cheeger, "C^α compactness for manifolds with Ricci curvature and injectivity radius bounded from below", *J. Diff. Geom.* **35** (1992), 265–281.

[Bando et al. 1989] S. Bando, A. Kasue, and H. Nakajima, "On a construction of coordinates at infinity on manifolds with fast curvature decay and maximal volume growth", *Invent. Math.* **97** (1989), 313–349.

[Cheeger 1970] J. Cheeger, "Finiteness theorems for Riemannian manifolds", *Am. J. Math.* **92** (1970), 61–74.

[Cheeger et al. 1992] J. Cheeger, K. Fukaya, and M. Gromov, "Nilpotent structures and invariant metrics on collapsed manifolds", *J. Amer. Math. Soc.* **5** (1992), 327–372.

[Cheeger et al. 1982] J. Cheeger, M. Gromov, and M. Taylor, "Finite propagation speed, kernel estimates for functions of the Laplace operator, and the geometry of complete Riemannian manifolds", *J. Diff. Geom.* **17** (1982), 15–53.

[DeTurck and Kazdan 1981] D. M. DeTurck and J. L. Kazdan, "Some regularity theorems in Riemannian geometry", *Ann. Sci. École Norm. Sup.* (4) **14** (1981), 249–260.

[Edwards and Kirby 1971] R. Edwards and R. Kirby, "Deformations of spaces of imbeddings", *Ann. Math.* (2) **93** (1971), 63–88.

[Fukaya 1990] K. Fukaya, "Hausdorff convergence of riemannian manifolds, and its applications", pp. 143–238 in Recent topics in differential and analytic geometry, edited by T. Ochiai, Adv. Stud. Pure Math. **18**-I, Kinokuniya, Tokyo, and Academic Press, Boston, 1990.

[Gao 1990] L. Gao, "Convergence of riemannian manifolds: Ricci and $L^{n/2}$ curvature pinching, *J. Diff. Geo.* **32** (1990), 349–381.

[Gilbarg and Trudinger 1983] D. Gilbarg and N. Trudinger, *Elliptic partial differential equations of second order*, 2nd ed., Grundlehren der math. Wiss. **224**, Springer, Berlin, 1983.

[Greene and Wu 1988] R. E. Greene and H. Wu, "Lipschitz convergence of Riemannian manifolds", *Pacific J. Math.* **131** (1988), 119–141.

[Gromov 1981a] M. Gromov, "Groups of polynomial growth and expanding maps", *Publ. Math. IHES* **53** (1981), 183–215.

[Gromov 1981b] M. Gromov, *Structures métriques pour les variétés riemanniennes*, edited by J. Lafontaine and P. Pansu, CEDIC, Paris, 1981.

[Jost and Karcher 1982] J. Jost and H. Karcher, "Geometrische methoder zur Gewinnung von a-priori-Schranken für harmonische Abbildungen", *Manuscripta Math.* **40** (1982), 27–77.

[Karcher 1989] H. Karcher, "Riemannian comparison constructions", pp. 170–222 in *Global differential geometry*, edited by S.-S. Chern, MAA Stud. Math. **27**, Math. Assoc. America, Washington, DC, 1989.

[Peters 1987] S. Peters, "Convergence of Riemannian manifolds", *Comp. Math.* **62** (1987), 3–16.

[Petersen 1993] P. Petersen, "Gromov–Hausdorff convergence of metric spaces", pp. 489–504 in *Differential geometry: Riemannian geometry* (Los Angeles, 1990), edited by R. Greene and S. T. Yau, Proc. Symp. Pure Math. **54** (3), Amer. Math. Soc., Providence, 1993.

[Petersen and Zhu 1994] P. Petersen and S.-H. Zhu, "$U(2)$-invariant four dimensional Einstein manifolds", preprint, Dartmouth University, 1994.

[Yang 1992] D. Yang, "Convergence of Riemannian manifolds with integral bounds on curvature I", *Ann. Sci. École Norm. Sup.* **25** (1992), 77–105.

PETER PETERSEN
DEPARTMENT OF MATHEMATICS
UNIVERSITY OF CALIFORNIA
LOS ANGELES, CA 90095

Comparison Geometry
MSRI Publications
Volume 30, 1997

Applications of Quasigeodesics
and Gradient Curves

ANTON PETRUNIN

ABSTRACT. This paper gathers together some applications of quasigeodesic
and gradient curves. After a discussion of extremal subsets, we give a proof
of the Gluing Theorem for multidimensional Alexandrov spaces, and a proof
of the Radius Sphere Theorem.

This paper can be considered as a continuation of [Perelman and Petrunin
1994]. It gathers together some applications of quasigeodesic and gradient curves.
The first section considers extremal subsets; in the second section we prove the
Gluing Theorem for multidimensional Alexandrov spaces; in the third we give
another proof of the Radius Sphere Theorem. Our terminology and notation
are those of [Perelman and Petrunin 1994] and [Burago et al. 1992]. We usually
formulate the results for general Alexandrov space, but for simplicity give proofs
only for nonnegative curvature.

NOTATION. We denote by M a complete n-dimensional Alexandrov space of
curvature $\geq k$. As in [Burago et al. 1992], we denote by p'_q the direction at q of
a shortest path to p. If H is a subset of M and $p, q \in H$, we denote by $|pq|_H$
the distance between p and q in the intrinsic metric of H. Finally, if X is a
metric space with metric ρ, we denote by X/c denote the space X with metric
ρ/c; where no confusion will arise, we may use the same notation for points in
X and their images in X/c.

1. Intrinsic Metric of Extremal Subsets

The notion of an extremal subset was introduced in [Perelman and Petrunin
1993, 1.1], and has turned out to be very important for the geometry of Alexandrov spaces. It gives a natural stratification of an Alexandrov space into open
topological manifolds. Also, as is shown in recent results of G Perelman, extremal
subsets in some sense account for the singular behavior of collapse. Therefore

---

This material is part of the author's Ph.D. thesis [Petrunin 1995].

the intrinsic metric of such subsets turns out to be important. Moreover, there is hope that extremal subsets with intrinsic metric will give a way to approach the idea of multidimensional generalized spaces with bounded integral curvature.

In this section we give a new proof of the generalized Lieberman lemma, prove a kind of "stability" property for extremal subsets, and prove the first variation formula for the intrinsic metric of extremal subsets. The Lieberman lemma can be understood as a totally quasigeodesic property of extremal subsets and therefore offers some hope that extremal subsets with the intrinsic metric might be Alexandrov spaces with the same curvature bound; at the end of this section we give a counterexample to this conjecture for extremal subsets with codimension at least 3. This question is still open for codimension one (i.e., for a boundary) and for codimension two.

THEOREM 1.1 (GENERALIZED LIEBERMAN LEMMA). *Any shortest path in the intrinsic metric of an extremal subset $F \subset M$ is a quasigeodesic in M.*

The first proof of this theorem was given in [Perelman and Petrunin 1993, 5.3].

PROOF. Assume γ is a shortest path in the length metric of some extremal subset F. Suppose γ is not a quasigeodesic. Then there is a point p such that the development $\tilde{\gamma}(t)$ from p is not convex in any neighborhood of some t_0. Now for any $\varepsilon > 0$ it is easy to find a "rounded" curve $\tilde{\delta}(t)$ such that $\tilde{\delta}(t) = \tilde{\gamma}(t)$ if $|t - t_0| > \varepsilon$, length$(\tilde{\delta}) < $ length$(\tilde{\gamma}) = $ length(γ), and for any t the points $\tilde{p}$, $\tilde{\gamma}(t)$, and $\tilde{\delta}(t)$ are collinear in this order.

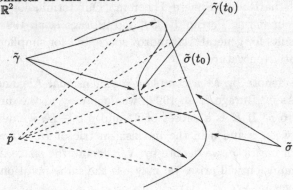

Now consider the curve in M given by

$$\delta(t) = \alpha_{\gamma(t)} \circ \rho_t^{-1}(|\tilde{p}\,\tilde{\delta}(t)|),$$

where $\alpha_{\gamma(t)} : [0, \infty) \to M$ is the dist$_p$-gradient curve that goes through $\gamma(t)$ such that $\alpha_{\gamma(t)}|_{[0,|p\gamma(t)|]}$ is a shortest path, and where ρ_t is its reparametrization, as in [Perelman and Petrunin 1994, 3.3(1)].

By [Perelman and Petrunin 1994, Theorem 6.3(a)], which states that if such a gradient curve starts at a point of an extremal subset F then it is contained in F, we obtain $\delta \subset F$. From [Perelman and Petrunin 1994, 3.3.3] (expansion along gradient curves is not more than in the model space)

$$\text{length}(\delta) \leq \text{length}(\tilde{\delta}) < \text{length}(\tilde{\gamma}) = \text{length}(\gamma).$$

Therefore γ is not a shortest path in F. $\square$

THEOREM 1.2. *Let M_n converge to M in the Gromov–Hausdorff topology with-out collapse (that is, $\dim M_n = \dim M$), and let $F_n \subset M_n$ be extremal subsets. Assume $F_n \to F \subset M$ as subsets. Then $F_n \xrightarrow{GH} F$ as length metric spaces with intrinsic metrics induced from M_n and M.*

PROOF. Let x and y lie in an extremal subset G. By the equivalence of the intrinsic metric of an extremal subset and the metric of the ambient space [Perelman and Petrunin 1993, 3.2(2)], we have for any open subset U in M an $\varepsilon = \varepsilon(\text{Vol}_n(U), \text{Diam}(U)) > 0$ such that $|xy|_G \leq \varepsilon^{-1}|xy|$ if $x, y \in U$. (The dependence on $\text{Vol}_n(U)$ and $\text{Diam}(U)$ can be easily obtained from the proof).

Consider $p, q \in F$ and $p_n, q_n \in F_n$ such that $p_n \to p$ and $q_n \to q$. It is easy to see that $|pq|_F \leq \liminf_{n\to\infty} |p_n q_n|_{F_n}$. Therefore we need to show only that $|pq|_F \geq \limsup_{n\to\infty} |p_n q_n|_{F_n}$. Set $\|pq\| = \limsup_{n\to\infty} |p_n q_n|_{F_n}$; this is easily seen to be a metric. From the previous paragraph, $\|pq\|$ does not depend on the choice of sequences $\{p_n\}$ and $\{q_n\}$, and we have $\|pq\| < \varepsilon^{-1}|pq|$, because from above ε can be found uniformly for all M_n in the absence of collapse.

Let $\gamma : [a, b] \to F$ be a shortest path in F between p and q parametrized by arclength. Assume $|pq|_F < \|pq\|$. Then, from [Busemann 1958, 5.14], for some $t_0 \in [a, b]$ and $\varepsilon > 0$ there is a sequence $t_i \to t_0\pm$ such that

$$\|\gamma(t_0)\gamma(t_i)\| \geq (1 + \varepsilon)|t_i - t_0|.$$

Setting $r = \gamma(t_0)$ and $s = \gamma(t_i)$, take sequences $r_n, s_n \in F_n$ such that $r_n \to r$ and $s_n \to s$. Let γ_i in F be the limit curve to the shortest paths between r_n and s_n in F_n. By [Perelman and Petrunin 1994, 2.3(3)] and the generalized Lieberman lemma, γ_i is a quasigeodesic between $\gamma(t_0)$ and $\gamma(t_i)$. From above, $\text{length}(\gamma_i) \geq (1 + \varepsilon)|t_i - t_0|$. Now consider the limit $(M/|t_0 - t_i|, r) \to (C_r, 0)$. Consider the curve in C_r given by

$$\gamma_*(t) = \lim_{i\to\infty} \left(\frac{\gamma_i}{|t_0 - t_i|}\right)(t\,|t_0 - t_i|) \in \frac{M}{|t_0 - t_i|},$$

where $(\gamma_i/|t_0 - t_i|)$ denotes the image of γ_i in $M/|t_0 - t_i|$. Then γ_* is a quasi-geodesic between 0 and the tangent vector $\gamma^\pm(t_0)$ which has length not less then $1 + \varepsilon$. This is a contradiction since $|\gamma^\pm(t_0)| = 1$ by [Perelman and Petrunin 1994, 2.3(2)]. $\square$

REMARK 1.3. The author does not know a counterexample for the following conjecture: *Let $M_n \xrightarrow{GH} M$, with $\dim M_n \leq C < \infty$, and let $F_n \subset M_n$ be extremal subsets. Assume that $F_n \to F \subset M$ as subsets and that $F_n \xrightarrow{GH} \bar{F}$. Then there is a discrete group of isometries G on $\bar{F}$ such that $F = \bar{F}/G$.*

As an example, consider the collapse of spaces with boundary $M_i \xrightarrow{GH} M$ such that $\dim M = \dim M_i - 1$. Then $\partial M_i \to M$ as subsets and $\partial M_i \xrightarrow{GH} \tilde{M}$, where $\tilde{M}$ is the double of M.

Now let M be an Alexandrov space and $F \subset M$ be an extremal subset. By the generalized Lieberman lemma, every shortest path in the length metric of F is a quasigeodesic as a curve in M, and every quasigeodesic at every point has directions of exit and entrance [Perelman and Petrunin 1994, 2.1(b) and 2.3(2)]. Thus if p and q lie in F we can define $q^\circ (= q_p^\circ)$ as the set of all directions of entrance in $\Sigma_p(F)$ of shortest paths between p and q in the length metric of F. It is easy to see that q° is compact.

THEOREM 1.4 (THE FIRST VARIATION FORMULA). *Let F be an extremal subset of the Alexandrov space M. Let $p, q \in F$, and let $\xi(t)$ be a curve in F starting from p in direction $\xi_0' \in \Sigma_p(F)$. Assume that $|p\,\xi(t)| = t + o(t)$. Then*

$$|\xi(t)\,q|_F = |pq|_F - \cos|\xi_0' q^\circ|_{\Sigma_p(F)}\, t + o(t).$$

PROOF. To prove this we have to prove two inequalities:

$$|\xi(t)q|_F \le |pq|_F - \cos|\xi_0' q^\circ|_{\Sigma_p(F)}\, t + o(t), \tag{1.1}$$

$$|\xi(t)q|_F \ge |pq|_F - \cos|\xi_0' q^\circ|_{\Sigma_p(F)}\, t + o(t). \tag{1.2}$$

PROOF OF (1.1). Take some $R \gg 1$. Set $\alpha = |\xi_0' q^\circ|_{\Sigma_p(F)}$ and $|pq|_F = l$. Take $\eta \in q^\circ$ such that $\alpha = |\xi_0' q^\circ|_{\Sigma_p(F)} = |\xi_0'\eta|$ and let $\gamma : [0, l] \to F$ be a shortest path between p and q in F such that $\gamma(0) = p$ and $\gamma^+(0) = \eta$. Then, by the triangle inequality,

$$|\xi(t)\,q|_F \le l - Rt + |\xi(t)\,\gamma(Rt)|_F.$$

The cosine rule gives us

$$|\xi_0'\, R\eta|_{C_p(F)} = \sqrt{R^2 + 1 - 2R\cos\alpha}.$$

Now, using Theorem 1.2, for the limit $(M/t, p) \to C_p$, we obtain

$$\lim_{t \to 0} |\xi(t)\gamma(Rt)|_F/t = |\xi_0'\, R\eta|_{C_p(F)}.$$

Therefore

$$|\xi(t)\,q|_F \le l - Rt + t\sqrt{R^2 + 1 - 2R\cos\alpha} + o(t)$$

$$\le l - \cos\alpha\, t + \frac{t}{R-1} + o(t).$$

When $R \to \infty$ we obtain

$$|\xi(t)q|_F \le |pq|_F - \cos|\xi_0' q^\circ|_{\Sigma_p(F)}\, t + o(t). \qquad \square$$

LEMMA 1.5. *Let $C = C(\Sigma)$ be a cone with curvature ≥ 0 (so the curvature of Σ is ≥ 1). Let γ be a quasigeodesic in C not passing through the vertex o. Then the projection of γ on Σ parametrized by the arclength is a quasigeodesic in Σ and the development of γ in the plane with respect to the vertex of C is a straight line.*

PROOF. To prove the second part of this lemma we have to prove that

$$(|\gamma(t)|^2)'' = 2.$$

In order to prove that $(|\gamma(t)|^2)'' \leq 2$, it is enough to consider the development of γ with respect to the vertex o of the cone. We prove that $(|\gamma(t)|^2)'' \geq 2$. Consider the Busemann function for $\theta \in \Sigma$, namely,

$$f_\theta = \lim_{\lambda \to \infty} (\text{dist}_{\lambda\theta} - \lambda).$$

The condition of convexity of the development with respect to $\lambda\theta$ gives the concavity of the function $f_\theta \circ \gamma(t)$ for every quasigeodesic γ in C. Using this for $\theta = \gamma(t)/|\gamma(t)|$ we get the needed inequality.

Therefore if γ^* is the projection of γ on Σ, we can choose a unique arclength parameter x on γ^* such that

$$\text{pr}\big(\gamma(c\tan x + d)\big) = \gamma^*(x)$$

for some constants $c > 0$ and d; without loss of generality we can set $d = 0$.

Now we have to prove that the development of γ^* in a standard sphere with respect to any $\theta \in \Sigma$ is convex, i.e., that $\cos(|\theta\gamma^*(x)|)'' + \cos(|\theta\gamma^*(x)|) \geq 0$. By [Perelman and Petrunin 1994, 1.7] it is enough to prove this only for $|\theta\gamma^*(x)| < \pi/2$. It is easy to see that

$$\cos(|\theta\gamma^*(x)|) = -\frac{f_\theta(\gamma(c\tan x))}{|\gamma(c\tan x)|}.$$

Then direct calculation gives what we need, because $f_\theta \circ \gamma$ is convex and because $|\gamma(c\tan x)| = c/\cos x$. $\square$

PROOF OF (1.2). Assume that (1.2) is false. Then one can find a sequence $\{t_i\}$, $t_i \to 0^+$, such that

$$|\xi(t_i)q|_F < |pq|_F - \cos|\xi_0'q^\circ|_{\Sigma_p(F)}\, t_i - \varepsilon\, t_i$$

for some fixed $\varepsilon > 0$.

Assume $|pq|_F = l$ and $|\xi(t_i)q|_F = l_i$. Let $\gamma_i : [0, l_i] \to F$ be the shortest paths between $\xi(t_i)$ and q in F such that $\gamma(0) = \xi(t_i)$. We can pass to a subsequence of $\{\gamma_i\}$ such that the shortest paths γ_i approach some shortest path $\gamma : [0, l] \to F$ between q and p. Let $\eta \in q^\circ$ be the direction of this shortest path γ. By Theorem 1.1, γ_i and γ are quasigeodesics.

Now consider the Gromov–Hausdorff limit $(M/t_i, p) \xrightarrow{GH} C_p$, and pass to a subsequence again, so that there exists $\hat{\gamma} : [0, \infty) \to C_p$ satisfying

$$\hat{\gamma}(t) = \lim_{i \to \infty} (\gamma_i/t_i)(tt_i) \in M/t_i,$$

where (γ_i/t_i) denotes the image of γ_i in M/t_i.

By [Perelman and Petrunin 1994, 2.3(3)], $\hat{\gamma}$ is a quasigeodesic in C_p, and it is easy to see that $\hat{\gamma}(0) \in \Sigma_p \subset C_p$.

We define the direction at infinity of the curve $\hat{\gamma}$ in C_p by

$$\lim_{t \to \infty} \frac{\hat{\gamma}(t)}{|o\,\hat{\gamma}(t)|}.$$

By Lemma 1.5 this is well defined for quasigeodesics.

We claim that the direction at infinity of $\hat{\gamma}$ is η. Indeed, let θ be the direction of $\hat{\gamma}$ at infinity. By the cosine rule we obtain, for $R \gg 1$,

$$
\begin{aligned}
|\hat{\gamma}(2R)|^2 &= \lim_{i \to \infty} \left(|p\,\gamma_i(2Rt_i)|/t_i\right)^2 \\
&\leq \lim_{i \to \infty} \left(|p\,\gamma_i(Rt_i)|^2 + (Rt_i)^2 - 2Rt_i|p\,\gamma_i(Rt_i)| \cos \angle(\gamma_i^+(Rt_i), p'_{\gamma_i(Rt_i)})\right)/t_i^2 \\
&= |\hat{\gamma}(R)|^2 + R^2 - 2R\,|\hat{\gamma}(R)| \lim_{i \to \infty} \cos \angle(\gamma_i^+(Rt_i), p'_{\gamma_i(Rt_i)}).
\end{aligned}
$$

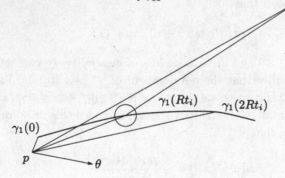

Now, by Lemma 1.5, we have for some β

$$
\begin{aligned}
\lim_{i \to \infty} \angle(\gamma_i^+(Rt_i), p'_{\gamma_i(Rt_i)}) &\geq \arccos \frac{|\hat{\gamma}(R)|^2 + R^2 - |\hat{\gamma}(2R)|^2}{2R\,|\hat{\gamma}(R)|} \\
&= \arccos \frac{((R^2 + 1 - 2R\cos\beta) + R^2 - (4R^2 + 1 - 4R\cos\beta))}{2R\sqrt{R^2 + 1 - 2R\cos\beta}} \\
&= \arccos\left(-\sqrt{\frac{R^2 - 2R\cos\beta + \cos^2\beta}{R^2 - 2R\cos\beta + 1}}\right) \geq \arccos\left(-\sqrt{1 - \frac{1}{(R-1)^2}}\right) \\
&\geq \arccos(-1 + 1/(R-1)^2) > \pi(1 - 1/R).
\end{aligned}
$$

Taking $r_k \to p$ such that $(r_k)'_p \to \theta$, we have

$$\lim_{i \to \infty} \tilde{\angle} p\,\gamma_i(Rt_i)\,r_k \geq \pi - \angle(\hat{\gamma}(R), (r_k)'_p) > \pi - \pi/R - \angle(\theta, (r_k)'_p).$$

The latter inequality is a corollary of Lemma 1.5, since $\sin \angle(\hat{\gamma}(R), \theta) \leq 1/R$. Therefore, since the perimeter of any triangle in the space of directions is at most 2π, we get

$$\lim_{i \to \infty} \angle(\gamma_i^+(Rt_i), (r_k)'_{\gamma_i(Rt_i)}) \leq 2\pi - \lim_{i \to \infty} \tilde{\angle} p\, \gamma_i(Rt_i)\, r_k - \lim_{i \to \infty} \angle(\gamma_i^+(Rt_i), p'_{\gamma_i(Rt_i)})$$

$$\leq \pi/R + \angle(\theta, (r_k)'_p) + \pi/R.$$

Using [Perelman and Petrunin 1994, 1.4(G2)] for γ_i with respect to the points r_k and starting at $\gamma_i(Rt_i)$, we obtain the estimates

$$|r_k \gamma(|pr_k|)| = \lim_{i \to \infty} |r_k\, \gamma_i(Rt_i + |\gamma_i(Rt_i)r_k|)|$$

$$\leq \lim_{i \to \infty} |\gamma_i(Rt_i)r_k| \lim_{i \to \infty} \angle(\gamma_i^+(Rt_i), (r_k)'_{\gamma_i(Rt_i)})$$

$$\leq |pr_k|\,(2\pi/R + \angle(\theta, (r_k)'_p)).$$

This means that η is $2\pi/R$-close to θ. Sending R to infinity we obtain $\theta = \eta$.

Now fix $R \gg 1$ and divide γ_i into two pieces using a parameter value $x_i \in [0, l_i)$ such that $|p\gamma_i(x_i)| = Rt_i$. We estimate the length of each part separately.

By Theorem 1.2 the length of the first part $|q\gamma_i(x_i)|_F$ is possible to estimate from the triangle inequality:

$$|q\gamma_i(x_i)|_F \geq |pq|_F - |p\gamma_i(x_i)|_F = |pq|_F - Rt_i + o(t_i).$$

The length of the second part is estimated using the fact that the limit of lengths of quasigeodesics is the length of the limit quasigeodesic [Perelman and Petrunin 1994, 2.2, 2.3(3)]. Therefore

$$|\xi(t_i)\gamma_i(x_i)|_F/t_i \to \text{length}(\hat{\gamma} \cap B_R(o) \subset C_p).$$

By Lemma 1.5 the last expression can be estimated from below as

$$R - \cos \angle(q^\circ, \xi_0') - C/R.$$

This estimate is easily deduced from the following diagram in the plane of the development $\tilde{\hat{\gamma}}$ of $\hat{\gamma}$ from o. Here α is the angle at $\tilde{o}$ subtended by $\tilde{\hat{\gamma}}$. Clearly α is not less than $\angle(q^\circ, \xi_0')$.

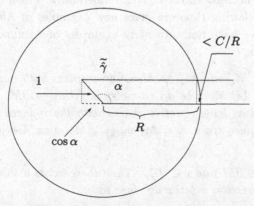

From these two estimates we obtain

$$|\xi(t_i)q|_F \geq |pq|_F - \cos|\xi_0'q^\circ|_{\Sigma_p(F)}\, t_i - C/R\, t_i + o(t_i),$$

which for $C/R < \varepsilon$ contradicts the assumption. This completes the proof of (1.2) and of the first variation formula. $\qquad\square$

A counterexample. In [Perelman and Petrunin 1993, 6.1] we conjectured that the intrinsic metric of a primitive extremal subset has curvature bounded from below. Here we show a counterexample to this conjecture for $\operatorname{codim} F \geq 3$. Therefore this question is still open for $\operatorname{codim} F = 1$ (i.e., for a boundary) and for $\operatorname{codim} F = 2$. Sergei Buyalo [1976] has settled the first of these questions affirmatively for a "smooth" Alexandrov space, i.e., for a convex subset in a Riemannian manifold with curvature bounded from below.

Consider a right simplex $\operatorname{conv}\{a_1 a_2 a_3 a_4 a_5\}$ in a standard S^4 such that $|a_i a_j| = \pi/2$ for $i \neq j$. Assume $a_5 = a_0$, take some $\varepsilon > 0$, and consider the closed broken geodesic

$$F = a_0^+\, a_1^-\, a_1^+\, a_2^-\, a_2^+\, a_3^-\, a_3^+\, a_4^-\, a_4^+\, a_5^-\, a_0^+,$$

where $a_i^\pm$ is the point on the geodesic $a_i a_{i\pm 1}$ such that $|a_i a_i^\pm| = \varepsilon$. Let $\Sigma = \operatorname{conv}\{F\}$. Then direct calculation shows that F is a primitive extremal subset of Σ and that, for ε sufficiently small, $\operatorname{length}(F) > 2\pi$. In particular, $C(F)$ is an extremal subset of $C(\Sigma)$, which has a singular point of negative curvature.

2. The Gluing Theorem

The Gluing Theorem for the two dimensional case is due to A. D. Alexandrov (see [Pogorelov 1973, § 11], for example). Later Perelman [1991, 5.2] proved the Doubling Theorem for multidimensional Alexandrov spaces; this is a special case of the theorem formulated below. The original Alexandrov's Theorem had a lot of applications to the bending of convex surfaces with boundary, which are currently impossible to generalize to the multidimensional case, because they are supported by the Theorem about convex embeddings [Pogorelov 1973, Sect6–7]. Formally, the following theorem gives new examples of Alexandrov spaces, but unfortunately we have not too many examples of Alexandrov spaces with isometric boundaries.

THEOREM 2.1. *Let M_1 and M_2 be Alexandrov spaces with nonempty boundary and curvature $\geq k$. Let there be an isometry* is : $\partial M_1 \to \partial M_2$, *where ∂M_1 and ∂M_2 are considered as length-metric spaces with the induced metric from M_1 and M_2. Then the glued space $X = M_1 \cup_{\mathrm{is}(x)=x} M_2$ is an Alexandrov space with curvature $\geq k$.*

LEMMA 2.2. *Let $p \in \partial M$ and $\eta \in \partial\Sigma_p$. Then there exists a shortest path in ∂M starting at p in a direction arbitrarily close to η.*

PROOF. Let $N = \partial M$. The boundary is an extremal subset and therefore we can use notation $q^\circ (= q_p^\circ)$ for the set of all directions of entrance in $\Sigma_p(N)$ of shortest paths between p and q in the length metric of N.

Choose a sequence of points $q_n \in N$ such that $q_n \to p$ and $\angle(q_n', \eta) \to 0$ (where $q_n' = (q_n)_p'$ is the direction at p of the shortest path pq). Assume that $\angle(\eta q_n^\circ) \geq \varepsilon$ for all n. Pass to a subsequence such that $\lim_{n\to\infty} \angle(\theta\, q_n^\circ) \to 0$ for some direction θ.

Find a point $r \in M$ such that $\angle(r', \theta) < \varepsilon/6$. Let $\{r_n\}$ be points on the shortest path pr such that $|pr_n| = |pq_n|$. Since the shortest path from p to q_n in N is a quasigeodesic (see Theorem 1.1), we conclude by using [Perelman and Petrunin 1994, 1.4(G2), 1.5] that $|r_n q_n| < (\varepsilon/5)|pq_n|$ for n sufficiently large, hence that $\lim_{n\to\infty} \angle(q_n', r') < \varepsilon/3$ for $\varepsilon \leq \pi/4$. Therefore

$$\lim_{n\to\infty} \angle(q_n', \theta) < \varepsilon/2.$$

We obtain a contradiction because $\lim_{n\to\infty} q_n' = \eta$ and $\angle(\eta, \theta) \geq \varepsilon$. $\qquad\square$

The rest of this section will be devoted to the proof of Theorem 2.1. Let $N = M_1 \cap M_2 = \partial M_i \subset X$.

DEFINITION 2.3. The m-predistance $|pq|_m$ between points p and q in X is the minimal length of broken geodesics with vertices $p = p_0, p_1, \ldots, p_{k+1} = q$, where $k \leq m$, $p_l p_{l\pm 1}$ is a shortest path that lies completely in one of M_i for every $l \in \{1, 2, \ldots, k\}$, and p_l lies in N. A broken geodesic that realizes this minimum is called an m-shortest path.

REMARK 2.4. It is easy to see that $|pq|_m \geq |pq|_{m+1} \geq |pq|$, $\lim_{m\to\infty} |pq|_m = |pq|$,

$$\begin{aligned} |pq|_m + |qr|_l &\geq |pr|_{m+l} \quad &&\text{if } q \in X \setminus N, \\ |pq|_m + |qr|_l &\geq |pr|_{m+l+1} \quad &&\text{if } q \in N . \end{aligned} \tag{2.1}$$

For every interior vertex $p = p_l$, $l \in \{1, 2, \ldots, k\}$, of an m-shortest path, we can define directions of exit and entrance ξ_i as directions in $\Sigma_p(M_i)$ of shortest paths in M_i.

By Theorem 1.2 the isometry is : $\partial M_1 \to \partial M_2 = N$ gives an isometry is$_p'$: $\partial \Sigma_p(M_1) \to \partial \Sigma_p(M_2) = \Sigma_p(N)$ and is$_p$: $\partial C_p(M_1) \to \partial C_p(M_2) = C_p(N)$. Set

$$\Sigma_p^\#(X) := \Sigma_p(M_1) \cup_{\text{is}_p'(x)=x} \Sigma_p(M_2),$$
$$C_p^\#(X) := C(\Sigma_p^\#(X)) = C_p(M_1) \cup_{\text{is}_p(x)=x} C_p(M_2).$$

From the induction hypothesis, $\Sigma_p^\#(X)$ will be an Alexandrov space with curvature ≥ 1, and therefore $C_p^\#(X)$ will be a cone with curvature ≥ 0.

NOTATION. If K_1 and K_2 are two compact metric spaces, we say that $K_1 \leq K_2$ if there is a noncontracting map $m : K_1 \to K_2$. If (L_1, p_1) and (L_2, p_2) are two locally compact metric spaces with base points, we say that $(L_1, p_1) \leq (L_2, p_2)$ if for any $R > 0$ there is a noncontracting map $m : B_R(p_1) \to B_R(p_2)$.

We will write $\limsup_{i\to\infty} K_i \leq K$ if for any Hausdorff subsequence $K_{i_k} \xrightarrow{GH} K'$ we have $K' \leq K$. Similarly one can write $\liminf_{i\to\infty} K_i \geq K$. We write $\limsup_{i\to\infty}(L_i, p_i) \leq (L, p)$ if for any $R > 0$ we have $\limsup_{i\to\infty} B_R(p_i) \leq B_R(p)$ (compare with [Burago et al. 1992, 7.13]).

PROOF. Proof of Theorem 2.1 As a base we can take the classical Gluing Theorem of A. D. Alexandrov in dimension 2 [Pogorelov 1973, §11] Assume we have already proved Theorem 2.1 for dimensions less than n.

LEMMA 2.5. *For any point $p \in N$ we have $\limsup_{\delta\to0}(X/\delta, p) \leq (C_p^\#(X), o)$.*

It is easy to see that as a corollary of Theorem 2.1 we will actually have equality in this theorem, instead of inequality.

PROOF OF LEMMA 2.5. Consider the gradient-exponential maps [Perelman and Petrunin 1994, 3.5] $\mathrm{gexp}_1 : C_p(M_1) \to M_1$ and $\mathrm{gexp}_2 : C_p(M_2) \to M_2$. By [Perelman and Petrunin 1994, 6.4(a)], we have $\exp_i(C_p(N)) \subset N$. We construct an exponential map $\exp : C_p^\#(X) \to X$ by setting

$$\exp(v) = \begin{cases} \mathrm{gexp}_1(v) & \text{for } v \in C_p(M_1) \subset C_p^\#(X), \\ \mathrm{gexp}_2(v) & \text{for } v \notin C_p(M_1). \end{cases}$$

Define $\exp_\delta : C_p^\#(X) \to X/\delta$ by $\exp_\delta(v) = i_\delta \circ \exp \circ(v\delta)$, where $i_\delta : X \to X/\delta$ is the canonical mapping.

Let $x = x_0, x_1, \ldots, x_k, x_{k+1} = y$ be vertices of an m-shortest path in $C_p^\#(X)$. It is easy to see that $|x_l x_{l+1}| \geq |\exp_\delta(x_l)\exp_\delta(x_{l+1})| + o(\delta)/\delta$. Therefore for the m-predistance in $C_p^\#(X)$ we have $|xy|_m \geq |\exp_\delta(x)\exp_\delta(y)| + o(\delta)/\delta$. Now $|xy| = \lim_{m\to\infty}|xy|_m$ for any $x, y \in C_p^\#(X)$. Hence

$$\lim_{\delta\to0} |\exp_\delta(x)\exp_\delta(y)| \leq \lim_{m\to\infty} |xy|_m = |xy|.$$

Now in order to complete the proof we need to verify that

$$\lim_{\delta\to0} \exp_\delta^{-1}(B_R(p) \subset X/\delta) \subset B_R(o) \subset C_p^\#(X).$$

for any $R > 0$, or, equivalently, that $\lim_{\delta\to0} |p\exp_\delta(x)| \geq |x|$ for any $x \in C_p^\#(X)$.

Assume otherwise. Therefore we can find $x \in C_p^\#(X)$ and a sequence $\delta_n \to 0$ such that for some $\varepsilon > 0$ we have

$$|p\exp_{\delta_n}(x)| \leq (1 - \varepsilon)|x|.$$

Consider shortest paths $p\exp_{\delta_n}(x) \subset X/\delta_n$ for all n. No subsequence lies completely in M_i/δ_n for fixed i. Let $y_n \in N/\delta_n \subset X/\delta_n$ be the closest point of N/δ_n to $\exp_{\delta_n}(x)$ on $p\exp_{\delta_n}(x)$. Pass to a subsequence of $\{\delta_n\}$ such that $\exp_{\delta_n}^{-1}(y_n) \to x^*$. By [Perelman 1991, 4.7], $x^* \in C(\Sigma_p(N)) = C(\partial M_i)$ and

$$\lim_{\delta_n\to0} |\exp_{\delta_n}(x)\exp_{\delta_n}(x^*)| = |xx^*|$$

(because a shortest path $\exp_{\delta_n}(x)y_n$ completely lies in one of the M_i and because $|y_n \exp_{\delta_n}(x^*)| = o(\delta_n)/\delta_n)$. Therefore $|p\exp_{\delta_n}(x^*)| \leq (1-\varepsilon)|x^*|$ for n sufficiently large, By Lemma 1.5, a limit of shortest paths in N/δ_n between p and $\exp_{\delta_n}(x^*)$ (which is a quasigeodesic by the generalized Lieberman lemma, Theorem 1.1) is a shortest path ox^* in $C_p(M_i)$. Because limits preserve lengths of quasigeodesics [Perelman and Petrunin 1994, 2.3(3)], we have

$$\lim_{n\to\infty} |p\exp_{\delta_n}(x^*)|_{N/\delta} = |x^*|.$$

Hence for n sufficiently large we get

$$|p\exp_{\delta_n}(x^*)| \leq (1-\varepsilon)|p\exp_{\delta_n}(x^*)|_{N/\delta}.$$

Therefore we can find a segment $s_n r_n$ on a shortest path $p\exp_{\delta_n}(x^*)$ that completely lies in one of the M_i/δ_n, such that $s_n, r_n \in N/\delta_n$ and

$$|s_n r_n|_{M_i} \leq (1-\varepsilon)(|pr_n|_N - |ps_n|_N) \tag{2.2}$$

(where we use the same notation for points in N and N/δ).

We can easily pass to a subsequence such that $\lim_{n\to\infty} |ps_n|_N/|pr_n|_N = c$. for some $0 \leq c \leq 1$.

Now we consider two cases, $c \neq 1$ and $c = 1$.

Suppose $c \neq 1$, and consider limit $(M_i/|pr_n|_N, p) \xrightarrow{GH} C_p(M_i)$. Pass to a subsequence such that $s_n \to s$ and $r_n \to r$. The boundary N is an extremal subset; therefore, by Theorem 1.2, $(N/|pr_n|_N, p) \xrightarrow{GH} C_p(N)$ as length-metric spaces. Hence

$$\lim_{n\to\infty} \frac{|s_n r_n|_{M_i}}{|pr_n|_N} = |sr| \geq |r| - |s| = |r|_{C(N)} - |s|_{C(N)} = 1 - \lim_{n\to\infty} \frac{|ps_n|_N}{|pr_n|_N},$$

contradicting (2.2).

Suppose instead that $c = 1$. Pass to a subsequence such that there exists a limit $(M_i/|s_n r_n|_{M_i}, s_n) \xrightarrow{GH} (M_s, s)$. (We remark that M_s need not be the tangent cone.) Set $N_s = \partial M_s$. By Theorem 1.2 we have

$$(N/|s_n r_n|_{M_i}, s_n) \xrightarrow{GH} (N_s, s).$$

Let $f_n : N/|s_n r_n|_{M_i} \to R$ be functions defined by

$$f_n(x) = |px|_{N/|s_n r_n|_{M_i}} - |ps_n|_{N/|s_n r_n|_{M_i}}.$$

Pass to a subsequence such that there exists a limit $f : N_s \to R$, $f = \lim_{n\to\infty} f_n$.

It is easy to see that M_s can be represented as a product $R \times M_s'$ such that $f(x) \leq pr_R(x)$, where pr_R is the projection $M_s \to R$. Indeed a sequence of quasigeodesics that prolong shortest paths ps_n in N easily goes to a straight line in M_s, so by the Toponogov splitting theorem we have such a representation. Therefore N_s is split as well, $N_s = R \times N_s'$.

Let σ_n be a shortest path in N between p and s_n, parametrized by distance from s_n, and let σ be a limit of $\{\sigma_n/|r_n s_n|_{M_i}\}$. By the triangle inequality, for any

$T > 0$ we have $|xp|_N - |s_np| \leq |x\sigma_n(|s_nr_n|T)| - |s_nr_n|T$. As a limit we obtain that $f(x) \leq |x\sigma(T)| - T$. For $T \to \infty$ the right side goes to the Busemann function of σ which coincides with pr_R.

Pass to a subsequence such that there is a limit as $r_n \to r$. We obtain

$$1 = |rs| \geq pr_R(r) \geq f(r) = \lim_{n\to\infty} (|pr_n|_N - |ps_n|_N)/|r_ns_n|_{M_i},$$

again contradicting (2.2). This concludes the proof of the lemma. $\square$

LEMMA 2.6. *The directions of exit and entrance (ξ_i) of any m-shortest path at every interior vertex $p = p_l$, for $l \in \{1, 2, \ldots, k \leq m\}$ (see Definition 2.3), are opposite in $C_p^\#(X)$ (that is, $|\xi_1\xi_2| = 2|\xi_1| = 2|\xi_2|$; see [Perelman and Petrunin 1994, 2.1]).*

PROOF. Let $\dot{\xi}_i \in \Sigma_p(M_i)$ be directions of exit/entrance of the m-shortest path at the interior vertex p. We first prove that $|\xi_1\nu|_0 + |\xi_2\nu|_0 = \pi$ for any $\nu \in \Sigma_p(N) \subset \Sigma_p^\#(X)$. Here the left side is the sum of two 0-distances in the glued space $\Sigma_p^\#(X)$, each of which, by Definition 2.3, is measured in one of the $\Sigma_p(M_i)$. Assume we have proved the lemma for dim $< n$, and let dim $\Sigma_p^\#(X) = n$. From the first variation formula we obtain

$$f(\nu) := |\xi_1\nu|_0 + |\nu\xi_2|_0 \geq \pi$$

for any $\nu \in \Sigma_p(N)$. Assume $\bar{\nu}$ is the minimum point in $\Sigma_p(N)$ of the last function. Thus, $\xi_1\bar{\nu}\xi_2$ is a 1-shortest path. Let γ be a shortest path in $\Sigma_p(N)$ such that $\gamma(0) = \bar{\nu}$ with arbitrary initial data $\gamma^+(0) = \eta$. Assume $f(\bar{\nu}) > \pi$. By the induction assumption, $|(\xi_1)'_{\bar{\nu}}\eta|_0 + |\eta(\xi_2)'_{\bar{\nu}}|_0 = \pi$. By the generalized Lieberman lemma, Theorem 1.1, γ is a quasigeodesic as a curve in $\Sigma_p(M_1)$ and $\Sigma_p(M_2)$. By [Perelman and Petrunin 1994, 1.4(G1)], the condition $f(\bar{\nu}) > \pi$ implies $(f \circ \gamma)(x) < (f \circ \gamma)(0) = f(\bar{\nu})$ for sufficiently small x. This contradicts the assumption that f has a minimum at $\bar{\nu}$.

Therefore $f(\bar{\nu}) = \pi$. Take any shortest path γ in $\Sigma_p(N)$ such that $\gamma(0) = \bar{\nu}$. Then γ is a quasigeodesic for $\Sigma_p(M_1)$ and $\Sigma_p(M_2)$. Set

$$g(\nu) := \cos|\xi_1\nu|_0 + \cos|\nu\xi_2|_0$$

for $\nu \in \Sigma_p(N)$. By the preceding arguments, $g(\bar{\nu}) = g \circ \gamma(0) = 0$, $(g \circ \gamma)'(0) = 0$ and $g \circ \gamma \leq 0$. By [Perelman and Petrunin 1994, 1.3(L2)], $(g \circ \gamma)'' + g \circ \gamma \geq 0$. Therefore $(g \circ \gamma)'' \geq 0$ and so $g \circ \gamma \equiv 0$; in particular for any ν, $g(\nu) = 0$. Therefore $f \equiv \pi$, that is, $|\xi_1\nu|_0 + |\xi_2\nu|_0 = \pi$ as claimed.

In order to prove that ξ_1 and ξ_2 are opposite, it is enough to show that $2|\xi_1| = 2|\xi_2| = |\xi_1\xi_2|$ holds in $C_p^\#(X)$, or equivalently that $|\xi_1\xi_2| = \pi$ holds in $\Sigma_p^\#(X)$. If this is false, there is m such that $|\xi_1\xi_2|_m < \pi$ in $\Sigma_p^\#(X)$. Let θ be the closest vertex to ξ_1 of the m-shortest path $\xi_1\xi_2$. By the preceding discussion, there is a 1-shortest path through θ of length π. Therefore we have two distinct directions at θ which are opposite to $(\xi_1)'_\theta$, a contradiction to the fact that $\Sigma_p^\#$ is an Alexandrov space. This completes the proof of the lemma. $\square$

COROLLARY 2.7. *Let $\xi_i \in \Sigma_p(M_i)$ be directions of exit/entrance of an m-shortest path at an interior vertex. For any $\eta \in \Sigma_p(M_i)$ there is a unique $\eta^* \in \Sigma_p(N)$ such that*

$$|\xi_1\eta|_0 + |\eta\eta^*|_0 + |\eta^*\xi_2|_0 = \pi$$

or

$$|\xi_1\eta^*|_0 + |\eta^*\eta|_0 + |\eta\xi_2|_0 = \pi.$$

PROOF. Suppose $\eta \in \Sigma_p(M_1)$. Consider the 1-shortest path $\eta\xi_2$. Applying Lemma 2.6 to $\Sigma_p^\#(X)$ we see that the directions at the vertex are opposite; therefore this 1-shortest path is a part of a 1-shortest path $\xi_1\xi_2$. □

LEMMA 2.8. *Let $\gamma : [a, b] \to X$ be a quasigeodesic in one of the int M_i or a shortest path in the length metric of N. Then*

$$\rho_k(|p\gamma(t)|_m)'' + k\rho_k(|p\gamma(t)|_m) \leq 1$$

for any $p \in X$

For the definition of ρ_k see [Perelman and Petrunin 1994, 1.4(L2)].

PROOF. We consider the case $k = 0$; we must show that $(|p\gamma(t)|_m^2)'' \leq 2$.
    This is true for $m = 0$ because

$$|pq|_0 = \begin{cases} |pq|_{M_i} & \text{if } p \in M_i, q \in \text{int } M_i \text{ or } q \in M_i, p \in \text{int } M_i, \\ \min_i |pq|_{M_i} & \text{if } p, q \in N, \\ \infty & \text{otherwise.} \end{cases}$$

(Recall that a shortest path in N is a quasigeodesic in both M_i by the generalized Lieberman Lemma).

Suppose the claim is true for all $l < m$ and false for m. Then the standard idea shows that in this case there exists $t_0 \in (a, b)$ and $\varepsilon > 0$ such that for $|t - t_0| < \varepsilon$

$$|p\gamma(t)|_m^2 \geq |p\gamma(t_0)|_m^2 - A(t - t_0) + (t - t_0)^2 + \varepsilon(t - t_0)^2,$$

for some constant A.

Assume $t_0 = 0$. Set $q = \gamma(0)$ and let $p = p_0 p_1 \ldots p_k p_{k+1} = q$ be an m-shortest path. Take a sequence $t_j \to 0$ such that the sequence $((\gamma(t_j)'_{p_k})^*$ (as in Corollary 2.5) goes to some direction $\nu \in \Sigma_{p_k}(N)$. Using Lemma 2.2 we can find a shortest path γ_k in N which goes from p_k in a direction arbitrarily close to ν.

In the following proof one might get lost in calculations and lose the main idea. If we assume that all $((\gamma(t_j)'_{p_k})^*$ coincide with ν and there is a shortest path (in the intrinsic metric of N) that goes in this direction, one can ignore the residue terms below.

Set $\alpha = \angle((p_k)'_q, \gamma^+(0))$, $\beta = \angle(q'_{p_k}, \gamma_k^+(0))$, $\beta_j = \angle(\gamma_k^+(0)\,(\gamma(t_j))'_{p_k})$, $\theta_j = \angle((\gamma(t_j))'_{p_k}\, q'_{p_k})$, and $\delta = \angle(\gamma_k^+, \nu)$, as in the figure below.

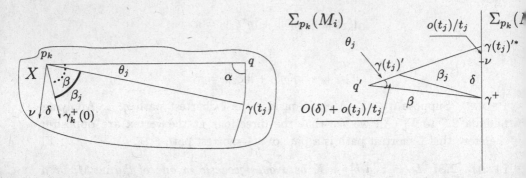

It is easy to see that

$$\theta_j \geq \frac{t_j \sin \alpha}{|p_k q|_0} + o(t_j).$$

We can assume that $q'_{p_k} \notin \Sigma_{p_k}(N)$; otherwise our m-shortest path lies completely in N. By the cosine rule applied to the triangle $\triangle q'_{p_k}(\gamma(t_j))'_{p_k}\gamma_k^+(0)$, we have

$$\beta - \beta_j \geq \left(1 + o(\delta) + o(t_j)/t_j\right)\theta_j \geq t_j\left(\frac{\sin\alpha}{|p_k q|_0} + o(\delta)\right) + o(t_j).$$

Hence

$$\cos(\beta - \beta_j) \leq 1 - \frac{t_j^2 \sin^2\alpha}{2\,|pq|_0^2} + o(\delta)t_j^2 + o(t_j^2).$$

From the induction assumption and Lemma 2.6 we have

$$|p\,\gamma_k(\tau)|_{m-1}^2 \leq |p\,p_k|_{m-1}^2 + 2\tau|p\,p_k|_{m-1}\cos\beta + \tau^2.$$

Because γ_k is a quasigeodesic for both of the M_i, we obtain

$$|\gamma(t_j)\gamma_k(\tau)|_0^2 \leq |\gamma(t_j)\,p_k|_0^2 - 2\cos\beta_j\,\tau\,|\gamma(t_j)\,p_k|_0 + \tau^2,$$

where these distances are measured in a fixed M_i.

Therefore, using (2.1) and the previous two inequalities, we have

$$|p\,\gamma(t_j)|_m^2 \leq \min_\tau (|p\,\gamma_k(\tau)|_{m-1} + |\gamma(t_j)\gamma_k(\tau)|_0)^2$$

$$\leq \min_\tau (|AB(\tau)| + |B(\tau)C|)^2$$

$$= |AC|^2 = |p\,p_k|_{m-1}^2 + |\gamma(t_j)\,p_k|_0^2 + 2\,|p\,p_k|_{m-1}|\gamma(t_j)\,p_k|_0\cos(\beta - \beta_j),$$

where $A, B(\tau)$ and C are as shown in the following diagram in the plane:

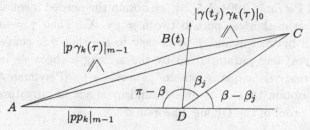

Because γ is either a quasigeodesic in one of the M_i, or a shortest path in N and therefore a quasigeodesic in both of the M_i (see Theorem 1.1), we conclude that

$$|p_k\,\gamma(t_j)|_0^2 \le |p_k q|_0^2 + t_j^2 - 2t_j|p_k q|_0 \cos\alpha$$

and so

$$|p_k\,\gamma(t_j)|_0 \le |p_k q|_0 - t_j \cos\alpha + \frac{t_j^2 \sin^2\alpha}{2|p_k q|_0} + o(t_j^2).$$

Hence

$$|p\,\gamma(t_j)|_m^2 \le |p\,p_k|_{m-1}^2 + |q\,p_k|_0^2 + t_j^2 - 2t_j|q\,p_k|_0\cos\alpha$$

$$+ 2|p\,p_k|_{m-1}\left(|p_k q|_0 - t_j\cos\alpha + \frac{t_j^2\sin^2\alpha}{2|p_k q|_0} + o(t_j^2)\right)$$

$$\times \left(1 - \frac{t_j^2\sin^2\alpha}{2|p_k q|_0^2} + t_j^2 o(\delta) + o(t_j^2)\right)$$

$$\le (|p\,p_k|_{m-1} + |p_k q|_0)^2 - 2t_j(|p\,p_k|_{m-1} + |p_k q|_0)\cos\alpha + t_j^2 + t_j^2 o(\delta) + o(t_j^2)$$

$$= |p\,q|_m^2 - 2t_j|p\,q|_m\cos\alpha + t_j^2 + t_j^2 o(\delta) + o(t_j^2).$$

This inequality for two sequences $t_j \to 0^+$ and $t_j \to 0^-$ contradicts our assumption for sufficiently small δ. $\square$

We continue the proof of Theorem 2.1, showing that every m-shortest path is a k-quasigeodesic. Indeed, using [Perelman and Petrunin 1994, 1.4(L2), 1.5], we only need to verify that $\rho_k(|\gamma(t)p|)'' \le 1 - k\rho_k(|\gamma(t)p|)$. Now $|\gamma(t)p| = \lim_{n\to\infty}|\gamma(t)p|_n$, and using Lemma 2.8 and [Perelman and Petrunin 1994, 1.3(4)] we obtain the needed inequality for all $t \ne t_l$ (where $\gamma(t_l) = p_l$).

Let σ be a shortest path between an arbitrary point x and $\gamma(t_l)$, parametrized by distance from $\gamma(t_l)$. By Lemmas 2.5 and 2.6 we conclude that, for fixed ε,

$$|\sigma(T)\gamma(t_l + T\varepsilon)| + |\sigma(T)\gamma(t_l - T\varepsilon)| \le 2T + CT\varepsilon^2 + o(T).$$

Therefore

$$\operatorname{dist}_x \circ \gamma(t_l + T\varepsilon) + \operatorname{dist}_x \circ \gamma(t_l - T\varepsilon) \le 2\operatorname{dist}_x \circ \gamma(t_l) + CT\varepsilon^2 + o(T).$$

Therefore, for $T \to 0$,

$$(\operatorname{dist}_x \circ \gamma)^+(t_l) \le (\operatorname{dist}_p \circ \gamma)^-(t_l) + C\varepsilon.$$

Hence, for $\varepsilon \to 0$, we obtain $(\mathrm{dist}_x \circ \gamma)^+(t_l) \le (\mathrm{dist}_x \circ \gamma)^-(t_l)$. From this, using [Perelman and Petrunin 1994, 1.3(2)], we obtain the needed inequality for any t.

Let γ_m be an m-shortest path between $p, q \in X$. Then $\gamma = \lim_{m \to \infty} \gamma_m$ is a shortest path between p and q. It is easy to see that γ is convex (as a limit of convex curves) and parametrized by the arclength (because $\mathrm{length}(\gamma_m) \to \mathrm{length}(\gamma)$); hence γ is a quasigeodesic. Therefore by [Perelman and Petrunin 1994, 1.6] we obtain that X is an Alexandrov space of curvature $\ge k$. This completes the proof of the Gluing Theorem. $\qquad\square$

3. The Radius Sphere Theorem

Theorem 3.2 below was proved independently by Karsten Grove and Peter Petersen [Grove and Petersen 1993]. Another proof follows immediately from [Perelman and Petrunin 1993, 1.2, 1.4.1]. The following proof is only a good demonstration of how beautiful quasigeodesics are.

PROPOSITION 3.1. *Let Σ be an Alexandrov space of curvature ≥ 1, with radius greater than $\pi/2$. Then for any $p \in \Sigma$ the space of directions Σ_p has a radius greater than $\pi/2$.*

PROOF. Assume that Σ_p has radius $\le \pi/2$, and let $\xi \in \Sigma_p$ be a direction such that $\mathrm{clos}\, B_\xi(\pi/2) = \Sigma_p$. Take a quasigeodesic of length $\pi/2$ starting at p in the direction ξ. Then the other endpoint q of this quasigeodesic satisfies $\mathrm{clos}\, B_q(\pi/2) = \Sigma$. (Indeed, for any point $r \in \Sigma$ we have $\angle rpq \le \pi/2$; therefore $|rq| \le \pi/2$ by the comparison inequality [Perelman and Petrunin 1994, 1.4(G2)]. This contradicts our assumption that Σ has radius $> \pi/2$. $\qquad\square$

THEOREM 3.2. *Let Σ be an Alexandrov space of curvature ≥ 1, with radius $> \pi/2$. Then Σ is homeomorphic to the sphere S^n.*

PROOF. Assume we have proved the theorem for $\dim \Sigma < n$. We now prove it for $\dim \Sigma = n$.

Let xy be a diameter of Σ. Let z be a critical point of dist_x. Then $\widetilde{\angle} xzy \le \angle xzy \le \pi/2$. By assumption $|xz|, |zy|, \pi/2 \le |xy|$. Therefore the last inequality can hold only for $z = y$. Therefore dist_x has no critical points but x and y. By [Perelman 1994], Σ is homeomorphic to $S(\Sigma_x)$. By Proposition 3.1 we have $\mathrm{Rad}(\Sigma_x) > \pi/2$. Hence by the induction assumption Σ_x is homeomorphic to S^{n-1}. Therefore Σ is homeomorphic to S^n. $\qquad\square$

References

[Buyalo 1976] S. V. Buyalo, "Кратчайшие на выпуклых гиперповерхностях риманова пространства" [Shortest lines on convex hypersurfaces of a Riemannian space], *Zap. Nauchn. Sem. Leningrad. Otdel. Mat. Inst. Steklov* **66** (1976), 114–131.

[Busemann 1958] H. Busemann, *Convex Surfaces*, Tracts in Pure and Appl. Math. **6**, Interscience, New York, 1958.

[Burago et al. 1992] Y. Burago, M. Gromov, and G. Perelman, "A. D. Aleksandrov spaces with curvatures bounded below", *Uspehi Mat. Nauk* **47**:2 (1992), 3–51, 222; translation in *Russian Math. Surveys* **47**:2 (1992), 1–58.

[Grove and Petersen 1993] K. Grove and P. Petersen, "A radius sphere theorem", *Invent. Math.* **112** (1993), 577–583.

[Liberman 1941] I. M. Liberman, "Geodesics on convex surfaces" (in Russian), Doklady Akademii Nauk **32**:5 (1941), 310–313.

[Perelman 1994] G. Perelman, "Elements of Morse theory on Alexandrov spaces" (Russian), *Algebra i analiz* **5**:1 (1993), 232–241; translation in *St. Petersburg Math. J.* **5**:1 (1994), 207–214.

[Perelman 1991] G. Ya. Perelman, "Alexandrov's spaces with curvatures bounded below, II", preprint, 1991.

[Perelman and Petrunin 1993] G. Ya. Perelman and A. M. Petrunin, "Extremal subsets in Aleksandrov spaces and the generalized Liberman theorem" (Russian), *Algebra i Analiz* **5**:1 (1993), 242–256; translation in *St. Petersburg Math. J.* **5**:1 (1994), 215–227.

[Perelman and Petrunin 1994] G. Ya. Perelman and A. M. Petrunin, "Quasigeodesics and gradient curves in Alexandrov spaces", preprint, University of California at Berkeley, 1994.

[Petrunin 1995] A. Petrunin, "Quasigeodesics in multidimensional Alexandrov spaces", Ph.D. thesis, Steklov Institute at St. Petersburg and University of Illinois at Urbana–Champaign, 1995.

[Pogorelov 1973] A. V. Pogorelov, *Extrinsic geometry of convex surfaces*, Transl. Math. Monographs **35**, American Mathematical Society, Providence, RI, 1973.

ANTON PETRUNIN
ST. PETERSBURG BRANCH
V. A. STEKLOV INSTITUTE OF MATHEMATICS (POMI)
RUSSIAN ACADEMY OF SCIENCES
FONTANKA 27
ST. PETERSBURG, 191011
RUSSIA

Comparison Geometry
MSRI Publications
Volume 30, 1997

The Comparison Geometry of Ricci Curvature

SHUNHUI ZHU

ABSTRACT. We survey comparison results that assume a bound on the manifold's Ricci curvature.

1. Introduction

This is an extended version of the talk I gave at the Comparison Geometry Workshop at MSRI in the fall of 1993, giving a relatively up-to-date account of the results and techniques in the comparison geometry of Ricci curvature, an area that has experienced tremendous progress in the past five years.

The term "comparison geometry" had its origin in connection with the success of the Rauch comparison theorem and its more powerful global version, the Toponogov comparison theorem. The comparison geometry of *sectional curvature* represents many ingenious applications of these theorems and produced many beautiful results, such as the $\frac{1}{4}$-pinched Sphere Theorem [Berger 1960; Klingenberg 1961], the Soul Theorem [Cheeger and Gromoll 1972], the Generalized Sphere Theorem [Grove and Shiohama 1977], The Compactness Theorem [Cheeger 1967; Gromov 1981c], the Betti Number Theorem [Gromov 1981a], and the Homotopy Finiteness Theorem [Grove and Petersen 1988], just to name a few. The comparison geometry of Ricci curvature started as isolated attempts to generalize results about sectional curvature to the much weaker condition on Ricci curvature. Starting around 1987, many examples were constructed to demonstrate the difference between sectional curvature and Ricci curvature; in particular, Toponogov's theorem was shown not to hold for Ricci curvature. At the same time, many new tools and techniques were developed to generalize results about sectional curvature to Ricci curvature. We will attempt to present the highlights of this progress.

As often is the case, a survey paper becomes outdated before it goes to press. The same can be said about this one. In the past year, many beautiful results

Supported in part by an NSF grant.

were obtained in this area, mainly by T. Colding and J. Cheeger [Colding 1996a; 1996b; 1995; Cheeger and Colding 1995] (see also Colding's article on pages 83–98 of this volume). These results are not included here. To compensate for this, we have tried to give complete proofs of the results we discuss; these results now constitute the "standard" part in the comparison theory of Ricci curvature. We benefited a lot from a course taught by J. Cheeger at Stony Brook in 1988 and from [Cheeger 1991]. Thanks also go to the participants of a topic course I gave at Dartmouth in the winter of 1995.

2. The Main Comparison Theorem through Weitzenböck Formula

The relation between curvature and the geometry is traditionally introduced through the second variation of arclength, as was first used by Myers. We chose to introduce it from the Weitzenböck formula, which gives a uniform starting point for many applications, including the recent results of Colding.

For a smooth function f, we define its gradient, Hessian, and Laplacian by

$$\langle \nabla f, X \rangle = X(f), \quad \text{Hess}\, f(X, Y) = \langle \nabla_X(\nabla f), Y \rangle, \quad \triangle f = \text{tr}(\text{Hess}\, f).$$

For a bilinear form A, we write $|A|^2 = \text{tr}(AA^t)$.

THEOREM 2.1 (THE WEITZENBÖCK FORMULA). *Let (M^n, g) be a complete Riemannian manifold. Then, for any function $f \in C^3(M)$, we have*

$$\tfrac{1}{2} \triangle |\nabla f|^2 = |\text{Hess}\, f|^2 + \langle \nabla f, \nabla(\triangle f) \rangle + \text{Ric}(\nabla f, \nabla f)$$

pointwise.

PROOF. Fix a point $p \in M$. Let $\{X_i\}_1^n$ be a local orthonormal frame field such that

$$\langle X_i, X_j \rangle = \delta_{ij}, \quad \nabla_{X_i} X_j(p) = 0.$$

Computation at p gives

$$\tfrac{1}{2} \triangle |\nabla f|^2 = \tfrac{1}{2} \sum_i X_i X_i \langle \nabla f, \nabla f \rangle = \sum_i X_i \langle \nabla_{X_i} \nabla f, \nabla f \rangle = \sum_i X_i \text{Hess}(f)(X_i, \nabla f)$$

$$= \sum_i X_i \text{Hess}(f)(\nabla f, X_i) \quad \text{(Hessian is symmetric)}$$

$$= \sum_i X_i \langle \nabla_{\nabla f}(\nabla f), X_i \rangle = \sum_i \langle \nabla_{X_i} \nabla_{\nabla f}(\nabla f), X_i \rangle + \langle \nabla_{\nabla f}(\nabla f), \nabla_{X_i} X_i \rangle$$

$$= \sum_i \langle \nabla_{X_i} \nabla_{\nabla f}(\nabla f), X_i \rangle \quad \text{(the other term vanishes at } p\text{)}$$

$$= \sum_i \langle R(X_i, \nabla f) \nabla f, X_i \rangle + \sum_i \langle \nabla_{\nabla f} \nabla_{X_i} \nabla f, X_i \rangle + \sum_i \langle \nabla_{[X_i, \nabla f]} \nabla f, X_i \rangle.$$

The first term is by definition $\mathrm{Ric}(\nabla f, \nabla f)$; the second term is

$$\sum_i (\nabla f)\langle \nabla_{X_i}\nabla f, X_i\rangle - \langle \nabla_{X_i}\nabla f, \nabla_{\nabla f}X_i\rangle = (\nabla f)\sum_i \langle \nabla_{X_i}\nabla f, X_i\rangle - 0 \ \ (\text{at } p)$$

$$= (\nabla f)\Delta f = \langle \nabla f, \nabla(\Delta f)\rangle,$$

and the third term is

$$\sum_i \mathrm{Hess}(f)([X_i, \nabla f], X_i) = \sum_i \mathrm{Hess}(f)(\nabla_{X_i}\nabla f - \nabla_{\nabla f}X_i, X_i)$$

$$= \sum_i \mathrm{Hess}(f)(\nabla_{X_i}\nabla f, X_i) - \mathrm{Hess}(f)(\nabla_{\nabla f}X_i, X_i)$$

$$= \sum_i \mathrm{Hess}(f)(\nabla_{X_i}\nabla f, X_i) - 0 \ \ (\text{at } p)$$

$$= \sum_i \mathrm{Hess}(f)(X_i, \nabla_{X_i}\nabla f)$$

$$= \sum_i \langle \nabla_{X_i}\nabla f, \nabla_{X_i}\nabla f\rangle = |\mathrm{Hess}(f)|^2.$$

The theorem follows. $\qquad\qquad\qquad\qquad\qquad\qquad\qquad\qquad\qquad\qquad\qquad\qquad$ $\square$

The power of this formula is that we have the freedom to choose the function f. Most of the results of comparison geometry are obtained by choosing f to be the distance function, the eigenfunction, and the displacement function, among others.

We will consider the distance function. Fix a point p, and let $r(x) = d(p, x)$ be the distance from p to x. This defines a Lipschitz function on the manifold, smooth except on the cut locus of p. It also satisfies $|\nabla r| = 1$ where it is smooth. In geodesic polar coordinates at p, we have $\nabla r = \partial/\partial r$. Let $m(r)$ denote the mean curvature of the geodesic sphere at p with outer normal N, i.e., if $\{e_1, \ldots, e_{n-1}\}$ be an orthonormal basis for the geodesic sphere, let

$$m(r) = \sum_{i=1}^{n-1} \langle \nabla_{e_i} N, e_i\rangle.$$

In geodesic polar coordinates, the volume element can be written as

$$d\,\mathrm{vol} = dr \wedge A_\omega(r)\,d\omega$$

where $d\omega$ is the volume form on the standard S^{n-1}. In what follows, we will suppress the dependence of $A_\omega(r)$ on ω for notational convenience. With these notations, we are now ready to state our main result of this section.

THEOREM 2.2 (MAIN COMPARISON THEOREM). *Let (M^n, g) be complete, and assume $\mathrm{Ric}(M) \geq (n-1)H$. Outside the cut locus of p, we have:*
(1) VOLUME ELEMENT COMPARISON:

$$\frac{A(r)}{A^H(r)} \text{ is nonincreasing along radial geodesics.}$$

(2) LAPLACIAN COMPARISON:

$$\Delta r \leq \Delta^H r.$$

(3) MEAN CURVATURE COMPARISON:

$$m(r) \leq m^H(r).$$

(Here quantities with a superscript H are the counterparts in the simply connected space form of constant curvature H.) Furthermore, equality holds if and only if all radial sectional curvatures are equal to H.

PROOF. We will prove the second part, the Laplacian comparison. The first and third parts follow from a lemma that we will prove momentarily.

Let $f(x) = r(x)$ in Theorem 2.1 and note that $|\nabla r| = 1$. We obtain, outside the cut locus of p,

$$|\operatorname{Hess} r|^2 + \frac{\partial}{\partial r}(\Delta r) + \operatorname{Ric}\left(\frac{\partial}{\partial r}, \frac{\partial}{\partial r}\right) = 0.$$

Let $\lambda_1, \ldots, \lambda_n$ be the eigenvalues of $\operatorname{Hess} r$. Since the exponential function is a radial isometry, one of the eigenvalues, say λ_1, is zero. By the Cauchy–Schwarz inequality, we have

$$|\operatorname{Hess}(r)|^2 = \lambda_2^2 + \cdots + \lambda_n^2 \geq \frac{(\lambda_2 + \cdots + \lambda_n)^2}{n-1} = \frac{\operatorname{tr}^2(\operatorname{Hess}(f))}{n-1} = \frac{(\Delta r)^2}{n-1}.$$

Thus, if $\operatorname{Ric} \geq (n-1)H$, then

$$\frac{(\Delta r)^2}{n-1} + \frac{\partial}{\partial r}(\Delta r) + (n-1)H \leq 0.$$

Let $u = (n-1)/\Delta r$. Then

$$\frac{u'}{1 + Hu^2} \geq 1.$$

Note that $\Delta r \to (n-1)/r$ when $r \to 0$; thus $u \to r$. Integrating the above inequality gives

$$\Delta r \leq \Delta^H r = \begin{cases} (n-1)\sqrt{H} \cot \sqrt{H}r & \text{for } H > 0, \\ (n-1)/r & \text{for } H = 0, \\ (n-1)\sqrt{-H} \coth \sqrt{-H}r & \text{for } H < 0. \end{cases}$$

We now discuss the equality case. If equality holds at r_0, then for any $r \leq r_0$, all the inequalities in the above argument become equalities. In particular, the $n-1$ eigenvalues of $\operatorname{Hess}(r)$ are equal to $\sqrt{N} \cot \sqrt{H}r$ (to simplify the notation, we assume $H > 0$. For $H \leq 0$, replace cot by coth.) Let X_i for $i = 2, 3, \ldots, n$, be the orthonormal eigenvectors of $\operatorname{Hess}(r)$ at r; thus

$$\nabla_{X_i} \frac{\partial}{\partial r} = \sqrt{H} \cot \sqrt{H}r X_i.$$

Extend X_i in such a way that $[X_i, \partial/\partial r] = 0$ at r, then

$$
\begin{aligned}
\sec\left(X_i, \frac{\partial}{\partial r}\right) &= -\left\langle \nabla_{\partial/\partial r}\nabla_{X_i}\frac{\partial}{\partial r}, X_i\right\rangle = -\langle \nabla_{\partial/\partial r}(\sqrt{H}\cot\sqrt{H}r)X_i, X_i\rangle \\
&= H\csc^2\sqrt{H}r - \langle\nabla_{\partial/\partial r}X_i, X_i\rangle \\
&= H\csc^2\sqrt{H}r - \sqrt{H}\cot\sqrt{H}r\left\langle\nabla_{X_i}\frac{\partial}{\partial r}, X_i\right\rangle \\
&= H\csc^2\sqrt{H}r - (\sqrt{H}\cot\sqrt{H}r)^2 = H. \qquad\qquad \square
\end{aligned}
$$

We now give a more geometric interpretation of $\triangle r$, which proves parts (1) and (3) of Theorem 2.2.

LEMMA 2.3. *Given a complete Riemannian manifold* (M^n, g) *and a point* $p \in M$, *we have* $\triangle r = m(r)$ *and* $m(r) = A'(r)/A(r)$.

PROOF. By definition,

$$
\begin{aligned}
\triangle r = \operatorname{tr}(\operatorname{Hess} r) &= \sum_{i=1}^{n-1}\langle\nabla_{e_i}(\nabla r), e_i\rangle + \langle\nabla_N(\nabla r), N\rangle \\
&= \sum_{i=1}^{n-1}\langle\nabla_{e_i}N, e_i\rangle + \langle\nabla_N N, N\rangle = \sum_{i=1}^{n-1}\langle\nabla_{e_i}N, e_i\rangle = m(r).
\end{aligned}
$$

This proves the first equality. For the second, consider the map $\phi : T_pM \to M$ defined by $\phi(v) = \exp_p(rv)$. Let $\{v_1, \ldots, v_{n-1}\}$ be an orthonormal basis for the unit sphere in T_pM. Then

$$
\begin{aligned}
A(r) &= d\operatorname{vol}\left(\frac{\partial}{\partial r}, \phi(v_1), \ldots, \phi(v_{n-1})\right) \\
&= d\operatorname{vol}\left(\frac{\partial}{\partial r}, d\exp_p(rv_1), \ldots, d\exp_p(rv_{n-1})\right) \\
&= J_1(r) \wedge J_2(r) \wedge \cdots \wedge J_{n-1}(r),
\end{aligned}
$$

where $J_i(r) = d\exp_p(rv_i)$. Fix r_0. We have

$$
\frac{A'(r_0)}{A(r_0)} = \frac{\sum_{i=1}^{n-1}J_1(r_0) \wedge \cdots \wedge J_i'(r_0) \wedge \cdots \wedge J_{n-1}(r_0)}{J_1(r_0) \wedge J_2(r_0) \wedge \cdots \wedge J_{n-1}(r_0)}.
$$

Let $\bar{J}_1(r), \ldots, \bar{J}_{n-1}(r)$ be linear combinations (with constant coefficients) of the $J_i(r)$'s such that $\bar{J}_1(r_0), \ldots, \bar{J}_{n-1}(r_0)$ form an orthonormal basis. Then

$$
\begin{aligned}
\frac{A'(r_0)}{A(r_0)} &= \frac{\sum_{i=1}^{n-1}J_1(r_0) \wedge \cdots \wedge J_i'(r_0) \wedge \cdots \wedge J_{n-1}(r_0)}{J_1(r_0) \wedge J_2(r_0) \wedge \cdots \wedge J_{n-1}(r_0)} \\
&= \frac{\sum_{i=1}^{n-1}\bar{J}_1(r_0) \wedge \cdots \wedge \bar{J}_i'(r_0) \wedge \cdots \wedge \bar{J}_{n-1}(r_0)}{\bar{J}_1(r_0) \wedge \bar{J}_2(r_0) \wedge \cdots \wedge \bar{J}_{n-1}(r_0)} \\
&= \sum_{i=1}^{n-1}\bar{J}_1(r_0) \wedge \cdots \wedge \bar{J}_i'(r_0) \wedge \cdots \wedge \bar{J}_{n-1}(r_0) = \sum_{i=1}^{n-1}\langle\bar{J}_i'(r_0), \bar{J}_i(r_0)\rangle.
\end{aligned}
$$

Let $f_i(t,s) = \exp_p(sv_i + t\vec{n})$. Then

$$J_i(r_0) = d\exp_p(r_0 v_i) = \frac{\partial}{\partial s}\Big|_{s=0} f_i(t,s)$$

and

$$J_i'(r_0) = \frac{\partial}{\partial t}\Big|_{t=r_0} \frac{\partial}{\partial s}\Big|_{s=0} f_i(t,s) = \frac{\partial}{\partial s}\Big|_{s=0} \frac{\partial}{\partial t}\Big|_{t=r_0} f_i(t,s) = \nabla_{J_i(r_0)} N$$

Therefore, we also have $\bar{J}_i'(r_0) = \nabla_{\bar{J}_i(r_0)} N$. Thus

$$\frac{A'(r_0)}{A(r_0)} = \sum_{i=1}^{n-1} \langle \nabla_{\bar{J}_i(r_0)} N, \bar{J}_i(r_0) \rangle = m(r_0). \qquad \square$$

3. Volume Comparison and Its Applications

The applications of the volume comparison theorem are numerous; we will divide them into several sections. A common feature is that all results in this section are about the fundamental group and the first Betti number.

Bishop–Gromov volume comparison and its direct applications. For most applications of the volume comparison theorem, an integrated form is used. One can integrate the inequality in Theorem 2.2 along radial directions and along a subset of the unit sphere at p. Then:

THEOREM 3.1. *Let $r \leq R$, $s \leq S$, $r \leq s$, $R \leq S$, and let Γ be any measurable subset of S_p^{n-1}. Let $A_{r,R}^\Gamma(p)$ be the set of $x \in M$ such that $r \leq r(x) \leq R$ and any minimal geodesic γ from p to x satisfies $\dot{\gamma}(0) \in \Gamma$. Then*

$$\frac{\mathrm{vol}(A_{s,S}^\Gamma(p))}{\mathrm{vol}(A_{r,R}^\Gamma(p))} \geq \frac{\mathrm{vol}^H(A_{s,S}^\Gamma)}{\mathrm{vol}^H(A_{r,R}^\Gamma)},$$

with equality if and only if the radial curvatures are all equal to H.

REMARK. The strength of this theorem is that now the balls do not have to lie inside the cut locus; hence it is a global result.

We will give a detailed proof of this theorem, since it does not seem to be in the literature.

LEMMA 3.2. *Let f, g be two positive functions defined over $[0, +\infty)$. If f/g is nonincreasing, then for any $R > r > 0$, $S > s > 0$, $r > s$, $R > S$, we have*

$$\frac{\int_r^R f(t)\,dt}{\int_s^S f(t)\,dt} \leq \frac{\int_r^R g(t)\,dt}{\int_s^S g(t)\,dt}.$$

PROOF. It suffices to show that the function

$$F(x,y) = \frac{\int_x^y f(t)\,dt}{\int_x^y g(t)\,dt}$$

satisfies

$$\frac{\partial F}{\partial x} \leq 0, \quad \frac{\partial F}{\partial y} \leq 0.$$

In fact, if this is true, then

$$\frac{\int_r^R f(t)\,dt}{\int_r^R g(t)\,dt} \leq \frac{\int_r^S f(t)\,dt}{\int_r^S g(t)\,dt} \leq \frac{\int_s^S f(t)\,dt}{\int_s^S g(t)\,dt},$$

which is what we want.

We now compute

$$\frac{\partial F}{\partial y} = \frac{1}{(\int_x^y g(t)\,dt)^2} \left(f(y) \int_x^y g(t)\,dt - g(y) \int_x^y f(t)\,dt \right)$$

$$= \frac{g(y)\int_x^y g(t)\,dt}{(\int_x^y g(t)\,dt)^2} \left(\frac{f(y)}{g(y)} - \frac{\int_x^y f(t)\,dt}{\int_x^y g(t)\,dt} \right).$$

But

$$\frac{f(t)}{g(t)} \geq \frac{f(y)}{g(y)} \quad \text{for } x \leq t \leq y.$$

Thus

$$\int_x^y f(t)\,dt \geq \int_x^y \frac{f(y)}{g(y)} \cdot g(t)\,dt = \frac{f(y)}{g(y)} \int_x^y g(t)\,dt,$$

that is,

$$\frac{f(y)}{g(y)} \leq \frac{\int_x^y f(t)\,dt}{\int_x^y g(t)\,dt},$$

which implies $\partial F/\partial y \leq 0$. □

PROOF OF THEOREM 3.1. Just as in lemma 3.2, we only need to show that

$$\frac{\mathrm{vol}(A_{x,y}^\Gamma)}{\mathrm{vol}(A^H(x,y))}$$

is nonincreasing.

Note that

$$\mathrm{vol}(A_{x,y}^\Gamma) = \int_\Gamma d\omega \int_{\min\{x,\mathrm{cut}(\omega)\}}^{\min\{y,\mathrm{cut}(\omega)\}} A(r,\omega)\,dr,$$

where $\mathrm{cut}(\omega)$ is the distance to the cut locus in the direction $\omega \in S_p^{n-1}$.

Since $A(r,\omega)/A^H(r)$ is nonincreasing for any ω and $r < \mathrm{cut}(\omega)$, Lemma 3.2 implies that, for $z \geq y$,

$$\frac{\int_{\min\{x,\mathrm{cut}(\omega)\}}^{\min\{y,\mathrm{cut}(\omega)\}} A(r,\omega)\,dr}{\int_{\min\{x,\mathrm{cut}(\omega)\}}^{\min\{y,\mathrm{cut}(\omega)\}} A^H(r)\,dr} \geq \frac{\int_{\min\{x,\mathrm{cut}(\omega)\}}^{\min\{z,\mathrm{cut}(\omega)\}} A(r,\omega)\,dr}{\int_{\min\{x,\mathrm{cut}(\omega)\}}^{\min\{z,\mathrm{cut}(\omega)\}} A^H(r)\,dr},$$

that is,

$$
\begin{aligned}
\int_{\min\{x,\mathrm{cut}(\omega)\}}^{\min\{y,\mathrm{cut}(\omega)\}} A(r,\omega)\,dr &\geq \frac{\int_{\min\{x,\mathrm{cut}(\omega)\}}^{\min\{y,\mathrm{cut}(\omega)\}} A^H(r)\,dr}{\int_{\min\{x,\mathrm{cut}(\omega)\}}^{\min\{z,\mathrm{cut}(\omega)\}} A^H(r)\,dr} \cdot \int_{\min\{x,\mathrm{cut}(\omega)\}}^{\min\{z,\mathrm{cut}(\omega)\}} A(r,\omega)\,dr \\
&\geq \frac{\int_{x}^{\min\{y,\mathrm{cut}(\omega)\}} A^H(r)\,dr}{\int_{x}^{\min\{z,\mathrm{cut}(\omega)\}} A^H(r)\,dr} \cdot \int_{\min\{x,\mathrm{cut}(\omega)\}}^{\min\{z,\mathrm{cut}(\omega)\}} A(r,\omega)\,dr \\
&\geq \frac{\int_{x}^{y} A^H(r)\,dr}{\int_{x}^{z} A^H(r)\,dr} \cdot \int_{\min\{x,\mathrm{cut}(\omega)\}}^{\min\{z,\mathrm{cut}(\omega)\}} A(r,\omega)\,dr,
\end{aligned}
$$

where the last inequality follows by considering the three possibilities $\mathrm{cut}(\omega) \leq y \leq z$, $y \leq \mathrm{cut}(\omega) \leq z$, and $y \leq z \leq \mathrm{cut}(\omega)$. The inequality before that uses the fact that $\int_{x}^{a} A^H(r)\,dr / \int_{x}^{b} A^H(r)\,dr$ is nonincreasing when $a < b$. Integrate the above over Γ, and we get

$$
\mathrm{vol}(A_{x,y}^{\Gamma}) \geq \frac{\int_{x}^{y} A^H(r)\,dr}{\int_{x}^{z} A^H(r)\,dr} \cdot \mathrm{vol}(A_{x,z}^{\Gamma}) = \frac{\mathrm{vol}(A^H(x,y))}{\mathrm{vol}(A^H(x,z))} \cdot \mathrm{vol}(A_{x,z}^{\Gamma}).
$$

The equality part follows from the equality discussion in Theorem 2.2. $\qquad\square$

By taking $s = r = 0$ and $\Gamma = S^{n-1}$, one gets the following frequently used corollary.

COROLLARY 3.3. (1) GROMOV'S RELATIVE VOLUME COMPARISON THEOREM:

$$
\frac{\mathrm{vol}(B_p(r))}{\mathrm{vol}(B_p(R))} \geq \frac{\mathrm{vol}(B^H(r))}{\mathrm{vol}(B^H(R))}.
$$

(2) BISHOP VOLUME COMPARISON THEOREM:

$$
\mathrm{vol}(B_p(r)) \leq \mathrm{vol}(B^H(r)).
$$

In both cases, equality holds if and only if $B_p(r)$ is isometric to $B^H(r)$.

We will give two applications of this result.

THEOREM 3.4. Assume $\mathrm{Ric} \geq (n-1)H > 0$.
(1) MYERS' THEOREM [1935]: $\mathrm{diam}(M) \leq \pi/\sqrt{H}$, and $\pi_1(M)$ is finite.
(2) CHENG'S MAXIMAL DIAMETER SPHERE THEOREM [1975]: If in addition $\mathrm{diam}(M) = \pi/\sqrt{H}$, then M^n is isometric to $S^n(H)$.

PROOF. Without loss of generality, we will assume $H = 1$.

(1) The classical proof of Myers' theorem is through second variation of geodesics, but one can easily see that it also follows from volume comparison. We will now use Theorem 2.2(2) to prove this.

Let p, q be such that $d(p, q) > \pi$, and let γ be a minimal geodesic from p to q. Since γ is minimal, $\gamma(\pi)$ is outside the cut locus of p; thus d_p is smooth at $\gamma(\pi)$, and $\triangle d_p \leq (n-1)\cot d_p$ by Theorem 2.2. Let $d_p \to \pi$ from the left. Then

$\triangle d_p \leq -\infty$. This is a contradiction, since the left-hand side is a finite number. Therefore the diameter of M is at most π.

Using the same argument for the universal cover of M, we conclude that the universal cover also has diameter less than π, and thus is compact. It then follows that $\pi_1(M)$ is finite.

(2) Cheng's original proof used an eigenvalue comparison theorem; we will use the volume comparison theorem Corollary 3.3 to give a more geometric argument. The first such proof in print seems to be in [Shiohama 1983].

Let $p, q \in M$ be such that $d(p, q) = \pi$. Consider two balls $B_p(\frac{\pi}{2})$ and $B_p(\frac{\pi}{2})$. If the interiors of the two balls intersect, then there is a point x in the intersection such that $d(x, p) < \frac{\pi}{2}$ and $d(x, q) < \frac{\pi}{2}$; therefore $d(p, q) \leq d(x, p) + d(x, q) < \pi$, a contradiction. Thus the two balls do not intersect in the interior. It follows that

$$\mathrm{vol}(M) \geq \mathrm{vol}(B_p(\tfrac{\pi}{2})) + \mathrm{vol}(B_q(\tfrac{\pi}{2}))$$

$$\geq \mathrm{vol}(B_p(\pi)) \cdot \frac{\mathrm{vol}(B^1(\frac{\pi}{2}))}{\mathrm{vol}(B^1(\pi))} + \mathrm{vol}(B_q(\pi)) \cdot \frac{\mathrm{vol}(B^1(\frac{\pi}{2}))}{\mathrm{vol}(B^1(\pi))}$$

$$= \mathrm{vol}(M) \cdot \tfrac{1}{2} + \mathrm{vol}(M) \cdot \tfrac{1}{2} = \mathrm{vol}(M).$$

Thus all inequalities are equalities. In particular, equality holds in the volume comparison. Therefore M has constant curvature 1, and $\mathrm{vol}(B_p(\frac{\pi}{2})) = \mathrm{vol}(B^1(\frac{\pi}{2}))$. It then follows M is simply connected and therefore isometric to $S^n(1)$. $\square$

REMARK. Myers' theorem is almost the only statement we can make about the fundamental group of manifolds with positive Ricci curvature. One conclusion one can draw from it is that the connected sum of two non-simply connected manifolds does not support any metric with positive Ricci curvature. The question remains open for the connected sum of simply connected manifolds.

We now turn to the second application of the relative volume comparison theorem, which was originally proved by analytic methods by Calabi and Yau.

THEOREM 3.5 [Yau 1976]. *If M^n is complete and noncompact with nonnegative Ricci curvature, then $\mathrm{vol}(B_p(r)) \geq cr$ for some $c > 0$.*

REMARK. This, together with Bishop's theorem, gives the growth of the volume of geodesic balls in noncompact manifolds with nonnegative Ricci curvature as $cr \leq \mathrm{vol}(B_p(r)) \leq \omega_n r^n$, where ω_n is the volume of the n-dimensional unit disc.

PROOF. Since M is noncompact, there is a ray γ with $\gamma(0) = p$. By the relative volume comparison theorem for an annulus, we have

$$\frac{\mathrm{vol}(B_{\gamma(t)}(t-1))}{\mathrm{vol}(A_{\gamma(t)}(t-1, t+1))} \geq \frac{\omega_n(t-1)^n}{\omega_n(t+1)^n - \omega_n(t-1)^n} = c(n)t;$$

therefore

$$\mathrm{vol}(B_{\gamma(t)}(t-1)) \geq c(n)\,\mathrm{vol}(A_{\gamma(t)}(t-1, t+1))t \geq c(n)\,\mathrm{vol}(B_p(1))t = c(M)t,$$

and then $\mathrm{vol}(B_p(t)) \geq \mathrm{vol}(B_{\gamma(t/2)}(t/2 - 1)) \geq ct$. $\square$

Packing and Gromov's precompactness theorem. One of the most useful consequences of a lower bound on Ricci curvature is the following.

LEMMA 3.6 (PACKING LEMMA [GROMOV 1981c]). *Let M^n be such that* $\mathrm{Ric} \geq (n-1)H$. *Given r, $\varepsilon > 0$, and $p \in M$, there is a covering of $B_p(r)$ by balls $B_{p_i}(\varepsilon)$, where $p_i \in B_r(p)$, such that the number of balls satisfies $N \leq C_1(n, Hr^2, r/\varepsilon)$ and the multiplicity of the covering is bounded by $C_2(n, H\varepsilon^2)$.*

PROOF. Take a maximal set of points $p_i \in B_p(r - \varepsilon/2)$ such that $\mathrm{dist}(p_i, p_j) \geq \varepsilon/2$ for $i \neq j$. It then follows that $B_{p_i}(\varepsilon/4) \cap B_{p_j}(\varepsilon/4) = \varnothing$. Therefore

$$
\begin{aligned}
N &\leq \frac{\mathrm{vol}(B_p(r))}{\min_i \mathrm{vol}(B_{p_i}(\varepsilon/4))} \\
&= \frac{\mathrm{vol}(B_p(r))}{\mathrm{vol}(B_{p_0}(\varepsilon/4))} \quad \text{for some } p_0 \\
&\leq \frac{\mathrm{vol}(B_{p_0}(2r))}{\mathrm{vol}(B_{p_0}(\varepsilon/4))} \quad \text{since } B_p(r) \subset B_{p_0}(2r) \\
&\leq \frac{\mathrm{vol}(B^H(2r))}{\mathrm{vol}(B^H(\varepsilon/4))} = C_1(n, Hr^2, r/\varepsilon).
\end{aligned}
$$

Next, if $B_{p_i}(\varepsilon/4) \cap B_{p_0}(\varepsilon/4) \neq \varnothing$, then $\mathrm{dist}(p_i, p_0) \leq 2\varepsilon$. Then the disjointness of $B_{p_i}(\varepsilon/4)$ and $B_{p_j}(\varepsilon/4)$ implies

$$
\begin{aligned}
\text{multiplicity} &\leq \frac{\mathrm{vol}(B_{p_0}(2\varepsilon + \varepsilon))}{\min_i \mathrm{vol}(B_{p_i}(\varepsilon/4))} \\
&= \frac{\mathrm{vol}(B_{p_0}(3\varepsilon))}{\mathrm{vol}(B_{p_1}(\varepsilon/4))} \quad \text{for some } p_1 \\
&\leq \frac{\mathrm{vol}(B_{p_0}(5\varepsilon))}{\mathrm{vol}(B_{p_1}(\varepsilon/4))} \quad \text{since } B_{p_0}(3\varepsilon) \subset B_{p_1}(5\varepsilon) \\
&\leq \frac{\mathrm{vol}(B^H(5\varepsilon))}{\mathrm{vol}(B^H(\varepsilon/4))} = C_2(n, H\varepsilon^2).
\end{aligned}
$$ $\square$

It is easy to construct examples to see this is not true if one drops the curvature condition—for example, by connecting two spheres with many thin tunnels.

The packing lemma says that under the assumption of Ricci curvature bounded below, there are only finitely many local intersection patterns on a fixed scale. This is made precise by introducing the Hausdorff distance between metric spaces, which induces a very coarse topology on the space of all compact metric spaces.

DEFINITION 3.7. Let X, Y be two compact metric spaces. A map $\phi : X \to Y$ is called an ε-Hausdorff approximation if the ε-neighborhood of $\phi(X)$ is equal to Y, and $|d(x_1, x_2) - d(\phi(x_1), \phi(x_2))| < \varepsilon$, for any $x_1, x_2 \in X$.

We can then define the Hausdorff distance between two compact metric spaces X, Y, denoted by $d_H(X, Y)$, to be the infimum of all ε such that there is a ε-Hausdorff approximation from X to Y and vice versa.

THEOREM 3.8 (GROMOV'S PRECOMPACTNESS THEOREM [1981c]). *The set of n-dimensional Riemannian manifolds satisfying* $\mathrm{Ric} \geq (n-1)H$ *and* $\mathrm{diam} \leq D$ *is precompact with respect to the Hausdorff topology.*

PROOF. By the packing lemma, for any M^n satisfying the conditions and any $j > 0$, there is a subset $x_1, x_2, \ldots, x_{N_j}$ with $N_j \leq N(n, D, H, j)$, which is $1/j$ dense and has diameter less than D. Thus, for any manifold under consideration, we have a sequence

$$M^1 \subset M^{1/2} \subset M^{1/3} \subset \cdots$$

such that $M^{1/j}$ is $(1/j)$-dense in M, and $|M^{1/j}| \leq N_1 + N_2 + \cdots + N_j$.

Let M_α be an infinite sequence. Since each $\{M_\alpha^1\}$ has N_1 elements and has bounded diameter, the precompactness of bounded sets in $\mathbb{R}^{N_1}$ implies the existence of a subsequence, still denoted by M_α^1, whose elements have pairwise distance less than 1. Thus the corresponding manifolds $M_{\alpha,1}$ have pairwise Hausdorff distance less than 1. For each $\{M_{\alpha,1}\}$, we can again find a subsequence $M_\alpha^{1/2}$ with pairwise distance less than $\frac{1}{2}$, and therefore a subsequence of manifolds $M_{\alpha,1/2}$, with pairwise Hausdorff distance less than $\frac{1}{2}$. Proceeding in this fashion, and using the diagonal argument, we will get a convergent subsequence. This proves that the set is precompact. $\square$

Growth of fundamental groups. Volume comparison is most often used to get information about the fundamental group. The key idea is that the condition on Ricci curvature is local, and so can be lifted to the universal covering space, and control over the volume will give control over the relative size of fundamental domains. In this section, we will begin this study by examining the size of the fundamental group, as measured by its growth.

Let G be a finitely generated group, $G = \langle g_1, g_2, \ldots, g_k \rangle$, and define the r-neighborhood with respect to the set of generators $g = \{g_1, g_2, \ldots, g_k\}$ as

$$U_g(r) = \{g \in G \mid g \text{ is a word of length} \leq r\} = \{g \in G \mid g = g_1^{i_1} \cdots g_k^{i_k}, \Sigma|i_j| \leq r\}.$$

DEFINITION 3.9. G has *polynomial growth* if there exists a set of generators g and a positive number s such that $|U_g(r)| \leq r^s$ for r large. G has *exponential growth* if there exists a set of generators g and a positive number c such that $|U_g(r)| \geq e^{cr}$ for r large.

In this definition, we may take for g *any* set of generators. Indeed, if $g = \{g_1, g_2, \ldots, g_k\}$ and $h = \{h_1, h_2, \ldots, h_l\}$ are two sets of generators, there are constants r_0, s_0 such that $h_i \in U_g(r_0)$ and $g_i \in U_h(s_0)$, so that $U_h(s) \subset U_g(r_0 s)$ and $U_g(s) \subset U_h(s_0 s)$. Therefore, if G is of polynomial growth because g satisfies an inequality as in the definition, h also satisfies such an inequality; and likewise for exponential growth. In fact, this shows that the quantity

$$\mathrm{ord}(G) = \liminf \frac{\ln |U_g(r)|}{r}$$

does not depend on g; we call it the *growth order* of G.

It is easy to check that the free abelian groups $\mathbb{Z}^k$ have polynomial growth of order k. In general, the order is related to the degree of commutativity of the group. One of the most striking results is the following characterization of groups with polynomial growth.

THEOREM 3.10 [Gromov 1981b]. *A group has polynomial growth if and only if it is almost nilpotent, that is, contains a nilpotent subgroup of finite index.*

REMARK. It was shown by Milnor [1968b] that solvable groups have exponential growth unless they are polycyclic.

THEOREM 3.11 [Milnor 1968a]. *If M^n is a complete n-dimensional manifold with $\mathrm{Ric} \geq 0$, then any finitely generated subgroup of $\pi_1(M)$ has polynomial growth of order at most n.*

PROOF. Take the universal cover $(\tilde{M}, \tilde{p}) \to (M, p)$ with the pullback metric. We identify $\pi_1(M)$ with the group of deck transformations of $\tilde{M}$. Let $H = \langle g_1, \ldots, g_k \rangle$ be a finitely generated subgroup of G. Then g_i can each be represented by a loop σ_i at p with length l_i. Let $\tilde{\sigma}_i$ be the liftings of σ_i at $\tilde{p}$; then, as deck transformations, we have $g_i(\tilde{p}) = \tilde{\sigma}_i(l_i)$. Let

$$\varepsilon < \min\{l_1, \ldots, l_k\}, \quad l = \max\{l_1, \ldots, l_k\}.$$

Then, for any distinct $h_1, h_2 \in H$, we have $h_1(B_{\tilde{p}}(\varepsilon)) \cap h_2(B_{\tilde{p}}(\varepsilon)) = \varnothing$, and $\bigcup_{h \in U(r)} h(B_{\tilde{p}}(\varepsilon)) \subset B_{\tilde{p}}(rl + \varepsilon)$. Thus

$$|U(r)| \cdot \mathrm{vol}(B_{\tilde{p}}(\varepsilon)) = \sum_{h \in U(r)} \mathrm{vol}(h(B_{\tilde{p}}(\varepsilon)) \leq \mathrm{vol}(B_{\tilde{p}}(rl + \varepsilon)),$$

and also

$$|U(r)| \leq \frac{\mathrm{vol}(B_{\tilde{p}}(rl + \varepsilon))}{\mathrm{vol}(B_{\tilde{p}}(\varepsilon))} = \frac{\omega_n}{\mathrm{vol}(B_{\tilde{p}}(\varepsilon))}(rl + \varepsilon)^n \leq cr^n. \qquad \square$$

REMARK. If the sectional curvature is nonpositive, one can also show that $\pi_1(M)$ has exponential growth. Note however, that $\mathrm{Ric} \leq 0$ is not enough.

Theorems 3.11 and 3.12 together show that, for manifolds M with $\mathrm{Ric} \geq 0$, any finitely generated subgroup of $\pi_1(M)$ is almost nilpotent. We can ask whether the converse is true. It was shown by Wei [1995] that any torsion-free nilpotent group is the fundamental group of some manifold with positive Ricci curvature, although the growth rate of the examples are far from optimal.

We end this section with a conjecture of Milnor [1968a]:

CONJECTURE 3.12. *If M is complete with $\mathrm{Ric}(M) \geq 0$, then $\pi_1(M)$ is finitely generated.*

Short loops and the first Betti number. Because of the relation between the fundamental group and the first Betti number given by the Hurewicz Theorem [Whitehead 1978], Ricci curvature can also give control on the first Betti number. This can also be seen through the Bochner technique. In this section, we will prove the following theorem.

THEOREM 3.13 [Gromov 1981c; Gallot 1983]. *If M^n is such that* $\mathrm{Ric} \geq (n-1)H$ *and* $\mathrm{diam}(M) \leq d$, *then* $b_1(M, R) \leq c(n, Hd^2)$ *and* $\lim_{Hd^2 \to 0} c(n, Hd^2) = n$.

REMARK. If Ricci curvature is replaced by sectional curvature, then the celebrated Betti number theorem of Gromov [1981a] shows that all higher Betti numbers can be bounded by sectional curvature and diameter. It was shown by Sha and Yang [1989a] that such an estimate is not true for Ricci curvature. Also note that in the above theorem, we used homology with real coefficients. The Betti number theorem of [Gromov 1981a] works for coefficients in any field. It is not known whether Theorem 3.13 is true for finite fields.

REMARK. The second part of the theorem should be compared with earlier results. Recall that Bochner showed that if $\mathrm{Ric} > 0$ then $b_1(M, \mathbb{R}) = 0$. Using an extension of Bochner's techniques, Gallot proved that if $\mathrm{Ric} \geq 0$ then $b_1(M, \mathbb{R}) \leq n$. For the Bochner Technique, see [Wu 1988; Berard 1988]. We will present the geometric proof due to Gromov [1981c].

PROOF. We first show that it is sufficient to prove that there is a finite cover $\hat{M} \to M$ such that $\pi_1(\hat{M})$ has at most $c(n, Hd^2)$ generators. In fact, if $G' = \pi_1(\hat{M}) = \langle \gamma_1, \ldots, \gamma_k \rangle$, and $G = \pi_1(M)$, then $|G/G'| = m < \infty$, i.e., there are $g_1, \ldots, g_m$ such that $g_i^m \in G'$ and G decomposes into left cosets as $G = g_1 G' \cup g_2 G' \cup \cdots \cup g_m G'$. Consider the Hurewicz map

$$G \overset{f}{\to} G/[G, G] \overset{i}{\to} H_1(M, \mathbb{R}).$$

Since $\{\gamma_1, \ldots, \gamma_k, g_1, \ldots, g_m\}$ generates G, $\{f(\gamma_1), \ldots, f(\gamma_k), f(g_1), \ldots, f(g_m)\}$ generates $G/[G, G]$. But

$$m \cdot (i \circ f)(g_i) = (i \circ f)(g_i^m) = (i \circ f)(h)$$

for some $h \in G'$, and $(i \circ f)(g_i)$ can be generated by $\{(i \circ f)(\gamma_i)\}$. Therefore the set $\{(i \circ f)(\gamma_1), \ldots, (i \circ f)(\gamma_k)\}$ generates $H_1(M, \mathbb{R})$. Thus, to bound $b_1(M, \mathbb{R})$, it is sufficient to bound the number of generators for $\pi_1(\hat{M})$.

We will now construct a finite cover $\hat{M}$ and, at the same time, give a set of generators whose size can be bounded by Ricci curvature and diameter. To this end, let $\tilde{\pi} : \tilde{M} \to M$ be the universal cover. Fix $\tilde{x}_0 \in \tilde{M}$ with $\tilde{\pi}(\tilde{x}_0) = x_0$ and $\varepsilon > 0$. Define $\|g\| = d(\tilde{x}_0, g(\tilde{x}_0))$. Take a maximal set of elements $\{g_1, \ldots, g_k\}$ of $\pi_1(M)$ such that $\|g_i\| \leq 2d + \varepsilon$, and $\|g_i g_j^{-1}\| \geq \varepsilon$, for $i \neq j$.

Let Γ be a subgroup of $\pi_1(M)$ generated by $\{g_i\}_1^k$, and let $\hat{\pi} : \hat{M} \to M$ be the covering of M with $\pi_1(\hat{M}) = \Gamma$. We need to show that $\hat{\pi}$ is a finite cover, and to give a bound for the number k.

To show $\hat{\pi}$ is finite, we will show that $\text{diam}(\hat{M}) \leq 2d + 2\varepsilon$. Let $\hat{x}_0 \in \hat{M}$ be such that $\hat{\pi}(\hat{x}_0) = x_0$. If $\text{diam}(\hat{M}) > 2d + 2\varepsilon$, then there is a point $\hat{z} \in \hat{M}$ with $d_{\hat{M}}(\hat{x}_0, \hat{z}) = d + \varepsilon$. But $d_M(x_0, \hat{\pi}(\hat{z})) \leq \text{diam}(M) = d$; therefore there is a deck transformation $\alpha \in \pi_1(M) \setminus \Gamma$ such that $d_{\hat{M}}(\hat{z}, \alpha \hat{x}_0) \leq d$. Then

$$d_{\hat{M}}(\hat{x}_0, \alpha \hat{x}_0) \geq d_{\hat{M}}(\hat{x}_0, \hat{z}) - d_{\hat{M}}(\hat{z}, \alpha \hat{x}_0) \geq \varepsilon,$$
$$d_{\hat{M}}(\hat{x}_0, \alpha \hat{x}_0) \leq d_{\hat{M}}(\hat{x}_0, \hat{z}) + d_{\hat{M}}(\hat{z}, \alpha \hat{x}_0) \leq 2d + \varepsilon.$$

Note there is a $\beta \in \pi_1(\hat{M}) = \Gamma$ such that

$$d_{\tilde{M}}(\Gamma \tilde{x}_0, \alpha \tilde{x}_0) = d_{\tilde{M}}(\beta \tilde{x}_0, \alpha \tilde{x}_0).$$

Therefore

$$\|\beta^{-1}\alpha\| = d_{\tilde{M}}(\tilde{x}_0, \beta^{-1}\alpha \tilde{x}_0) = d_{\tilde{M}}(\beta \tilde{x}_0, \alpha \tilde{x}_0)$$
$$= d_{\tilde{M}}(\Gamma \tilde{x}_0, \alpha \tilde{x}_0) = d_{\hat{M}}(\hat{x}_0, \alpha \hat{x}_0) \leq 2d + \varepsilon.$$

Furthermore, for any $g \in \pi_1(\hat{M}) = \Gamma$, we have

$$\|g^{-1}\beta^{-1}\alpha\| = d_{\tilde{M}}(\tilde{x}_0, g^{-1}\beta^{-1}\alpha \tilde{x}_0) = d_{\tilde{M}}(\beta g \tilde{x}_0, \alpha \tilde{x}_0)$$
$$\geq d_{\tilde{M}}(\Gamma \tilde{x}_0, \alpha \tilde{x}_0) = d_{\hat{M}}(\hat{x}_0, \alpha \hat{x}_0) \geq \varepsilon.$$

Thus $\beta^{-1}\alpha$ should also be in Γ, since Γ is maximal. But α is not in Γ, so this is a contradiction. Therefore $\text{diam}(\hat{M}) \leq 2d + 2\varepsilon$.

We now bound k. Since $\varepsilon \leq \|g_i^{-1}g_j\| = d_{\tilde{M}}(g_i^{-1}g_j \tilde{x}_0, \tilde{x}_0) = d_{\tilde{M}}(g_j \tilde{x}_0, g_i \tilde{x}_0)$, we have

$$B_{g_i \tilde{x}_0}^{\tilde{M}}(\varepsilon/2) \cap B_{g_j \tilde{x}_0}^{\tilde{M}}(\varepsilon/2) = \varnothing \quad \text{for } i \neq j$$

and

$$\bigcup_{i=1}^{k} B_{g_i \tilde{x}_0}^{\tilde{M}}(\varepsilon/2) \subset B_{\tilde{x}_0}^{\tilde{M}}(2d + 3\varepsilon/2).$$

Therefore, choosing $d = \varepsilon/2$, we get

$$k = \frac{\text{vol}(\bigcup_1^k B_{g_i \tilde{x}_0}^{\tilde{M}}(d))}{\text{vol}(B_{\tilde{x}_0}^{\tilde{M}}(d))} \leq \frac{\text{vol}(B_{\tilde{x}_0}^{\tilde{M}}(2d + 3d))}{\text{vol}(B_{\tilde{x}_0}^{\tilde{M}}(d))} \leq \frac{\text{vol}^H(B(2d + 3d))}{\text{vol}^H(B(d))} = c(n, Hd^2).$$

We are now going to refine the preceding argument by choosing longer loops to generate $H_1(M, \mathbb{R})$.

Again, let $\phi : G \to H_1(M, \mathbb{R})$ be the Hurewicz map, and let $\{g_1, \ldots, g_k\}$ be chosen as before, with $\varepsilon = 2d$. Then $\{\phi(g_1), \ldots, \phi(g_k)\}$ is a basis for $H_1(M, \mathbb{R})$. Let $\Gamma = \langle g_1, \ldots, g_k \rangle$. If Γ contains an element γ such that $\|\gamma\| < 2d$ and $\phi(\gamma) \neq 0$, then $\phi(\gamma)$ is of infinite order, and there exists an $m > 0$ such that $2d \leq \|\gamma\| \leq 4d$. Since

$$\phi(\gamma) = a_1 \phi(g_1) + \cdots + a_k \phi(g_k) \neq 0,$$

we can assume, without loss of generality, that $a_1 \neq 0$. Let $\Gamma_1 = \langle \gamma^m, g_2, \ldots, g_k \rangle$. Obviously, $\phi(\gamma^m), \phi(g_2), \ldots, \phi(g_k)$ form a basis for $H_1(M, \mathbb{R})$. Furthermore, $\gamma \notin \Gamma_1$. In fact, if

$$\gamma = b_1 \cdot (\gamma^m)^{k_1} \cdot b_2 \cdot (\gamma^m)^{k_2} \cdots (\gamma^m)^{k_l} \cdot b_{l+1},$$

where the b_i's are words in $\{g_2, \ldots, g_k\}$, then

$$\phi(\gamma) = \phi(b_1) + \phi(\gamma^{mk_1}) + \cdots + \phi(\gamma^{mk_l})$$
$$= \phi(b_1 b_2 \cdots b_{l+1}) + m(k_1 + k_2 + \cdots + k_l)\phi(\gamma),$$

that is,

$$\phi(b_1 b_2 \cdots b_{l+1}) = (1 - m(k_1 + k_2 + \cdots + k_l))\phi(\gamma).$$

Since $m \geq 2$, the coefficient is not zero; therefore

$$\phi(\gamma) = \frac{1}{1 - m(k_1 + k_2 + \cdots + k_l)}\phi(b_1 b_2 \cdots b_{l+1}) = a_2\phi(g_2) + \cdots + a_k\phi(g_k),$$

which contradicts $a_1 \neq 0$.

Thus, each time we have an element $\gamma \in \Gamma$ with $\|\gamma\| < 2d$ and $\phi(\gamma) \neq 0$, we can replace it by γ^m such that $\|\gamma^m\| \geq 2d$, and still have a basis. By repeating this process a finite number of times (since $\pi_1(M)$ contains finitely many γ with $\|\gamma\| < 2d$), we get a set, still denoted by $\{g_1, \ldots, g_k\}$, such that
(1) $\{\phi(g_1), \ldots, \phi(g_k)\}$ is a basis for $H_1(M, \mathbb{R})$;
(2) $2d \leq \|g_i\| \leq 4d$; and
(3) $\|g\| < 2d$ for every element $g \in \langle g_1, \ldots, g_k \rangle$ with $\phi(g) \neq 0$.
    Let

$$U(N) = \left\{ g \in H_1(M, \mathbb{R}) \,\middle|\, g = \sum r_i \phi(g_i), \sum |r_i| \leq N \right\}.$$

By (3), $\|g^{-1}h\| \geq 2d$, which implies that for all $g \in U(N)$ such that $\phi(g) \neq 0$, the balls $B_{g(\tilde{x}_0)}(d)$ are disjoint. Furthermore,

$$\bigcup_{g \in U(N)} B_{g(\tilde{x}_0)}(d) \subset B_{\tilde{x}_0}(2Nd + 4d).$$

By taking the volume on both sides, we get

$$|U(N)| = \frac{\mathrm{vol}(\bigcup_{g \in U(N)} B_{g(\tilde{x}_0)}(d))}{\mathrm{vol}(B_{\tilde{x}_0}(d))}$$
$$\leq \frac{\mathrm{vol}(B_{\tilde{x}_0}(2Nd + 4d))}{\mathrm{vol}(B_{\tilde{x}_0}(d))} \leq \frac{\mathrm{vol}^H(B(2Nd + 4d))}{\mathrm{vol}^H(B(d))}$$
$$= \frac{\int_0^{2Nd+4d} \frac{1}{\sqrt{H}} \sinh^{n-1} \sqrt{H} t \, dt}{\int_0^d \frac{1}{\sqrt{H}} \sinh^{n-1} \sqrt{H} t \, dt} = \frac{\int_0^{(2N+4)d\sqrt{H}} \sinh^{n-1} t \, dt}{\int_0^{d\sqrt{H}} \sinh^{n-1} t \, dt},$$

which is bounded by cN^n when Hd^2 is small. This implies that $b_1(M, \mathbb{R}) \leq n$. $\square$

Short geodesics and the fundamental group. We now consider a bigger class of manifolds, the set $\mathcal{M}$ of n-dimensional manifolds satisfying

$$\text{Ric} \geq -(n-1)\Lambda^2,$$

diam $\leq D$, and vol $\geq v > 0$. Note that the first condition will not yield any restriction on π_1, and the first two conditions allow infinitely many isomorphism classes of π_1 (as the example of lens spaces shows). In this section, we will prove the following theorem.

THEOREM 3.14 [Anderson 1990a]. *The class $\mathcal{M}$ contains only finitely many isomorphism classes of π_1.*

To control the size of the fundamental group is to count the number of fundamental domains, as was shown in the previous sections. For this to work, the fundamental domain should not be too thin, i.e., it should contain a geodesic ball of size bounded from below. This is controlled by the first systol, defined as

$$\text{sys}^1(M, g) = \inf\{\text{length}(\gamma) \mid \gamma \text{ is noncontractible}\}.$$

The following result gives an estimate of most noncontractible curves for the class $\mathcal{M}$.

LEMMA 3.15. *For any Λ, v, D, there exist positive numbers $N(n, \Lambda, v, D)$ and $L(n, \Lambda, v, D)$ such that if $M \in \mathcal{M}$ and $[\gamma] \in \pi_1(M)$ have order $\geq N$, then*

$$\text{length}(\gamma) \geq L.$$

PROOF. Let $\Gamma \subset \pi_1(M, x_0)$ be the subgroup generated by γ, so that $|\Gamma| \geq N$. Let $\pi : \tilde{M} \to M$ be the universal cover, and let $F \subset \tilde{M}$ be a fundamental domain with $\tilde{x}_0 \in F$. Then $B_{\tilde{x}_0}(r) \cap F$ is mapped isometrically by π onto $B_{x_0}(r)$, modulo a set of measure zero corresponding to the boundary of F. In particular, $\text{vol}(B_{\tilde{x}_0} \cap F) = \text{vol}(B_{x_0}(r))$.

Let $U(r) = \{g \in \Gamma \mid g = \gamma^i, |i| < r\}$. Note that $d_{\tilde{M}}(\gamma \tilde{x}_0, \tilde{x}_0) \leq \text{length}(\gamma) = l(\gamma)$; therefore $d_{\tilde{M}}(g\tilde{x}_0, \tilde{x}_0) \leq 2N \cdot l(\gamma)$, for any $g \in U(N)$. Then

$$\bigcup_{g \in U(N)} g(B_{\tilde{x}_0}(D) \cap F) \subset B_{\tilde{x}_0}(2N \cdot l(\gamma) + D).$$

Taking the volume, we obtain

$$(2N + 1)\,\text{vol}(M) \leq \text{vol}(B_{\tilde{x}_0}(2N \cdot l(\gamma) + D)) \leq \text{vol}(B^\Lambda(2N \cdot l(\gamma) + D)).$$

If $l(\gamma) < D/2N$, then,

$$N \leq \frac{\text{vol}(B_{\tilde{x}_0}(2N \cdot l(\gamma) + D))}{2\,\text{vol}(M)} < \frac{\text{vol}(B^\Lambda(2D))}{2v}.$$

Thus, if we set $N = N(n, \Lambda, D, v) = \lfloor \text{vol}(B^\Lambda(2D))/2v \rfloor + 1$, we will have

$$l(\gamma) \geq \frac{D}{2N} = \frac{Dv}{\text{vol}(B^\Lambda(2D))} = L(n, \Lambda, D, v).$$

□

REMARK. In this proof we only considered a subgroup generated by one loop. The same argument applies to subgroups generated by several loops. We thus can prove that the subgroup of $\pi_1(M)$ generated by loops of length $\leq L$ must have order $\leq N$.

REMARK. If the sectional curvature is bounded below by $-\Lambda^2$, one can bound the length of closed geodesics, which gives a lower bound on sys^1. This was proved by Cheeger [1970] in connection with his finiteness theorem.

LEMMA 3.16 [Gromov 1981c]. *For any compact manifold M with diameter D, one can choose a set of generators $\{g_1, \ldots, g_m\}$ of $\pi_1(M, p)$ and representative loops $\{\gamma_1, \ldots, \gamma_m\}$ such that length $(\gamma_i) \leq 2D$ and all relations are of the form $g_i g_j g_k^{-1} = 1$.*

REMARK. In this lemma, we do not have a bound on the number of generators. This number is in general very big, and thus such a representation is not efficient for other purposes.

PROOF. Fix a constant ε smaller than the injectivity radius. Choose a triangulation K of M such that any n-simplex lies in a ball of radius less than ε. Let $\{x_i\}$ be the vertices and $\{e_{ij}\}$ the edges. Since cut(p) has measure zero, we can assume that all x_i are not in cut(p). Let γ_1 be the minimal geodesic from p to x_i, and set $\sigma_{ij} = \gamma_i e_{ij} \gamma_j^{-1}$. Then $\sigma_{ij} \in \pi_1(M, p)$ and the length of $\sigma_{i,j}$ is less than $2D + \varepsilon$. Given any loop σ at p, we can deform σ to lie in the one-skeleton of K. Thus, σ can be written as a product of σ_{ij}s. This shows that the σ_{ij} generate $\pi_1(M, p)$.

If Δ_{ijk} is a two-simplex with vertices x_i, x_j, x_k, we have

$$\sigma_{ij}\sigma_{jk} = \sigma_{ik}.$$

If $\sigma = e$ is a relation with σ a product of σ_{ij}s, the homotopy can be represented by a collection of two simplices (e.g., take simplicial approximations of σ and the homotopy.) Therefore it can be generated by the above set of relations.

Let g_{ij} be a geodesic loop at p in the homotopy class of σ_{ij}. Since the set of lengths of such loops form a discrete set, we can choose δ small enough such that there are no loops with length in $[2D, 2D + \delta]$. Hence if we further require $\varepsilon < \delta$, each g_{ij} has length at most $2D$. □

PROOF OF THEOREM 3.14. By Lemma 3.5.2, we only need to bound the number of generators: in fact, if there are p generators as in Lemma 3.16, the relations can be chosen from among a finite number (p^3) of possibilities.

Let $\{g_1, \ldots, g_p\}$ be such a set of generators. Fix $\tilde{x}_0 \in \tilde{M}$ in the universal cover. Consider $B_{g_i \tilde{x}_0}(L/2)$ with L as in Lemma 3.14. Since length$(g_i) \leq 2D$, we have

$$\bigcup_{i=1}^{p} B_{g_i \tilde{x}_0}(L/2) \subset B_{\tilde{x}_0}(2D + L/2).$$

These balls can intersect, but if

$$B_{g_i \tilde{x}_0}(L/2) \cap B_{g_j \tilde{x}_0}(L/2) \neq \varnothing,$$

then $g_i g_j^{-1}$ is represented by a loop of length less than $2 \cdot L/2 = L$. By Lemma 3.14, there are at most N such loops. Therefore the balls $\{B_{g_i \tilde{x}_0}(L/2)\}$ have multiplicity bounded by N. Thus

$$\mathrm{vol}(B_{\tilde{x}_0}(2D + L/2)) \geq \mathrm{vol}\left(\bigcup_{i=1}^{p} B_{g_i \tilde{x}_0}(L/2) \right) \geq \mathrm{vol}\left(\bigcup B_{g_j \tilde{x}_0}(L/2) \right)$$

$$\geq \frac{p}{N} \, \mathrm{vol}(B_{x_0}(L/2)),$$

where the second union is taken over the balls that do not intersect. It then follows that

$$p \leq N \cdot \frac{\mathrm{vol}(B_{\tilde{x}_0}(2D + L/2))}{\mathrm{vol}(B_{x_0}(L/2))} \leq N \cdot \frac{\mathrm{vol}(B^\Lambda(2D + L/2))}{\mathrm{vol}(B^\Lambda(L/2))}. \qquad \square$$

4. Laplacian Comparison and Its Applications

Weak maximum principle and regularity. It is clear now that in order to apply any analysis to the distance function, we have to either restrict ourselves to the complement of the cut locus, or to extend the analysis to Lipschitz functions. In this section we will extend the maximum principle to this situation. Much of this section is adapted from [Cheeger 1991].

DEFINITION 4.1. A *lower barrier* (or *support function*) for a continuous function f at the point x_0 is a C^2 function g, defined in a neighborhood of x_0, such that $g(x_0) = f(x_0)$ and $g(x) \leq f(x)$ in the neighborhood.

DEFINITION 4.2. If f is continuous, we say that $\Delta f \geq a$ at x_0 *in the barrier sense* if, for any $\varepsilon > 0$, there is a barrier $f_{x_0, \varepsilon}$ of f at x_0 such that

$$\Delta f_{x_0, \varepsilon} \geq a - \varepsilon.$$

THEOREM 4.3 (WEAK MAXIMUM PRINCIPLE, HOPF–CALABI). *Let M be a connected Riemannian manifold and let $f \in C^0(M)$. Suppose that $\Delta f \geq 0$ in the barrier sense. Then f attains no weak local maximum value unless it is a constant function.*

PROOF. Let p be a weak local maximum, so that $f(p) \geq f(x)$ for all x_0 near p. Take a small normal coordinate ball $B_p(\delta)$, and assume that there exists a point $z \in \partial B_p(\delta)$ such that $f(p) > f(z)$. Then, by continuity, $f(p) > f(z')$ for $z' \in \partial B_p(\delta)$ sufficiently close to z. Choose a normal coordinate system $\{x_i\}$ such that $z = (\delta, 0, \dots, 0)$. Put $\phi(x) = x_1 - d(x_2^2 + \cdots + x_n^2)$, where d is a number so large that if $y \in \partial B_p(\delta)$ and $f(y) = f(p)$, then $\phi(y) < 0$. Note that

$$\nabla \phi = \frac{\partial}{\partial x^1} - \cdots \neq 0.$$

Put $\psi = e^{a\phi} - 1$. Then $\triangle\psi = (a^2|\nabla\phi|^2 + a\triangle\phi)e^{a\phi}$. Thus, for a large enough, $\triangle\psi > 0$. Moreover, $\psi(p) = 0$. Thus, for $\eta > 0$ sufficiently small,

$$(f + \eta\psi)|_{\partial B_p(\delta)} < f(p), \quad (f + \eta\psi)(p) = f(p).$$

Therefore $f + \eta\psi$ has an interior maximum at some point $q \in B_p(\delta)$.

If $f_{q,\varepsilon}$ is a barrier for f at q with $\triangle f_{q,\varepsilon} \geq -\varepsilon$, then $f_{q,\varepsilon} + \eta\psi$ is also a barrier for $f + \eta\psi$ at q. For ε sufficiently small, we have $\triangle(f_{q,\varepsilon} + \eta\psi) > 0$. Since $f + \eta\psi$ has a local maximum at q, and

$$f_{q,\varepsilon} + \eta\psi < f + \eta\psi, \quad (f_{q,\varepsilon} + \eta\psi)(q) = (f + \eta\psi)(q),$$

we find that $f_{q,\varepsilon} + \eta\psi$ has a local maximum at q. This is not possible because $\triangle(f_{q,\varepsilon} + \eta\psi) > 0$.

It follows that for all small δ, we have $f|_{\partial B_p(\delta)} = f(p)$. Since M is connected, this implies that f is constant. $\qquad\square$

The following regularity theorem is not necessary for the proof of later results, but it simplifies the proof of the Splitting Theorem considerably.

THEOREM 4.4 (REGULARITY). *If $\triangle f = 0$ in the barrier sense, then f is smooth.*

PROOF. Since regularity is a local property, this theorem follows from standard elliptic regularity [Gilbarg and Trudinger 1983, Theorem 6.17]. $\qquad\square$

The Splitting Theorem. Recall that a geodesic $\gamma : [0, +\infty) \to M$ is a *ray* if $d(0, \gamma(t)) = t$ for all $t > 0$. A geodesic $\gamma : (-\infty, +\infty) \to M$ is a *line* if $d(\gamma(s), \gamma(t)) = |s - t|$ for all t, s. It is easy to see that if M is noncompact, it contains a ray. If it has at least two ends, it contains a line.

The purpose of this section is to prove the Splitting Theorem of Cheeger and Gromoll [1971]. We will first provide some preliminary properties of the Busemann functions.

Let σ be a ray, and define $b_r^\sigma : M \to \mathbb{R}$ by $b_r^\sigma(x) = r - d(x, \sigma(r))$.

LEMMA 4.5. (1) b_r^σ *is increasing in r when x is fixed.*
(2) b_r^σ *is bounded by $d(x, \sigma(0))$.*
(3) *The family b_r^σ is uniformly continuous.*

PROOF. (1) By the triangle inequality, for any $r > s$,

$$d(x, \sigma(r)) - d(x, \sigma(s)) \leq d(\sigma(r), \sigma(s)) = r - s,$$

so that $b_r^\sigma(x) \geq b_s^\sigma(x)$. (2) By the triangle inequality,

$$b_r^\sigma(x) = r - d(x, \sigma(r)) = d(\sigma(0), \sigma(r)) - d(x, \sigma(r)) \leq d(\sigma(0), x).$$

(3) We have $|b_r^\sigma(x) - b_r^\sigma(y)| = |d(x, \sigma(r)) - d(y, \sigma(r))| \leq d(x, y)$; uniform continuity follows. $\qquad\square$

DEFINITION 4.6. The Busemann function associated to σ is defined as

$$b^{\sigma}(x) = \lim_{r \to \infty} b_r^{\sigma}(x) = \lim_{r \to \infty} (r - d(x, \sigma(r))).$$

LEMMA 4.7. If $\mathrm{Ric} \geq 0$ and σ is a ray, then $\triangle b^{\sigma} \geq 0$ in the barrier sense.

PROOF. Fix a point $p \in M$. We will construct a barrier for b^{σ} at p. For this, we first define the asymptote of σ. Take a sequence of points $t_i \to \infty$, and let δ_i be a minimizing geodesic from p to $\sigma(t_i)$. A subsequence of δ_i converges to a ray δ, which is called an asymptote of σ at p.

We now show that, for any $r > 0$, $b_r^{\delta}(x) + b^{\sigma}(p)$ is a barrier for b^{σ} at p.

In fact, since δ is a ray, $\delta(r)$ is not a cut point of p, hence p is not a cut point of $\delta(r)$, therefore b_r^{δ} is a smooth function near p. Furthermore, $(b_r^{\delta}(x) + b^{\sigma}(p))(p) = b^{\sigma}(p)$. We thus only need to prove that $b_r^{\delta}(x) + b^{\sigma}(p) \leq b^{\sigma}(x)$ near p.

For any $k > 0$, there exists a geodesic δ_k from p to $\sigma(t_k)$ such that

$$d(\delta(t), \delta_k(t)) \leq 1/k$$

for $t \in [0, t_k]$. Thus,

$$
\begin{aligned}
b_{t_k}^{\sigma}(x) - b_{t_k}^{\sigma}(p) &= d(p, \sigma(t_k)) - d(x, \sigma(t_k)) \\
&= d(p, \delta_k(r)) + d(\delta_k(r), \sigma(t_k)) - d(x, \sigma(t_k)) \\
&\geq r + (d(\delta(r), \sigma(t_k)) - 1/k) - d(x, \sigma(t_k)) \geq r - 1/k - d(\delta(r), x).
\end{aligned}
$$

As $k \to \infty$, we obtain $b^{\sigma}(x) - b^{\sigma}(p) \geq r - d(\delta(r), x) = b_r^{\delta}(x)$, that is, $b_r^{\delta}(x) + b^{\sigma}(p) \leq b^{\sigma}(x)$. This proves the claim.

Now we use the Laplacian comparison theorem to compute

$$\triangle(b_r^{\delta}(x) + b^{\sigma}(p)) = \triangle(r - d(\delta(r), x)) = -\triangle d(\delta(r), x) \geq -\frac{n-1}{d(\delta(r), x)} \geq -\varepsilon,$$

for $d(x, \delta(r))$ big enough. Thus $\triangle b^{\sigma} \geq 0$ in the barrier sense. $\square$

If σ is a line, then we have two rays σ^+ and σ^-, and thus also two Busemann functions b^+ and b^-.

LEMMA 4.8. If $\mathrm{Ric} \geq 0$, and σ is a line, then
(1) $b^+ + b^- = 0$, and b^+, b^- are smooth; and
(2) through every point in M, there is a unique line perpendicular to the set $V_0 = \{b^+ = 0\}$.

PROOF. (1) By lemma 4.7, $\triangle(b^+ + b^-) \geq 0$. The triangle inequality implies $b^+ + b^- \leq 0$. Obviously $(b^+ + b^-)(\sigma(0)) = 0$. By Theorem 4.3, $b^+ + b^- = 0$.

This, together with $\triangle b^+ \geq 0$ and $\triangle b^- \geq 0$, implies that $0 \leq \triangle b^+ = -\triangle b^- \leq 0$. Thus, $\triangle b^+ = \triangle b^- = 0$, and is therefore smooth by the regularity theorem.

(2) Fix a point $p \in M$. We have at least two asymptotes δ^+, δ^-. Note that

$$b^+(\delta^+(t)) = \lim_{r \to \infty} (r - d(\delta^+(t), \delta^+(r)))$$

$$= \lim_{r \to \infty} (r - d(\delta^+(0), \delta^+(r))) + \lim_{r \to \infty} (d(\delta^+(0), \delta^+(r)) - d(\delta^+(t), \delta^+(r)))$$

$$= b^+(\delta^+(0)) + t.$$

Similarly, $b^-(\delta^-(s)) = b^s(\delta^s(0)) + s$.

Thus,

$$d(\delta^+(t), \delta^-(s)) \geq b^+(\delta^+(t)) - b^+(\delta^-(s))$$

$$\geq b^+(\delta^+(t)) + b^-(\delta^-(s)) = b^+(\delta^+(0)) + t + b^-(\delta^-(0)) + s$$

$$= t + s + b^+(p) + b^-(p) = t + s.$$

Thus, δ^+, δ^- fit together to form a line. Since δ^+, δ^- are arbitrary asymptotes, they are unique.

Moreover, for any $y \in V_0$,

$$d(y, \delta^+(t)) \geq b^+(\delta^+(t)) - b^+(y) = b^+(\delta^+(0)) + t - b^+(y) = t.$$

Thus δ^+ is perpendicular to V_0. $\qquad\square$

THEOREM 4.9 (SPLITTING THEOREM [Cheeger and Gromoll 1971]). *If M has nonnegative Ricci curvature, and contains a line, then M is isometric to the product $\mathbb{R} \times V$, for some $(n-1)$-dimensional Riemannian manifold V.*

PROOF. We can now define a map $\phi : V_0 \times \mathbb{R} \to M$ as

$$\phi(v, t) = exp_v(t\dot\sigma(0)),$$

where V_0 is as in Lemma 4.8, and σ is the unique line passing through $v \in V_0$. We claim that ϕ is an isometry.

To see that ϕ is a bijection, note that, for any $x \in M$, there is a line γ such that $\gamma \perp V_0$, $\gamma(0) \in V_0$, and $x = \gamma(t_0)$. Thus $\phi(\gamma(0), t_0) = x$, which shows that ϕ is surjective. If $\phi(v_1, t_1) = \phi(v_2, t_2) = x$, then $t_1 = t_2 = d(x, V_0)$. Now since σ_1 and σ_2 are lines, they cannot intersect unless they are the same, i.e., $v_1 = v_2$. Therefore ϕ is injective. Since σ is a line, $\exp_v$ is a local diffeomorphism. This implies that ϕ is a diffeomorphism.

To see that ϕ is an isometry, we let $S(t) = (b^+)^{-1}(t)$. Then the mean curvature of $S(t)$ is $m(t) = \Delta b^+ = 0$. On the other hand, letting $f = b^+$ in Theorem 2.1 (note that b^+ is smooth by Lemma 4.8) gives

$$m'(t) = \text{Ric} + |\text{Hess}(b^+)|^2,$$

which, together with Ric ≥ 0, gives

$$|\text{Hess}(b^+)|^2 \leq 0.$$

Thus, $\text{Hess}(b^+) = 0$, and $S(t)$ is totally geodesic. Therefore, for any X tangent to V_0 and $N = \nabla b^+$,

$$R(N, X)N = \nabla_N \nabla_X N - \nabla_X \nabla_N N - \nabla_{[X,N]} N = \nabla_N \nabla_X N = 0.$$

Let $J(t) = \phi_*(X)$. Then J is a variational vector field of geodesics, and satisfies the Jacobi equation, which, with vanishing curvature, says that $J''(t) = 0$. Thus J is a constant, implying $|\phi_*(X)| = |X|$, and hence ϕ is an isometry.   □

From the proof of the Splitting Theorem, it seems natural to make the following conjecture.

CONJECTURE 4.10 (LOCAL SPLITTING). *Let M be complete, and let γ be a line. If M has nonnegative Ricci curvature in a neighborhood of γ, then a (smaller) neighborhood of γ splits as a product.*

The main difficulty in proving this conjecture is that in proving Lemma 4.7 and in using the Laplacian Comparison Theorem to conclude that $\triangle b \geq 0$, it is necessary that the asymptotes stay in the region where $\text{Ric} \geq 0$. In general, this is very hard to achieve. In [Cai et al. 1994], it was proved that if the Ricci curvature is nonnegative outside of a compact set, then any line in this region will cause a splitting. Furthermore, under the condition $0 \leq \sec \leq c$, the local splitting conjecture is true.

We now prove two corollaries on the fundamental group of manifolds with nonnegative Ricci curvature. They generalize Myers's theorem to manifolds with $\text{Ric} \geq 0$, and strengthen the result of Milnor (Theorems 3.11 and 3.4).

COROLLARY 4.11. *If M is compact with nonnegative Ricci curvature, there is a finite group F and a Bieberbach group B_k of $\mathbb{R}^n$ such that the sequence*

$$0 \to F \to \pi_1(M) \to B_k \to 0$$

is exact.

PROOF. By Theorem 4.9, we may write the universal covering space $\tilde{M}$ of M as $N \times \mathbb{R}^k$, where N contains no lines. Since isometries map lines to lines, the covering transformations $\Gamma = \pi_1(M)$ (actually all isometries) are of the form $(f, g)(x, y) = (f(x), g(y))$ where $f : N \to N$ and $g : \mathbb{R}^k \to \mathbb{R}^k$ are isometries. Let ρ be the projection of $\tilde{M}$ on the first factor in $N \times \mathbb{R}^k$, and let F be a compact fundamental domain for Γ, which exists because M is compact. Then the orbit $\rho(F)$ under $\rho(\Gamma)$ is all of N. We claim that N is compact. Otherwise there exists a ray γ and a sequence $g_i \in \rho(\Gamma)$ such that $g_i^{-1}(\gamma(i)) \in \rho(F)$. By the compactness of $\rho(F)$ we can find a subsequence, still denoted by g_i, such that $(g_i^{-1})_*(\gamma'(i))$ converges to a tangent vector v at some point $p \in \rho(F)$. If σ is a geodesic of N with $\sigma'(0) = v$, then σ is a line. This contradiction shows that N is compact.

Now let $\psi : (f, g) \to g$ be the projection of the isometry group of $\tilde{M}$ into the second factor, and consider the short exact sequence:

$$0 \to \ker \phi \to \pi_1(M) \to \operatorname{im} \phi \to 0.$$

Since $\ker \phi = (f, 1)$, and f acts discretely on N which is compact, $\ker \phi$ is thus finite. The image $\operatorname{im} \phi$ acts on $\mathbb{R}^k$ with compact quotients (since M is compact). Therefore $\operatorname{im} \phi$ is a Biebabach group. The corollary follows. $\qquad\square$

COROLLARY 4.12. *Let M^n be a complete Riemannian manifold with nonnegative Ricci curvature. Then any finitely generated subgroup of π_1 has polynomial growth of order $n - 1$ unless M is a compact flat manifold, in which case the order is n.*

PROOF. Let $\pi : \tilde{M} \to M$ be the universal cover of M with the pullback metric, and let $\tilde{M} = N \times \mathbb{R}^k$ be a splitting such that N does not contain any line. Fixing $(p, 0) \in N \times \mathbb{R}^k$, we will estimate the volume of $B_p(r) \cap \Gamma(p, 0)$ for r large. Note that any deck transformation in $\Gamma = \pi_1(M)$ is of the form (f, g), where f and g act as isometries of the two factors. We claim that there are no sequences (f_i, g_i) such that $f_i(p) = \gamma_i(t_i)$ for a geodesic γ_i, and that $\gamma_i|_{[t_i - i, t_i + i]}$ is minimal. In fact, if such a sequence exists, then $f_i^{-1} \circ \gamma_i|_{[-i, i]}$ is minimal with base point p, and a subsequence will converge to a line in N, contradicting the fact that N contains no lines. Thus there is a number $i_0 > 0$ such that, for any $(f, g) \in \Gamma$ and any geodesic γ from p to $f(p)$ with $f(p) = \gamma(t)$, the restriction $\gamma|_{[t - i_o, t + i_o]}$ is not minimizing. Denote by C the set of points $x \in N$ such that $x \in B_p(i_0)$ or any geodesic γ from p to $x = \gamma(t)$ does not minimize to $t + i_o$. We have shown that

$$\Gamma(p, 0) \in C \times \mathbb{R}^k.$$

Since C is the image under $\exp_p$ of a set contained in T_pN, which is the union of sections of annular regions with width i_0 and a ball of radius i_0, it follows from the volume comparison theorem (unless $\dim N = 0$, in which case M is flat) that

$$\operatorname{vol}(B_p(r) \cap C) \le c(i_0^{\dim N} + i_0 r^{\dim N - 1}) \le c r^{n-1}.$$

As in the proof of Theorem 3.11, let $H = \langle g_1, \ldots, g_k \rangle$ be a finitely generated subgroup of Γ. Then g_i can each be represented by a loop σ_i at $\pi(p, 0)$, with length l_i. Let $\tilde{\sigma}_i$ be the lifting of σ_i at $(p, 0)$; then $g_i(p, 0) = \tilde{\sigma}_i(l_i)$ as deck transformations. Let

$$\varepsilon < \min\{l_1, \ldots, l_k\}, \quad l = \max\{l_1, \ldots, l_k\}.$$

Then, for any distinct $h_1, h_2 \in H$, we have $h_1(B_{\tilde{p}}(\varepsilon)) \cap h_2(B_{\tilde{p}}(\varepsilon)) = \varnothing$ and

$$\bigcup_{h \in U(r)} h(B_{\tilde{p}}(\varepsilon)) \subset B_{(p,0)}(rl + \varepsilon) \cap C.$$

Thus

$$|U(r)| \cdot \mathrm{vol}(B_{\tilde{p}}(\varepsilon)) = \sum_{h \in U(r)} \mathrm{vol}(h(B_{\tilde{p}}(\varepsilon))) \leq \mathrm{vol}(B_{\tilde{p}}(rl + \varepsilon) \cap C),$$

and

$$|U(r)| \leq \frac{\mathrm{vol}(B_{\tilde{p}}(rl + \varepsilon) \cap C)}{\mathrm{vol}(B_{\tilde{p}}(\varepsilon))} = \frac{c}{\mathrm{vol}(B_{\tilde{p}}(\varepsilon))}(rl + \varepsilon)^{n-1} \leq cr^{n-1}. \qquad \square$$

Abresch and Gromoll's estimate of the excess function. Given two points $p, q \in M$, the *excess function* is defined as

$$e_{p,q}(x) = d(p, x) + d(x, q) - d(p, q).$$

(When no confusion can occur, we will drop the reference to p, q.) Thus, the excess function measures how much the triangle inequality fails to be an equality. Since the excess function is made up of distance functions, the properties below follow directly from those of the distance function.

LEMMA 4.13. (1) $e(x) \geq 0$;
(2) $\mathrm{dil}(e) \leq 2$, *where* $\mathrm{dil}(f) = \inf |f(x) - f(y)|/d(x, y)$;
(3) $e|_\gamma = 0$ *where* γ *is a minimizing geodesic from* p *to* q; *and*
(4) *if* $\mathrm{Ric} \geq 0$, *then*

$$\Delta e \leq (n - 1)\Big(\frac{1}{d(x, p)} + \frac{1}{d(x, q)}\Big)$$

in the barrier sense.

The following analytic lemma uses the Weak Maximum Principle, Theorem 4.3.

LEMMA 4.14. *Suppose* $\mathrm{Ric} \geq 0$, *and let* $u : B_y(R + \eta) \to \mathbb{R}^1$ *(for some* $\eta > 0$*) be a Lipschitz function satisfying*
(1) $u \geq 0$;
(2) $u(y_0) = 0$ *for some* $y_0 \in \overline{B_y(R)}$;
(3) $\mathrm{dil}(u) \leq a$; *and*
(4) $\Delta u \leq b$ *in the barrier sense.*
Then, for all $c \in (0, R)$, u *satisfies* $u(y) \leq a \cdot c + G(c)$, *where* G *is defined as*

$$G(x) = \frac{b}{2n}\Big(x^2 + \frac{2}{n-2}R^n x^{2-n} - \frac{n}{n-2}R^2\Big).$$

PROOF. Take $\varepsilon < \eta$, and define the function G using the value $R + \varepsilon$ in place of R. Since we can eventually let $\varepsilon \to 0$, it will suffice to prove the inequality in this case.

If d denotes the distance function to any point in the n-dimensional space form $S^n(H)$ of curvature H, then $G \circ d$ is the unique function on $S^n(H)$ satisfying
(i) $G \circ d(x) > 0$ for $0 < d(x) < R$,
(ii) G is decreasing for $0 < d(x) < R$,

(iii) $G(R) = 0$, and

(iv) $\Delta^H G \circ d = b$.

Now fix $c \in (0, R)$ and suppose the bound is false. Then $u(y) \geq a \cdot c + G(c)$, and it follows that

$$u|_{\partial B_y(c)} \geq u(y) - c \cdot \mathrm{dil}(u) \geq a \cdot c + G(c) - c \cdot a = G(c) = G|_{\partial B_y(c)},$$

and $u|_{\partial B_y(R)} \geq 0 = G|_{\partial B_y(R)}$. Thus, for ε small, we have $u|_{\partial A} \leq 0$ for the annulus $A = B_y(R) - B_y(\varepsilon)$. But $(G - u)(y_0) = G(y_0) > 0$; hence $G - u$ has a strict interior maximum in A. This violates the maximum principle since $\Delta(G - u) \geq 0$. $\square$

For convenience, we set $s(x) = \min(d(p, x), d(q, x))$ and define the height $h(x) = \mathrm{dist}(x, \gamma)$ for any fixed minimal geodesic γ, from p to q.

THEOREM 4.15 (EXCESS ESTIMATE [Abresch and Gromoll 1990]). *If* Ric ≥ 0 *and* $h(x) \leq s(x)/2$, *then*

$$e(x) \leq 8 \left(\frac{h(x)^n}{s} \right)^{1/(n-1)}.$$

PROOF. By Lemma 4.13, we can choose $a = 2, b = 4(n-1)/s(x)$, and $R = h(x)$ in Lemma 4.14, and let

$$c = \left(\frac{2h^n}{s} \right)^{1/(n-1)}.$$

Then

$$c \leq 2 \left(\frac{2h^n}{s} \right)^{1/(n-1)} + G(c)$$

$$= \frac{b}{2n} \left(\left(\frac{2h^n}{s} \right)^{2/(n-1)} - \frac{n}{n-2} h^2 + \frac{2}{n-2} h^n 2^{2-n} \left(\frac{2h^n}{s} \right)^{(2-n)/(n-1)} \right)$$

$$\leq 8 \left(\frac{h(x)^n}{s} \right)^{1/(n-1)}. \qquad \square$$

REMARK. In the above, we only stated the case where the Ricci curvature is nonnegative. It is easy to see that an estimate holds for general lower bounds on Ricci curvature, and takes the form

$$e(x) \leq E \left(\frac{h}{s} \right) \cdot h,$$

for some function E and $E(0) = 0$.

Critical points of the distance function and Toponogov's theorem.

One of the most useful theories of differential topology is Morse theory. The main idea is that, if $f : M \to \mathbb{R}$ is a smooth function, the topology of M is reflected by the critical points of f. In geometry, the classical application of Morse theory is the energy functional of the loop space. As we have seen, the distance function is a natural function closely tied to the geometry and topology of a manifold. In this section, we will try to develop a Morse theory for the distance function.

This theory was originally proposed by Grove–Shiohama [1977], and formally formulated by Gromov [1981a].

DEFINITION 4.16. A point x is a *critical point* of d_p if, for all vectors $v \in M_p$, there is a minimizing geodesic γ from x to p such that the angle $\angle(v, \dot{\gamma}(0)) \leq \pi/2$.

The main result we will use from Morse theory is the following theorem.

THEOREM 4.17 (ISOTOPY LEMMA). *If $r_1 < r_2 \leq +\infty$ and $\overline{B_p(r_2)} \setminus B_p(r_1)$ has no critical point, then this region is homeomorphic to $\partial B_p(r_1) \times [r_1, r_2]$.*

PROOF. If x is not critical, there is a vector $w_x \in M_x$ with $\angle(w_x, \dot{\gamma}(0)) < \pi/2$ for all minimizing geodesics γ from x to p. By continuity, we can extend w_x to a vector field W_x in an open neighborhood U_x of x, such that if $y \in U_x$ and σ is any minimizing geodesic from y to p, then $\angle(\dot{\sigma}(0), W_x(y)) < \pi/2$. Take a finite subcover of $\overline{B_p(r_2)} \setminus B_p(r_1)$ by sets U_{x_i} (locally finite if $r_2 = +\infty$), and a smooth partition of unity $\sum \phi_i = 1$. Let $W = \sum \phi_i W_{x_i}$. From the restriction on angles, it follows that W is nonvanishing. Let $\psi_x(t)$ be the integral curve of W through x, and let $\sigma_t(s)$ be a minimal geodesic from p to $\psi_x(t)$. The first variation formula gives

$$d(\psi_x(t_2)) - d(\psi_x(t_1)) = \int_{t_1}^{t_2} \frac{d}{dt} d(\psi_x(t)) = \int_{t_1}^{t_2} -\cos\angle(\dot{\sigma}_t(0), W(\psi_x(t))\,dt$$
$$\leq -\cos(\pi/2 - \varepsilon)(t_2 - t_1),$$

for some $\varepsilon > 0$. This implies that d is strictly decreasing along $\psi(t)$. It now follows that the flow along ψ gives a homeomorphism: $\phi : \partial B_{r_1} \times [r_1, r_2) \to M$ with $\phi(x, t) = \psi_x(t)$. $\qquad\square$

Since the definition of critical points requires a control on the angle, the standard application of this Morse theory requires the following result:

THEOREM 4.18 (TOPONOGOV COMPARISON THEOREM). *Let M^n be a complete Riemannian manifold with $\sec \geq H$.*
(1) *Let $\{\gamma_0, \gamma_1, \gamma_2\}$ be a triangle in M, and assume that all three geodesics are minimizing. Then there is a triangle $\bar{\gamma}_0, \bar{\gamma}_1, \bar{\gamma}_2$ in the two-dimensional space form $S^2(H)$ of curvature H with $\text{length}(\gamma_i) = \text{length}(\bar{\gamma}_i)$. Furthermore, if α_i is the angle opposite γ_i, then $\bar{\alpha}_i \leq \alpha_i$.*
(2) *Let $\{\gamma_1, \gamma_2, \alpha\}$ be a hinge in M, and assume that γ_1 and γ_2 are minimizing. Then there is a hinge $\bar{\gamma}_1, \bar{\gamma}_2$ in $S^2(H)$ with $\text{length}(\gamma_i) = \text{length}(\bar{\gamma}_i) = l_i$ and same angle α. Furthermore, $d(\gamma_1(l_1), \gamma_2(l_2)) \leq d^H(\bar{\gamma}_1(l_1), \bar{\gamma}_2(l_2))$.*

We have not stated this theorem in the strongest possible form. For a complete statement and proof, we refer the reader to [Cheeger and Ebin 1975].

To show typical applications for the Morse theory, and provide background for later results, we prove the following two well-known results.

THEOREM 4.19 (GROVE–SHIOHAMA DIAMETER SPHERE THEOREM [1977]). *Let M^n be a Riemannian manifold with sec $\geq H > 0$. If $\operatorname{diam}(M) > (\pi/2\sqrt{H})$, then M is homeomorphic to the sphere S^n.*

PROOF. Without loss of generality, we assume $H = 1$. Let $p, q \in M$ be such that $d(p, q) = \operatorname{diam}(M)$, a simple first variation argument shows that p, q are mutually critical. The theorem follows from the following claim.

Any $x \in M \setminus \{p, q\}$ is noncritical for d_p and d_q.

To prove the claim, assume x is a critical point of d_p. Let σ_1 be a minimal geodesic from q to x, with length l_1. Since x is critical, there is a minimal geodesic σ_2 from x to p with length l_2 such that $\angle(\dot{\sigma}_2(0), \dot{\sigma}_1(l_1)) \leq \pi/2$. Since p, q are mutual critical points, there are minimal geodesics $\sigma_3, \tilde{\sigma}_3$ from q to p such that $\angle(\dot{\sigma}_3(0), \dot{\sigma}_1(0)) \leq \pi/2$, and $\angle(-\dot{\tilde{\sigma}}_3(0), -\dot{\sigma}_2(l_2)) \leq \pi/2$. Let $\{\bar{\sigma}_1, \bar{\sigma}_1, \bar{\sigma}_1\}$ be the comparison triangle in $S^n(1)$. Applying Toponogov's theorem twice— once to the triangle $\{\sigma_1, \sigma_2, \sigma_3\}$, then to the triangle $\{\sigma_1, \sigma_2, \tilde{\sigma}_3\}$—we obtain $\bar{\alpha}_i \leq \pi/2$, for $i = 1, 2, 3$. It now follows from elementary spherical geometry that $\operatorname{length}(\bar{\sigma}_3) \leq \pi/2$. But then Toponogov's theorem implies $d(p, q) = \operatorname{length}(\sigma_3) \leq \operatorname{length}(\bar{\sigma}_3) \leq \pi/2$, a contradiction. $\square$

THEOREM 4.20. *If M is complete and has nonnegative sectional curvature, it has finite topological type, i.e., it is homeomorphic to the interior of a compact manifold with boundary.*

PROOF. We first prove that if x is a critical point of d_p, and y is such that $d(y) \geq \mu d(x)$, then for any minimal geodesic γ_1 of length l_1 from p to x and for any minimal geodesic γ_2 of length l_2 from p to y the angle $\theta = \angle(\dot{\gamma}_1(0), \dot{\gamma}_2(0))$ is at least $\cos^{-1}(1/\mu)$.

In fact, using part (2) of Toponogov's theorem, we have

$$l_3^2 = d(x, y)^2 \leq l_1^2 + l_2^2 - 2l_1 l_2 \cos\theta.$$

Choose an arbitrary minimal geodesic γ_3 from x to y. Since x is critical, there is a minimal geodesic $\bar{\gamma}_1$ from x to p such that $\angle(\dot{\bar{\gamma}}_1(0), \dot{\gamma}_3(0)) \leq \pi/2$. Applying Toponogov's theorem again to this triangle we get $l_2^2 \leq l_1^2 + l_3^2$. Combining these two inequalities gives the desired lower bound on θ.

We now claim that d_p does not have critical points outside a compact set. If not, we would have a sequence of critical points x_i such that $d(x_{i+1}) \geq \mu d(x_i)$, for any i. Let γ_i be a minimal geodesic from p to x_i. From what we just proved in the preceding paragraph, we have

$$\angle(\dot{\gamma}_i, \dot{\gamma}_j) \geq \cos^{-1}(1/\mu),$$

and setting $\mu = 2$ we get $\angle(\dot{\gamma}_i, \dot{\gamma}_j) \geq \pi/3$. This gives a covering of the unit sphere at p by an infinite number of balls of fixed size (corresponding to solid angles of $\pi/6$), no two of which intersect. This is not possible since S_p is compact. Therefore d_p has no critical point outside of a compact set. By the isotopy lemma, M has finite topological type. $\square$

REMARK. This result is a weaker version of the Soul Theorem [Cheeger and Gromoll 1972], which says any complete manifold with nonnegative sectional curvature is diffeomorphic to the normal bundle of a compact totally geodesic submanifold, called the *soul*. The argument given above is due to Gromov [1981a].

Diameter growth and topological finiteness. Given the results just seen, it is quite natural to ask whether the conclusion of Theorem 4.20 is still true if one replaces nonnegative sectional curvature by nonnegative Ricci curvature. The examples of Sha and Yang showed this not to be the case [Sha and Yang 1989a; 1989b; Anderson 1990a]. In this section, we will prove a topological finiteness theorem for Ricci curvature under some additional conditions, due to Abresch and Gromoll [1990]. The excess estimate of section 4.3 was originally designed for this purpose. It turned out to be useful for other applications, as we will see in next section.

DEFINITION 4.21. For $r > 0$, the open set $M \setminus \overline{B_p(r)}$ contains only finitely many unbounded components, and each such component has finitely many boundary components Σ_r. Define the diameter growth function $D(r, p)$ as the maximum diameter of the Σ_r, as measured in M.

THEOREM 4.22 [Abresch and Gromoll 1990]. *Let M^n be a complete Riemannian manifold of nonnegative Ricci curvature, sectional curvature bounded below by $H > -\infty$, and satisfying $D(r, p) = o(r^{1/n})$. Then d_p has no critical point outside of a compact set. In particular, M has finite topological type.*

Given the lower bound on the Ricci curvature, the excess estimate in the last section gives a upper bound for the excess function. To see the relevance of the lower sectional bound in the above theorem, we state the following lemma, which gives a lower bound for the excess.

LEMMA 4.23. *If M^n is complete with $\sec \geq -1$, and x a critical point of d_p, then for any $\varepsilon > 0$, there is a $\delta > 0$ such that if $d(q, x) \geq 1/\delta$, then*

$$e_{p,q}(x) \geq \ln\left(\frac{2}{1 + e^{-2d(p,x)}}\right) - \varepsilon.$$

PROOF. Take an arbitrary minimal geodesic γ from x to q. Since x is critical for d_p, there is a minimal geodesic σ from x to p such that $\angle(\dot\sigma(0), \dot\gamma)(0) \leq \frac{\pi}{2}$. By Toponogov's theorem, we have

$$\cosh d(p, q) \leq \cosh d(x, p) \cdot \cosh d(x.q).$$

When $d(x, q) \to \infty$, $d(p, q) \to \infty$ with $d(p, x)$ fixed, the above inequality becomes

$$\frac{e^{d(p,q)}}{2} \leq \cosh d(x, p)\frac{e^{d(x,q)}}{2},$$

and the lemma follows from the definition of the excess function. □

PROOF OF THEOREM 4.22. Given a boundary component Σ_r for a noncompact component of $M \setminus B_p(r)$, we can construct a ray γ such that $\gamma(t) \in U(r)$, for all $t > r$. Thus, if $x \in \Sigma_r$ is critical, then $D(r, p) = o(r^{1/n})$, and Theorem 4.15 implies $e(x) \to 0$. But Lemma 4.23 gives a positive lower bound for $e(x)$. This is not possible. Thus, for any $r > r_0$, no point of any set Σ_r is critical for d_p.

Now fix r_0, U_{r_0}, a boundary component Σ_{r_0}, and a ray γ with $\gamma(r_0) \in \Sigma_{r_0}$ and $\gamma(t) \in U_{r_0}$ for $t > r_0$. For each $t \geq r_0$, let Σ_t denote the boundary component of the unbounded component of $M \setminus \overline{B_p(t)}$ with $\gamma(t) \in \Sigma_t$. Using the isotopy lemma we can construct an embedding $\psi : (r_0, \infty) \times \Sigma_0 \to U_{r_0}$ such that $\psi((t, \Sigma_{r_0})) = \Sigma_t$. It follows easily that $\psi((r_0, \infty) \times \Sigma_{r_0}))$ is open and closed in U_{r_0}. Hence $\psi((r_0, \infty) \times \Sigma_{r_0}) = U_{r_0}$. $\qquad\square$

REMARK. Opinions are divided as to whether the lower bound on sectional curvature is necessary. For that matter, it is interesting to consider the finiteness question for complete manifolds with positive Ricci curvature and bounded diameter. The techniques in [Perelman 1994] may help to settle this question in the negative.

Because of the relation between Ricci curvature and the volume growth of geodesic balls, as given by Theorem 3.5, for example, another approach to obtaining a finiteness theorem is to put conditions on the volume growth of geodesic balls, e.g., to require $\text{vol}(B_p(r))$ to be close to $\omega_n r^n$ or close to cr. A positive result is the following theorem due to Perelman, which we will prove in the next section.

THEOREM 4.24 [Perelman 1994]. *For any $n > 0$, there is a positive number $\varepsilon(n)$ such that if M is a complete n-dimensional manifold with $\text{Ric} \geq 0$ and*

$$\text{vol}(B_p(r)) \geq (1 - \varepsilon)\omega_n r^n,$$

then M is contractible.

This is a topological stability result for the maximal volume growth condition. From [Perelman 1994], it seems possible to construct examples of complete manifolds of positive Ricci curvature with infinite topological type and satisfying the condition that $\text{vol}(B_p(r)) \geq cr^n$, for some positive constant c. But the detailed computation still needs to be completed. No such example is known about the case where the volume grows slowly.

These finiteness considerations are closely related to the attempt to generalize Cheeger's Finiteness Theorem [1970] and Grove–Petersen's Homotopy Finiteness Theorem [1988] to Ricci curvature. The latter theorem says that the class of n-dimensional manifolds of sectional curvature $\geq H$, volume $\geq v > 0$, and diameter $\leq d$ contains only finitely many homotopy types. The crucial step in the proof is to show that geodesic balls of a fixed (small) size have simple topology. When the sectional curvature condition is replaced by Ricci curvature, a small geodesic ball (when rescaled) will resemble a complete manifold with $\text{Ric} \geq 0$ and $\text{vol}(B_p(r)) \geq cr^n$. The above example of Perelman also shows that this finiteness conjecture is false in dimensions four and above. In dimension three, there is a homotopy finiteness theorem due to Zhu [1990].

Perelman's almost maximal volume sphere theorem. Because of the Grove–Shiohama diameter sphere theorem (Theorem 4.19), efforts were made to generalize the result to Ricci curvature. This turned out not to be possible in dimension four and above. In fact, Anderson [1990b] and Otsu [1991] constructed metrics on $CP^n \# CP^n$ and $S^n \times S^m$ that have Ricci curvature at least equal to the dimension minus one, and diameter arbitrarily close to π. In dimension three, it was shown recently by Shen and Zhu [1995] that the diameter sphere theorem for Ricci curvature is still true (with slightly bigger diameter requirement).

It follows from Bishop's Volume Comparison Theorem that if the volume of an n-dimensional manifold with Ric $\geq n - 1$ is close to that of the unit sphere, then the diameter is close to π. One can thus attempt to prove a sphere theorem with conditions on the Ricci curvature and volume. This result was proved by Perelman, using the excess estimate of Abresch and Gromoll (Theorem 4.15).

THEOREM 4.25 [Perelman 1995]. *For any $n > 0$, there is a positive number $\varepsilon(n)$, such that if M^n satisfies Ric $\geq n - 1$ and $\mathrm{vol}(M) \geq (1 - \varepsilon)\,\mathrm{vol}(S^n(1))$, then M is homeomorphic to S^n.*

REMARK. One needs only to prove that $\pi_k(M) = 0$ for all $k < n$, which implies that M is a homotopy sphere. Then the above theorem follows from the solution to Poincaré's conjecture when $n \geq 4$ (see [Smale 1961] and [Freedman 1982]), and from Hamilton's result [1982] when $n = 3$.

Theorem 4.24 can be considered as a noncompact version of Theorem 4.25. Both results are consequences of the next lemma.

MAIN LEMMA 4.26. *For any $C_2 > C_1 > 1$ and any integer $k \geq 0$, there is a constant $\delta = \delta_k(C_1, C_2, n) > 0$ such that, if M^n has Ricci curvature bounded below by $n - 1$ and satisfies $\mathrm{vol}(B_q(\rho)) \geq (1 - \delta)\,\mathrm{vol}(B^1(\rho))$ for any $B_q(\rho) \subset B_p(C_2 R)$ and $0 < R < \pi/C_2$, the following two k-parametrized properties hold:*

A(k) *Any continuous map $f : S^k = \partial D^{k+1} \to B_p(R)$ can be extended to a continuous map $g : D^{k+1} \to B_p(C_1 R)$.*

B(k) *Any continuous map $f : S^k \to M \setminus B_p(R)$ can be continuously deformed to a map $h : S^k \to M \setminus B_p(C_1 R)$.*

Theorem 4.25 follows from A(k) and B(k). Theorem 4.24 follows from A(k).

PROOF. We will work by induction on k. The induction step will be stated in Lemma 4.29 below, but first we need two other lemmas.

LEMMA 4.27. *If Ric $\geq n - 1$, there is a continuous function $E : \mathbb{R} \to \mathbb{R}^+$ such that*

$$e_{p,q}(x) \leq E\Big(\frac{h(x)}{d(p,x)}\Big) \cdot h(x),$$

assuming that $d(p, x) \leq d(q, x)$ and $h(x)$ is the height function, i.e., the distance from x to a minimal geodesic from p to q.

This is a restatement of Theorem 4.15.

LEMMA 4.28. *For any $C_2 > C_1 > 1$ and $\varepsilon > 0$, there is a positive constant $\gamma(C_1, C_2, \varepsilon, n)$ such that, if $\mathrm{Ric} \geq n-1$ and $\mathrm{vol}(B_p(C_2 R)) \geq (1-\gamma)\,\mathrm{vol}(B^1(C_2 R))$, for any $0 < R < \pi/C_2$, then for any $a \in B_p(R)$ there is a point $b \in M \setminus B_p(C_1 R)$ such that*

$$d(a, \overline{pb}) \leq \varepsilon R,$$

where $\overline{pb}$ denotes a minimal geodesic from p to b.

Thus, under the curvature and volume conditions, there are lots of thin and long triangles. The excess estimate in Lemma 4.27 is a good estimate only when applied to such thin and long triangles.

PROOF. Set $\Gamma = \{\dot{\sigma} \mid d(a, \sigma) \leq \varepsilon R\} \subset S_p^{n-1}$. Assume that, for all $v \in \Gamma$, the cut point in the direction of v is less than $C_1 R$. We will derive a contradiction. In fact, under this assumption, it follows that

$$\mathrm{vol}(B_p(C_2 R)) = \int_\Gamma \int_0^{\mathrm{cut}(v)} A_M(t)\, dt\, dv + \int_{S^{n-1} \setminus \Gamma} \int_0^{\min\{C_2 R, \mathrm{cut}(v)\}} A_M(t)\, dt\, dv$$

$$\leq \mathrm{vol}(\Gamma) \int_0^{C_1 R} A^1(t)\, dt + (\mathrm{vol}(S^{n-1}) - \mathrm{vol}(\Gamma)) \int_0^{C_2 R} A^1(t)\, dt$$

$$= -\mathrm{vol}(\Gamma) \int_{C_1 R}^{C_2 R} A^1(t)\, dt + \mathrm{vol}(S^{n-1}) \int_0^{C_2 R} A^1(t)\, dt$$

$$\leq -\mathrm{vol}(\Gamma) \int_{C_1 R}^{C_2 R} A^1(t)\, dt + \mathrm{vol}(B^1(C_2 R)).$$

(Note that $A^1(t) = \sin^{n-1} t$.) Thus

$$(1 - \gamma)\,\mathrm{vol}(B^1(C_2 R)) \leq -\mathrm{vol}(\Gamma) \int_{C_1 R}^{C_2 R} A^1(t)\, dt + \mathrm{vol}(B^1(C_2 R)),$$

which implies that

$$\mathrm{vol}(\Gamma) \leq \gamma \cdot \frac{\mathrm{vol}(B^1(C_2 R))}{\int_{C_1 R}^{C_2 R} A^1(t)\, dt}.$$

Also note that

$$\mathrm{vol}(B_a(\varepsilon R)) \leq \mathrm{vol}(A_{0, C_1 R}^\Gamma(p)) \leq \mathrm{vol}(\Gamma) \int_0^{C_1 R} A^1(t)\, dt$$

$$\leq \gamma \cdot \frac{\mathrm{vol}(B^1(C_2 R))}{\int_{C_1 R}^{C_2 R} A^1(t)\, dt} \cdot \int_0^{C_1 R} A^1(t)\, dt.$$

(For the definition of $A_{r,R}^{\Gamma}(p)$, see Theorem 3.1.) By the relative volume comparison, we have

$$\operatorname{vol}(B_a(\varepsilon R)) \geq \operatorname{vol}(B_a(R + C_2 R)) \cdot \frac{\int_0^{\varepsilon R} A^1(t)\,dt}{\int_0^{R+C_2 R} A^1(t)\,dt}$$

$$\geq \operatorname{vol}(B_p(C_2 R)) \cdot \frac{\int_0^{\varepsilon R} A^1(t)\,dt}{\int_0^{R+C_2 R} A^1(t)\,dt}$$

$$\geq (1 - \gamma)\operatorname{vol}(B^1(C_2 R)) \cdot \frac{\int_0^{\varepsilon R} A^1(t)\,dt}{\int_0^{R+C_2 R} A^1(t)\,dt}.$$

These two inequalities together imply that

$$(1 - \gamma)\operatorname{vol}(B^1(C_2 R)) \cdot \frac{\int_0^{\varepsilon R} A^1(t)\,dt}{\int_0^{R+C_2 R} A^1(t)\,dt} \leq \gamma \cdot \frac{\operatorname{vol}(B^1(C_2 R))}{\int_{C_1 R}^{C_2 R} A^1(t)\,dt} \cdot \int_0^{C_1 R} A^1(t)\,dt,$$

which gives a bound $\gamma > C(C_1, C_2, n, \varepsilon)$. This is a contradiction if we choose $\gamma(C_1, C_2, n, \varepsilon) = C$. $\qquad\square$

We now give an outline of the proof of $A(k)$, which, as we have mentioned, uses induction on k. As usual, we view D^{k+1} as $S^k \times [1, 0)$ plus a point. In order to extend a function $f : S^k \to B_p(R)$ to D^{k+1}, we need to be able to get a map $\tilde{f} : S^k \to B_p(\alpha R)$, for $\alpha < 1$; we can then add the final point by continuity. The usual requirement is that $\tilde{f}$ be homotopic to f, so the function g is obtained by extending gradually along the radial direction from $S^k \times \{1\}$ to $S^k \times \{0\}$. Perelman's idea is that, in fact, one does not have to require $\tilde{f}$ to be homotopic to f. Instead, one only needs $\tilde{f}$ to be close to f. Then f and $\tilde{f}$ give an extension to the k-skeleton of a cell decomposition K_0 of D^{k+1}, as in the figure.

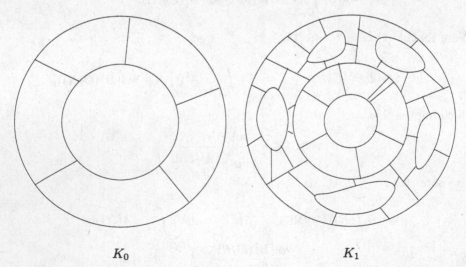

$K_0 \qquad\qquad\qquad\qquad\qquad K_1$

The requirement that $\tilde{f}$ and f be close guarantees that each k-cell of K_0 is mapped into a smaller ball. Now regard the boundary of each k-cell as S^k, and proceed in each such cell to extend the map to a finer cell decomposition K_1 of D^{k+1}, as in the second figure. Continuing in this fashion, one obtains the desired extension by adding infinitely many points (corresponding to the center of the k-cells of K_j when $j \to \infty$.) Thus, the main difficulty is to construct the map $\tilde{f}$ that maps into a smaller ball and is close to f. We will prove the existence of $\tilde{f}$ in the following lemma, which uses Lemmas 4.27 and 4.28, and the induction hypothesis $A(k-1)$. In the lemma, d_0 is a (small) positive number to be chosen later.

LEMMA 4.29 (INDUCTION LEMMA $P(k)$). *Assume $A(k-1)$ is true. Then, given any map $\phi : S^k \to B_q(\rho)$, and a fine triangulation T of S^k with $\mathrm{diam}(\phi(\Delta)) \leq d_0\rho$ for any $\Delta \in T$, there exists a continuous map $\tilde{\phi} : S^k \to B_q((1-d_0)\rho)$ such that*

$$\mathrm{diam}(\phi(\Delta) \cup \tilde{\phi}(\Delta)) \leq \beta\rho,$$

for any $\Delta \in T$, where $\beta = (4 + 2d_0/k)^{-k}(1 - 1/C_1)$.

PROOF. We will construct the map $\tilde{\phi}$ on the skeletons $\mathrm{skel}_i(T)$ of T, for $i = 0, 1, \ldots, k$. We proceed by induction on i.

When $i = 0$, for any $x \in \mathrm{skel}_0(T)$, let γ_x be a minimal geodesic from q to $\phi(x)$. Define $\tilde{\phi}(x) = \gamma_x((1 - 2d_0)\rho)$. Then $\tilde{\phi}(\Delta) \subset B_q((1 - 2d_0)\rho)$ and $\mathrm{diam}(\phi(\Delta) \cup \tilde{\phi}(\Delta)) \leq 10d_0\rho$ (we could use $2d_0\rho$ on the right-hand side for the latter inequality).

Assume that $\tilde{\phi}$ has already been defined on the i-skeleton skel_i, with $\tilde{\phi}(\Delta) \subset B_q((1-d_0(2-\frac{i}{k}))\rho)$ and $\mathrm{diam}(\phi(\Delta)\cup\tilde{\phi}(\Delta)) \leq 10d_i\rho$, where d_i will be determined later. We now construct $\tilde{\phi}$ on skel_{i+1}.

For any $\Delta \in \mathrm{skel}_{i+1}$, we can assume that $\phi(\Delta) \not\subset B_q((1 - 2d_0)\rho)$; otherwise we are done. By Lemma 4.28, there is a point $y_\Delta \in M \setminus B_q(C_1\rho)$ such that

$$d(\phi(\Delta), \overline{qy_\Delta}) \leq 2d_0\rho.$$

Let σ be a minimal geodesic from q to y_Δ, and let $q_\Delta = \sigma((1 - d_{i+1})\rho)$. Then, for any $x \in \partial\Delta$, we check that the triangle $\tilde{\phi}(x)q_\Delta q$ is thin and long:

$$d(\tilde{\phi}(x), q_\Delta \bar{y}_\Delta) \leq d(\phi(x), q_\Delta \bar{y}_\Delta) + \mathrm{diam}(\phi(\partial(\Delta)) \cup \tilde{\phi}(\partial\Delta)) \leq 2d_0\rho + 10d_i\rho \leq 20d_i\rho,$$

$$d(\tilde{\phi}(x), q_\Delta) \geq d(q_\Delta, \phi(\Delta)) - \mathrm{diam}(\phi(\partial\Delta) \cup \tilde{\phi}(\partial\Delta)) \geq d_{i+1}\rho - 10d_i\rho \geq \frac{d_{i+1}}{2}\rho,$$

$$d(\tilde{\phi}(x), y_\Delta) \geq d(y_\Delta, q) - d(\tilde{\phi}(x), q) \geq C_1\rho - (1 - d_0(2-i/k))\rho \geq \tfrac{1}{2}d_{i+1}\rho.$$

Here we have assumed that $d_{i+1} \geq 100d_i$ and $d_k \leq C_1 - 1$.

By Lemma 4.27, we have

$$d(\tilde{\phi}(x), y_\Delta) + d(\tilde{\phi}(x), q_\Delta) - d(q_\Delta, y_\Delta) \leq 20d_i\rho E\left(\frac{100d_i}{d_{i+1}}\right).$$

The triangle inequality gives

$$d(q_\Delta, y_\Delta) + \rho(1 - d_{i+1}) = d(q, y_\Delta) \le d(\tilde{\phi}(x), y_\Delta) + d(\tilde{\phi}(x), q).$$

Adding these two inequalities, we obtain

$$d(\tilde{\phi}(x), q_\Delta) \le d(\tilde{\phi}(x), q) + 20 d_i \rho E\left(\frac{100 d_i}{d_{i+1}}\right) - \rho(1 - d_{i+1})$$

$$\le (1 - d_0(2 - \frac{i}{k}))\rho + 20 d_i \rho E\left(\frac{100 d_i}{d_{i+1}}\right) - \rho(1 - d_{i+1})$$

$$\le \left(d_{i+1} - d_0\left(2 - \frac{2i+1}{2k}\right)\right)\rho,$$

where we need

$$20 d_i E\left(\frac{100 d_i}{d_{i+1}}\right) \le d_0/2k.$$

Since A(i) is true (note that $i \le k - 1$), we can use A(i) to extend the map (using $1 + d_0/2k$ as the factor C_1; we thus require $\delta_{i+1} < \delta_i(1 + d_0/2k, C_2)$). We get

$$\tilde{\phi}\left(\Delta, q_\Delta\right) \le \left(1 + \frac{d_0}{2k}\right)\left(d_{i+1} - d_0\left(2 - \frac{2i+1}{2k}\right)\right)\rho$$

$$\le \left(d_{i+1} - d_0\left(2 - \frac{i+1}{k}\right)\right)\rho.$$

Therefore

$$\mathrm{diam}(\phi(\Delta) \cup \tilde{\phi}(\Delta)) \le \mathrm{diam}(\phi(\partial\Delta) \cup \tilde{\phi}(\partial\Delta)) + \mathrm{diam}(\phi(\Delta)) + \mathrm{diam}(\tilde{\phi}(\Delta))$$

$$\le 10 d_i \rho + d_0 \rho + \left(d_{i+1} - d_0\left(2 - \frac{i+1}{k}\right)\right)\rho \le 10 d_{i+1}\rho \le \beta.$$

Here we require $10 d_k \le \beta$. Also,

$$d(\tilde{\phi}(\Delta), q) \le d(q, q_\Delta) + d(\tilde{\phi}(x), q_\Delta)$$

$$\le (1 - d_{i+1})\rho + \left(d_{i+1} - d_0\left(2 - \frac{i+1}{k}\right)\right)\rho \le (1 - d_0)\rho.$$

Thus we complete the proof of the Induction Lemma if we choose

$$\delta_k \le \min\{\gamma(C_1, C_2, d_0), \delta_i(1 + d_0/(2k), C_2) \text{ for } i = 0, 1, \ldots, k - 1\}. \qquad \Box$$

We now turn to the proof of Lemma 4.26 proper, by induction on k. The case $k = 0$ is obvious. Assume A($k - 1$) is true, i.e., any map $f : S^i \to B_p(R)$, for $i \le k - 1$, can be extended to $g : D^{i+1} \to B_p(\alpha R)$, with $\alpha = 1 + d_0/2k$. Now consider a map $f : S^k \to B_p(R)$ and view $S^k = \partial D^{k+1}$. Give a fine triangulation T of S^k, view $D^{k+1} = T \times (0, 1] \cup \{0\}$. We define a sequence of cell decompositions $K_j (j = 0, 1, \ldots)$ of D^{k+1} as discussed before Lemma 4.29. We only give the k-skeleton; the lower skeletons are naturally induced from T:

$$\mathrm{skel}_k(K_0) = \partial D^{k+1} = S^k$$

$$\mathrm{skel}_k(K_1) = (S^k \times \{\tfrac{1}{2}\}) \cup (S^k \times \{1\}) \cup (\mathrm{skel}_{k-1}(T) \times [\tfrac{1}{2}, 1])$$

Each k-simplex in K_1 can be considered as a map $\sigma : S^k \to B_p(R)$, and then can be further subdivided to define K_j inductively using the above formula for each k-cell of K_{j-1}. We then define a function f_j on K_j so that $f_0 = f$, $f_{j+1}|_{S^k \times \{1\}} = f_j$, $f_{j+1}|_{S^k \times \{\frac{1}{2}\}} = \tilde{f}_j$, by Lemma 4.29 $P(j)$, and $f_{j+1}|_{\mathrm{skel}_{k-1}(T) \times [\frac{1}{2}, 1]}$ (extension by induction hypothesis on $\mathrm{skel}_i(T) \times [\frac{1}{2}, 1]$, for $i = 0, 1, \ldots, k-1$).

We will let the desired extension be $g = \lim f_j$. To see that this gives a continuous function with the desired properties, we estimate the size of its image as follows.

If $j = 0$, we have $f_0 : \mathrm{skel}_k(K_0) \to B_p(R)$.

If $j = 1$, the map $f_1 : \mathrm{skel}_k(K_1) \to M$ satisfies, for any $\Delta \in \mathrm{skel}_k(K_1)$,

$$f_1|_{\Delta^k \times \{1\}} = f_0 \subset B_p(R),$$
$$f_1|_{\Delta^k \times \{\frac{1}{2}\}} = \tilde{f}_0 \subset B_p((1 - d_0)R);$$

therefore

$$\mathrm{diam}(f_1|_{\mathrm{skel}_k(K_1)}(\Delta)) \leq \mathrm{diam}(f_1(\Delta^k \times \{1\}) \cup f_1(\Delta^k \times \{\tfrac{1}{2}\})) + \mathrm{diam}(f_1|_{\partial \Delta \times [\frac{1}{2}, 1]})$$

$$\leq \beta R + \alpha^k \beta R = (1 + \alpha^k) \beta R$$

and

$$d(f_1|_{\mathrm{skel}_k(K_1)}, p) \leq d(f_1|_{\Delta^k \times \{\frac{1}{2}\}}) + \mathrm{diam}(f_1|_{\mathrm{skel}_k(K_1)}) \leq (1 - d_0)R + (1 + \alpha^k)\beta R,$$

where we denote by p_Δ the center of the ball containing $f_1(\partial \Delta)$.

We have spelled out these two cases to make it easier to see the general formula. For general j, we have

$$\mathrm{diam}(f_j|_{\mathrm{skel}_k(K_j)}(\Delta)) \leq ((1 + \alpha^k)\beta)^j R,$$

$$d(f_j|_{\mathrm{skel}_k(K_j)}(\Delta), p_\Delta)) \leq (1 - d_0)R + R \sum_{i=1}^{j} ((1 + \alpha^k)\beta)^i + R(1 - d_0) \sum_{i=1}^{j-1} ((1 + \alpha^k)\beta)^i$$

$$\leq (1 - d_0)R + 2R \sum_{i=1}^{j} ((1 + \alpha^k)\beta)^i \leq \frac{1 - C_1^{-1}}{2},$$

where we used the definition of α.

Therefore

$$\mathrm{diam}(f_j|_{\mathrm{skel}_k(K_j)}(\Delta)) \leq \left(\frac{1 - C_1^{-1}}{2} \right)^j R \to 0$$

$$d(f_j|_{\mathrm{skel}_k(K_j)}(\Delta), p_\Delta)) \leq \left((1 - d_0) + \frac{1 - C_1^{-1}}{1 - \frac{1}{2}(1 - C_1^{-1})} \right) R$$

$$\leq C_1 R.$$

Thus, each cell is mapped into a ball that is smaller by a factor bounded away from 1. Therefore when $j \to \infty$, the limit $g = \lim f_j$ is continuous. This concludes the proof of Lemma 4.26. $\square$

5. Mean Curvature Comparison and its Applications

Direct applications of the Mean Curvature Comparison Theorem as stated in Theorem 2.2(3) have not been numerous. On the other hand, we should point out that Ricci curvature naturally enters into the second variation for the area of hypersurfaces. This aspect of the relation between Ricci curvature and mean curvature was explored extensively in dimension three by Meeks, Schoen, and Yau, among others, as in the proof of the Positive Mass Conjecture by Schoen and Yau. In this section, we will only give one application of the mean curvature comparison, to the Diameter Sphere Theorem of Perelman. The result was also obtained by Colding by a different method.

Recall that the direct generalization of the Grove–Shiohama Diameter Sphere Theorem to Ricci curvature is not correct in dimensions four and above, as shown by the examples of Anderson and Otsu. In dimension three, it holds by a recent result of Shen and Zhu [1995]. In the preceding section we strengthened the condition to volume. Here we will keep the diameter condition, but add a condition on the sectional curvature.

THEOREM 5.1 [Perelman 1997]. *For any positive integer n and any number H, there exists a positive number $\varepsilon(n, H)$ such that if a n-dimensional manifold M^n satisfies $\mathrm{Ric} \geq n - 1$, $\sec \geq H$, and $\mathrm{diam} \geq \pi - \varepsilon$, then M is a twisted sphere.*

LEMMA 5.2. *For any $\delta > 0$ and any positive integer n, there is a positive number $\varepsilon(\delta, n)$ such that if $p, q \in M^n$ satisfy $\mathrm{Ric} \geq n - 1$ and $d(p, q) \geq \pi - \varepsilon$, then $e_{p,q}(x) \leq \delta$.*

PROOF. Let $e = e_{p,q}(x)$. The triangle inequality implies that the three geodesic balls $B_p(d(x, p) - e/2)$, $B_q(d(x, q) - e/2)$, and $B_x(e/2)$ have disjoint interiors. Therefore

$$\mathrm{vol}(M) \geq \mathrm{vol}(B_x(e/2)) + \mathrm{vol}(B_p(d(x, p) - e/2)) + \mathrm{vol}(B_q(d(x, q) - e/2))$$

$$\geq \mathrm{vol}(M) \left(\frac{\mathrm{vol}(B^1(e/2))}{\mathrm{vol}(B^1(d(p, q)))} + \frac{\mathrm{vol}(B^1(d(p, x) - e/2))}{\mathrm{vol}(B^1(d(p, q)))} + \frac{\mathrm{vol}(B^1(d(q, x) - e/2))}{\mathrm{vol}(B^1(d(p, q)))} \right).$$

Thus

$$\mathrm{vol}(d(p, q)) \geq \mathrm{vol}(B^1(e/2)) + \mathrm{vol}(B^1(d(p, x) - e/2)) + \mathrm{vol}(B^1(d(q, x) - e/2))$$

$$\geq \mathrm{vol}(B^1(e/2)) + 2\,\mathrm{vol}\left(B^1\left(\frac{(d(p, x) - e/2) + (d(q, x) - e/2)}{2} \right) \right)$$

$$= \mathrm{vol}(B^1(e/2)) + \mathrm{vol}\left(B^1\left(\frac{d(p, q)}{2} \right) \right),$$

where, for the second inequality, we used the fact that the volume of balls in $S^n(1)$ is a convex function of the radius. Therefore

$$\mathrm{vol}(B^1(e/2)) \leq \mathrm{vol}(B^1(d(p, q))) - 2\,\mathrm{vol}(B^1(\frac{d(p, q)}{2})).$$

The right-hand side approaches 0 when $\varepsilon \to 0$. This gives the desired bounds. $\square$

COROLLARY 5.3. *For any $\rho > 0$, any H, and any positive integer n, there is a positive number $\varepsilon(n, \rho, H)$ such that if $p, q \in M^n$ satisfy* Ric $\geq n - 1$, sec $\geq H$, *and $d(p, q) \geq \pi - \varepsilon$, then no x with $\min\{d(x, p), d(x, q)\} > \rho$ is not a critical point of d_p, d_q.*

PROOF. Without loss of generality, we will assume $H = -K^2$ is negative. Denoting by α the angle at x, and applying Toponogov's theorem to the triangle pqx, we obtain

$$\cosh K d(p, q) \leq \cosh K d(p, x) \cosh K d(x, q) - \sinh K d(p, x) \sinh K d(x, q) \cos \alpha.$$

Using the excess estimate of the previous lemma, we immediately conclude that $\cos \alpha \to -1$ as $\varepsilon \to 0$. $\square$

PROOF OF THEOREM 5.1. By the preceding discussion, we only need to consider points that are very close to p or q. Let m be a critical point of d_p such that $d_p(m) \leq \rho$. Define a function $g : M \to R$ by

$$g(x) = \min_{y \in [pm]} \{d(x, y) + (d(p, y) - d(m, y))^2 - d(p, m)^2\},$$

where $[pm]$ denotes the union of all minimal geodesics from p to m.

LEMMA 5.4. (1) $g(x) < d(x, m)$.
(2) *If $M(R) = \{x | g(x) \leq R\}$ and $R_1 = \inf\{r > 0 | M(R) \cap \overline{B_q(r)} \neq \varnothing\}$, then for any $x_0 \in M(R) \cap \overline{B_q(R_1)}$, we have $g(x_0) < d(x_0, p)$.*
(3) $|R + R_1 - \pi| < 2\rho$.

REMARK. This lemma implies that $M(R) \cap \overline{B_q(R_1)} \neq \{m, p\}$. This is the reason for considering the function g; we could just use the distance function to $[pm]$ if $[pm]$ forms a close geodesic.

PROOF. (1) Note that if we let $y = m$, then $d(x, y) + (d(p, y) - d(m, y))^2 - d(p, m)^2 = d(x, m)$. Thus $g(x) \leq d(x, m)$. Too see the strict inequality, we note that since m is a critical point, the angle at m is at most $\pi/2$. By Toponogov's theorem,

$$d(x, y) \leq \sqrt{d^2(x, m) + d^2(m, y)} = d(x, m) + O(d^2(m, y)).$$

Thus

$$d(x, y) + (d(p, y) - d(m, y))^2 - d(p, m)^2 = d(x, y) - 4d(p, y)d(y, m)$$
$$\leq d(x, m) + O(d^2(m, y)) - 4d(p, y)d(y, m)$$
$$< d(x, m).$$

(2) Since $d(q, m) \leq d(q, p) \leq d(q, x_0) + d(x_0, p) = R_1 + d(x_0, p)$, we have

$$d(m, B_q(R_1)) \leq d(x_0, p).$$

Thus, there exists a point $x_1 \in \partial B_q(R_1)$, such that $d(m, x_1) \leq d(x_0, p)$. This, together with the definition of R_1, implies that

$$g(x_0) \leq g(x_1) < d(x_1, m) \leq d(x_0, p).$$

(3) Take x_0 as in (2). Then

$$R + R_1 = g(x_0) + d(x_0, q) < d(x_0, q) + d(x_0, q) \leq d(p, q) \leq \pi.$$

Similarly, if $g(x_0)$ is realized at y_0, an interior point of some shortest line from p to m, then

$$R + R_1 = g(x_0) + d(x_0, q) = d(x_0, y_0) - 4d(p, y_0)d(y_0, m) + d(x_0, q)$$
$$\geq d(y_0, q) - \rho \geq \pi - 2\rho,$$

if we choose ε in Corollary 5.3 smaller than ρ. $\qquad \square$

LEMMA 5.5. *For any point $x_0 \in \partial M(R)$, let $\vec{n}$ be the unit normal vector pointing away from p. Then the mean curvature of $\partial M(R)$ can be estimated as*

$$m_{\vec{n}}(x_0) \leq (n-2)\coth\sqrt{K}R + \tanh\sqrt{K}R + 10.$$

PROOF. By Lemma 5.4, at x_0, the value of $g(x_0)$ is achieved at an interior point y_0 of some minimal geodesic γ from p to m. Then the function $g(x)$ is a perturbation of the distance function to the geodesic γ, whose mean curvature does not exceed $(n-2)\coth\sqrt{K}R + \tanh\sqrt{K}R$.

For convenience of notation, we parametrize γ so that $\gamma(0) = y_0$. Let σ be a minimal geodesic from y_0 to x_0 with $\sigma(t_0) = x_0$. Let $V(t)$ be the vector field along σ obtained by parallel translation of $\dot{\gamma}(0)$ along σ. Let v_1 be the projection of $V(t_0)$ in the tangent plane of $\partial M(R)$ (v_1 is not a unit vector). Let $\{v_2, \ldots, v_{n-1}\}$ be orthonormal tangent vectors of $\partial M(R)$ that are all perpendicular to v_1. By the first variation formula, the v_i are all perpendicular to $\dot{\sigma}(t_0)$, and the values of g along the directions v_i are all achieved at y_0. It now follows from the Hessian Comparison Theorem that $\langle \nabla_{v_i}\dot{\sigma}(t_0), v_i \rangle \leq \coth\sqrt{K}t_0$ for $i = 2, \ldots, n-1$.

We now consider the direction v_1. Let $J(t)$ be a Jacobian field along σ such that $J(0) = \dot{\gamma}(0)$ and $J(t_0) = v_1$. We decompose J into directions along σ and orthogonal to σ, then call the orthogonal component $W(t)$, i.e.,

$$J(t) = (at + b)\dot{\sigma}(t) + W(t).$$

In general, $W(t_0)$ is not a unit vector. Let $\bar{J}(t) = cJ(t)$ be such that the orthogonal component $\bar{W}$ of $\bar{J}$ at t_0 has unit length. Then the Hessian comparison theorem applied to this case implies

$$\langle \nabla_{\bar{W}(t_0)}\dot{\sigma}(t_0), \bar{W}(t_0) \rangle \leq \tanh\sqrt{K}t_0.$$

To get the desired estimate on the mean curvature of $\partial M(R)$, we need to estimate the numbers a, b, c in this decomposition, which in return depend on the angle $\angle(\dot{\sigma}(0), \dot{\gamma}(0))$.

Note that $g(x_0) = d(x_0, y_0) - 4d(y_0, p)d(y_0, m)$. The first variation formula gives

$$0 = -\cos \angle(\dot{\sigma}(0), \dot{\gamma}(0)) - 4(\rho - 2d(y_0, m));$$

thus

$$|\cos \angle(\dot{\sigma}(0), \dot{\gamma}(0))| \le 4\rho.$$

This will give a uniform bound for a, b, c. Note also that

$$|R - t_0| = 4d(y_0, p)d(y_0, m) \le 4\rho^2.$$

The lemma follows if we choose ρ small enough. □

We continue with the proof of Theorem 5.1. Choose $R > 0$ such that

$$(n-1)\cot R - (n-2)\sqrt{K}\coth\sqrt{K}R - \sqrt{K}\tanh\sqrt{K}R - 10 > 0.$$

Thus, R depends on n, K. Then choose ρ small enough so that Lemma 5.5 holds; this in return determines the number ε. For such an ε, by Corollary 5.3, there are no critical points at distance ρ from p and q. Assume there is a critical point m such that $(p, m) \le \rho$; we will derive a contradiction using the mean value comparison.

On the one hand, using the condition Ric $\ge n - 1$ and applying Theorem 2.2 to d_q, we conclude, letting $m^1(R_1)$ be the mean curvature of the sphere of radius R_1 in $S^n(1)$:

$$
\begin{aligned}
m_{\partial B_q(R_1)}(x_0) &\le m^1(R_1) \le (n-1)\cot R_1 \\
&\le (n-1)\cot(\pi - R - 2\rho) = -(n-1)\cot(R + 2\rho) \\
&\le -(n-1)\cot R + \kappa,
\end{aligned}
$$

where $\kappa = \kappa(n, \rho)$ and $\lim_{\rho \to 0} \kappa = 0$. On the other hand, Lemma 5.5 implies

$$m_{\partial M(R), N}(x_0) \ge -(n-2)\sqrt{K}\coth\sqrt{K}R - \sqrt{K}\tanh\sqrt{K}R - 10.$$

Denote by N the unit normal vector pointing away from q. Since $\partial M(R)$ and $\partial B_q(R_1)$ are tangent to each other at x_0, and $\partial M(R)$ lies outside (with respect to N) of $\partial B_q(R_1)$, we can write

$$m_{\partial M(R), N}(x_0) \le m_{\partial B_q(R_1)}(x_0),$$

which implies

$$(n-1)\cot R - (n-2)\sqrt{K}\coth\sqrt{K}R - \sqrt{K}\tanh\sqrt{K}R - 10 \le \kappa.$$

We get a contradiction if we choose ε so small that κ violates this inequality. □

References

[Abresch and Gromoll 1990] U. Abresch and D. Gromoll, "On complete manifolds with nonnegative Ricci curvature", *J. Amer. Math. Soc.* **3** (1990), 355–374.

[Anderson 1990a] M. T. Anderson, "Short geodesics and gravitational instantons", *J. Diff. Geom.* **31** (1990), 265–275.

[Anderson 1990b] M. T. Anderson, "Metrics of positive Ricci curvature with large diameter", *Manuscripta Math.* **68** (1990), 405–415.

[Berard 1988] P. Berard, "From vanishing theorems to estimating theorems: the Bochner technique revisited", *Bull. Amer. Math. Soc.* **19** (1988), 371–406.

[Berger 1960] M. Berger, "Les variétés riemanniennes $\frac{1}{4}$-pincées", *Ann. Scuola Norm. Sup. Pisa* **14** (1960), 161–170.

[Bochner 1946] S. Bochner, "Vector fields and Ricci curvature", *Bull. Amer. Math. Soc.* **52** (1946), 776–797.

[Cai et al. 1994] M. Cai, G. Galloway and Z. Liu, "Local splitting theorems for Riemannian manifolds", *Proc. Amer. Math. Soc.* **120** (1994), 1231–1239.

[Cheeger 1967] J. Cheeger, "Comparison and finiteness theorems for Riemannian manifolds", Ph.D. Thesis, Princeton University, 1967.

[Cheeger 1970] J. Cheeger, "Finiteness theorems for Riemannian manifolds", *Am. J. Math.* **92** (1970), 61–74.

[Cheeger 1991] J. Cheeger, "Critical points of distance functions and applications to geometry", pp. 1–38 in *Geometric topology: recent developments* (Montecatini Terme, 1990), edited by J. Cheeger et al., Lecture Notes in Math. **1504**, Springer, Berlin, 1991.

[Cheeger and Colding 1995] J. Cheeger and T. H. Colding, "Almost rigidity of warped products and the structure of spaces with Ricci curvature bounded below", *C. R. Acad. Sci. Paris* Sér. I **320** (1995), 353–357.

[Cheeger and Ebin 1975] J. Cheeger and D. Ebin, *Comparison theorems in Riemannian geometry*, North-Holland, Amsterdam, and Elsevier, New York, 1975.

[Cheeger and Gromoll 1971] J. Cheeger and D. Gromoll, "The splitting theorem for manifolds of nonnegative Ricci curvature", *J. Differential Geom.* **6** (1971), 119–129.

[Cheeger and Gromoll 1972] J. Cheeger and D. Gromoll, "On the structure of complete manifolds with nonnegative curvature", *Ann. of Math.* **96** (1972), 413–443.

[Cheng 1975] S. Y. Cheng, "Eigenvalue comparison theorems and its geometric applications", *Math. Z.* **143** (1975), 289–297.

[Colding 1995] T. H. Colding, "Stability and Ricci curvature", *C. R. Acad. Sci. Paris* Sér. I **320** (1995), 1343–1347.

[Colding 1996a] T. H. Colding, "Shape of manifolds with positive Ricci curvature", *Invent. Math.* **124** (1996), 175–191.

[Colding 1996b] T. H. Colding, "Large manifolds with positive Ricci curvature", *Invent. Math.* **124** (1996), 193–214.

[Freedman 1982] M. H. Freedman, "Topology of four-dimensional manifolds", *J. Diff. Geom.* **17** (1982), 357–453.

[Gallot 1983] S. Gallot, "A Sobolev inequality and some geometric applications", pp. 45–55 in *Spectra of Riemannian manifolds*, edited by M. Berger et al., Kaigai, Tokyo, 1983.

[Gilbarg and Trudinger 1983] D. Gilbarg and N. Trudinger, *Elliptic partial differential equations of second order*, 2nd ed., Grundlehren der math. Wiss. **224**, Springer, Berlin, 1983.

[Greene and Wu 1988] R. E. Greene and H. Wu, "Lipschitz convergence of Riemannian manifolds", *Pacific J. Math.* **131** (1988), 119–141.

[Gromov 1981a] M. Gromov, "Curvature, diameter, and Betti numbers", *Comment. Math. Helv.* **56** (1981), 53–78.

[Gromov 1981b] M. Gromov, "Groups of polynomial growth and expanding maps", *Publ. Math. IHES* **53** (1981), 183–215.

[Gromov 1981c] M. Gromov, *Structures métriques pour les variétés riemanniennes*, edited by J. Lafontaine and P. Pansu, CEDIC, Paris, 1981.

[Grove and Petersen 1988] K. Grove and P. Petersen, "Bounding homotopy types by geometry", *Ann. of Math.* (2) **128** (1988), 195–206.

[Grove and Shiohama 1977] K. Grove and K. Shiohama, "A generalized sphere theorem", *Ann. of Math.* (2) **106** (1977), 201–211.

[Hamilton 1982] R. S. Hamilton, "Three-manifolds with positive Ricci curvature", *J. Diff. Geom.* **17** (1982), 255–306.

[Klingenberg 1961] W. Klingenberg, "Über Mannigfaltigkeiten mit positiver Krümmung", *Comm. Math. Helv.* **35** (1961), 47–54.

[Milnor 1968a] J. Milnor, "A note on curvature and fundamental group", *J. Diff. Geom.* **2** (1968), 1–7.

[Milnor 1968b] J. Milnor, "Growth of finitely generated solvable groups", *J. Diff. Geom.* **2** (1968), 447–449.

[Myers 1935] S. Myers, "Riemannian manifolds in the large", *Duke Math. J.* **1** (1935), 39–49.

[Otsu 1991] Y. Otsu, "On manifolds of positive Ricci curvature with large diameter", *Math. Z.* **206** (1991), 255–264.

[Perelman 1994] G. Perelman, "Manifolds of positive Ricci curvature with almost maximal volume", *J. Amer. Math. Soc.* **7** (1994), 299–305.

[Perelman 1995] G. Perelman, "A diameter sphere theorem for manifolds of positive Ricci curvature", *Math. Z.* **218** (1995), 595–596.

[Perelman 1997] G. Perelman, "Construction of manifolds of positive Ricci curvature with big volume and large Betti numbers", pp. 157–163 in this volume.

[Peters 1987] S. Peters, "Convergence of Riemannian manifolds", *Compositio Math.* **62** (1987), 3–16.

[Schoen and Yau 1979] R. Schoen and S.-T. Yau, "On the proof of the positive mass conjecture in general relativity", *Comm. Math. Phys.* **65** (1979), 45–76.

[Schoen and Yau 1982] R. Schoen and S.-T. Yau, "Complete three-dimensional manifolds with positive Ricci curvature and scalar curvature", pp. 209–228 in

Seminar on Differential Geometry, edited by S.-T. Yau, Ann. of Math. Stud. **102**, Princeton Univ. Press, Princeton, NJ, 1982.

[Sha and Yang 1989a] J. Sha and D. Yang, "Examples of manifolds of positive Ricci curvature", *J. Diff. Geom.* **29** (1989), 95–103.

[Sha and Yang 1989b] J. Sha and D. Yang, "Positive Ricci curvature on the connected sums of $S^n \times S^{m}$", *J. Diff. Geom.* **33** (1991), 127–137.

[Shen and Zhu 1995] Y. Shen and S. Zhu, "Ricci curvature, minimal surfaces and sphere theorems", preprint, 1995. To appear in *J. Geom. Anal.*

[Shiohama 1983] K. Shiohama, "A sphere theorem for manifolds of positive Ricci curvature", *Trans. Amer. Math. Soc.* **275** (1983), 811–819.

[Smale 1961] S. Smale, "Generalized Poincaré's conjecture in dimensions greater than four", *Ann. of Math.* **74** (1961), 391–466.

[Wei 1995] G. Wei, "Aspects of positive curved spaces", Ph.D. Thesis, SUNY Stony Brook, 1989.

[Whitehead 1978] G. Whitehead, *Elements of homotopy theory*, Graduate Texts in Math. **61**, Springer, New York, 1978.

[Wu 1988] H.-H. Wu, *The Bochner technique in differential geometry*, Mathematical reports **3**, part 2, Harwood, Chur (Switzerland) and New York, 1988.

[Yau 1976] S.-T. Yau, "Some function-theoretic properties of complete Riemannian manifolds and their applications to geometry", *Indiana Math. J.* **25** (1976), 659–670.

[Zhu 1990] S. Zhu, "A finiteness theorem for Ricci curvature in dimension three", *Bull. Amer. Math. Soc.* **23** (1990), 423–426.

SHUNHUI ZHU
DEPARTMENT OF MATHEMATICS
DARTMOUTH COLLEGE
HANOVER, NH 03755